Lecture Notes in Electrical Engineering

Volume 158

For further volumes:
http://www.springer.com/series/7818

Nicola Adami · Andrea Cavallaro
Riccardo Leonardi · Pierangelo Migliorati
Editors

Analysis, Retrieval and Delivery of Multimedia Content

Springer

Editors
Nicola Adami
Department of Information Engineering
University of Brescia
Brescia
Italy

Riccardo Leonardi
Department of Information Engineering
University of Brescia
Brescia
Italy

Andrea Cavallaro
School of Electrical Engineering
and Computer Science
Queen Mary University of London
London
UK

Pierangelo Migliorati
Department of Information Engineering
University of Brescia
Brescia
Italy

ISSN 1876-1100 ISSN 1876-1119 (electronic)
ISBN 978-1-4614-3830-4 ISBN 978-1-4614-3831-1 (eBook)
DOI 10.1007/978-1-4614-3831-1
Springer New York Heidelberg Dordrecht London

Library of Congress Control Number: 2012942703

Printed on acid-free paper

Springer is part of Springer Science+Business Media (www.springer.com)

Contents

Preface

This book presents an extended version of a few selected papers originally submitted to the 11th International Workshop on Image Analysis for Multimedia Interactive Services, which took place in April 2010 in Desenzano del Garda, Brescia, Italy. This workshop is one of the main international events for the presentation and discussion of the latest technological advances in interactive multimedia services. The objective of the workshop is to bring together researchers and developers from academia and industry working in the areas of image, video, and audio applications, with a special focus on analysis.

The book is organized into five main sections, considering Multimedia Content Analysis, Motion and Activity Analysis, High-Level Descriptors and Video Retrieval, 3D and Multi-View, and Multimedia Delivery.

Part 1: Multimedia Content Analysis

Multimedia Content Analysis is of great relevance in the scenario of image analysis for multimedia interactive services. In this respect, it is very important to consider also the audio signal and caption text eventually superimposed on the considered images. Also, the objects displayed in the images could be very helpful in content analysis.

Panagiotis Sidiropoulos, Vasileios Mezaris, Ioannis Kompatsiaris, Hugo Meinedo, Miguel Bugalho, and Isabel Trancoso, in the book chapter "*On the use of audio events for improving video scene segmentation*" deal with the problem of automatic temporal segmentation of a video into elementary semantic units known as scenes. The novelty lies in the use of high-level audio information, in the form of audio events, for the improvement of scene segmentation performance. More specifically, the proposed technique is built upon a recently proposed audio-visual scene segmentation approach that involves the construction of multiple scene transition graphs (STGs) that separately exploit information coming from different modalities. In the extension of the latter approach presented in this chapter, audio event detection results are introduced to the definition of an audio-based scene transition graph, while a visual-based scene transition graph is also defined independently. The results of these two types of STGs are subsequently combined. The results of the application of the proposed technique to broadcast videos demonstrate the usefulness of audio events for scene segmentation and highlight the importance of introducing additional high-level information to the scene segmentation algorithms.

The important problem of caption text extraction is addressed in the chapter "*Region-based caption text extraction*", authored by Miriam León, Veronica Vilaplana, Antoni Gasull, and Ferran Marques. The authors present a method for caption text detection that takes advantage of texture and geometric features to detect the caption text. Texture features are estimated using wavelet analysis and mainly applied for *text candidate spotting*. In turn, *text characteristics verification* relies on geometric features, which are estimated exploiting the region-based image model. Analysis of the region hierarchy provides the final caption text objects. The final step of *consistency analysis for output* is performed by a binarization algorithm that robustly estimates the thresholds on the caption text area of support.

Image classification is a challenging task in computer vision. For e.g., fully understanding real-world images may involve both scene and object recognition. Many approaches have been proposed to extract meaningful descriptors from images and classifying them in a supervised learning framework. In the chapter "*K-nn boosting prototype learning for object classification*", Paolo Piro, Michel Barlaud, Richard Nock, and Frank Nielsen, revisit the classic k-nearest neighbors classification rule, which has shown to be very effective when dealing with local image descriptors. However, *k-nn* still features some major drawbacks, mainly due to the uniform voting among the nearest prototypes in the feature space. In this chapter, the authors propose therefore a generalization of the classic knn rule in a supervised learning (boosting) framework. Namely, they redefine the voting rule as a strong classifier that linearly combines predictions from the k closest prototypes. In order to induce this classifier, they propose a novel learning algorithm, *MLNN* (Multiclass Leveraged Nearest Neighbors), which gives a simple procedure for performing prototype selection very efficiently. Experiments carried out first on object classification using 12 categories of objects, then on scene recognition, using 15 real-world categories, show significant improvement over classic *K-nn* in terms of classification performances.

Part 2: Motion and Activity Analysis

Motion and activity information plays certainly a crucial role in content-based video analysis and retrieval. In this context the problem of automatic tracking of moving object in a video have been extensively studied in the literature and also in this book.

In the book chapter titled "*Semi-automatic object tracking in video sequences by extension of the MRSST algorithm*", Marko Esche, Mustafa Karaman, and Thomas Sikora investigate a new approach for segmentation of real-world objects in video sequences. While some amount of user interaction is still necessary for most algorithms in this field, in order for them to produce adequate results, these can be reduced making use of certain properties of graph-based image segmentation algorithms. Based on one of these algorithms a framework is proposed that tracks individual foreground objects through arbitrary video sequences and partly automates the necessary corrections required from the user. Experimental results suggest that the proposed algorithm performs well on both low- and high-resolution video sequences and can even, to a certain extent, cope with motion blur and gradual object deformations.

The problem of tracking a non-rigid object in an uncalibrated static multi-camera environment is considered in "*A multi-resolution particle filter tracking with a dual consistency check for model update in a multi-camera environment*", where Yifan Zhou, Jenny Benois-Pineau, and Henri Nicolas present a novel tracking method with a multi-resolution approach and a dual model check. The proposed method is based on particle filtering using color features. The major contributions of the method are: multi-resolution tracking to handle strong and non-biased object motion by short-term particle filters; stratified model consistency check by Kolmogorov-Smirnov test, and object trajectory-based view corresponding deformation in a multi-camera environment.

An interesting application of trajectories analysis in a surveillance scenario is proposed by Mattia Daldoss, Nicola Piotto, Nicola Conci, and Francesco G. B. De Natale in the book chapter "*Activity detection using regular expressions*". The authors propose a novel method to analyze trajectories in surveillance scenarios by means of Context-Free Grammars (CFGs). Given a training corpus of trajectories associated to a set of actions, a preliminary processing phase is carried out to characterize the paths as sequences of symbols. This representation turns the numerical representation of the coordinates into a syntactical description of the activity structure, which is successively adopted to identify different behaviors through the CFG models. Such a modeling is the basis for the classification and matching of new trajectories versus the learned templates and it is carried out through a parsing engine that enables the online recognition of human activities. An additional module is provided to recover parsing errors (i.e., insertion, deletion, or substitution of symbols) and update the activity models previously learned. The proposed system has been validated in indoor, in an assisted living context, demonstrating good capabilities in recognizing activity patterns in different configurations, and in particular in presence of noise in the acquired trajectories, or in case of concatenated and nested actions.

Katharina Quast, and André Kaup, in "*Shape adaptive mean shift object tracking using gaussian mixture models*" propose a new object tracking algorithm based on a combination of the mean shift and Gaussian mixture models (GMMs), named GMM-SAMT. GMM-SAMT stands for Gaussian mixture model-based shape adaptive mean shift tracking. Instead of a symmetrical kernel like in traditional mean shift tracking, GMM-SAMT uses an asymmetric shape adapted kernel which is retrieved from an object mask. During the mean shift iterations the kernel scale is altered according to the object scale, providing an initial adaptation of the object shape. The final shape of the kernel is then obtained by segmenting the area inside and around the adapted kernel into object and non-object segments using Gaussian mixture models.

Part 3: High-Level Descriptors and Video Retrieval

In the context of content-based video retrieval the high-level descriptors are clearly of great relevance. This topic is covered in this part of the book.

Seunghan Han, Bonjung Koo, Andreas Hutter, and Walter Stechele in "*Forensic reasoning upon pre-obtained surveillance metadata using uncertain spatiotemporal rules and subjective logic*" present an approach to modeling uncertain contextual rules using subjective logic for forensic visual surveillance. Unlike traditional real-time visual surveillance, forensic analysis of visual surveillance data requires matching of high level contextual cues with observed evidential metadata where both the specification of the context and the metadata suffer from uncertainties. To address this aspect, there has been work on the use of declarative logic formalisms to represent and reason about contextual knowledge, and on the use of different uncertainty handling formalisms. In such approaches, uncertainty attachment to logical rules and facts are crucial. However, there are often cases that the truth value of rule itself is also uncertain thereby, uncertainty attachment to rule itself should be rather functional. '*The more X then the more Y*' type of knowledge is one of the examples. To enable such type of rule modeling, in this chapter, the authors propose a reputational subjective opinion function upon logic programming, which is similar to fuzzy membership function but can also take into account uncertainty of membership value itself. Then they further adopt subjective logic's fusion operator to accumulate the acquired opinions over time. To verify the proposed approach, the authors present a preliminary experimental case study on reasoning likelihood of being a good witness that uses metadata extracted by a person tracker and evaluates the relationship between the tracked persons. The case study is further extended to demonstrate more complex forensic reasoning by considering additional contextual rules.

Nowadays, multimedia data is produced and consumed at an ever-increasing rate. Similar to this trend, diverse storage approaches for multimedia data have been introduced. These observations lead to the fact that distributed and heterogeneous multimedia repositories exist, whereas an easy and unified access to the stored multimedia data is not given. In this respect, Florian Stegmaier, Mario Döller, Harald Kosch, Andreas Hutter, and Thomas Riegel in "*AIR: architecture for interoperable retrieval on distributed and heterogeneous multimedia repositories*" present an architecture, named AIR, that offers the aforementioned retrieval possibilities. To ensure interoperability, AIR makes use of recently issued standards, namely the MPEG Query Format (multimedia query language) and the JPSearch transformation rules (metadata interoperability).

In the final chapter of this section, the detection of high-level concepts in video is considered. More specifically, Vasileios Mezaris, Anastasios Dimou, and Ioannis Kompatsiaris propose in "*Local invariant feature tracks for high-level video feature extraction*" the use of feature tracks for the detection of high-level features (concepts) in video. Extending previous work on local interest point detection and description in images, feature tracks are defined as sets of local interest points that are found in different frames of a video shot and exhibit spatio-temporal and visual continuity, thus defining a trajectory in the 2D + Time space. These tracks jointly capture the spatial attributes of 2D local regions and their corresponding long-term motion. The extraction of feature tracks and the selection and representation of an appropriate subset of them allow the generation of a Bag-of-Spatiotemporal-Words model for the shot, which facilitates capturing the dynamics of video content. Experimental evaluation of the proposed approach on two challenging datasets (TRECVID 2007, TRECVID 2010) highlights how the selection, representation, and use of such feature tracks enhance the results of traditional keyframe-based concept detection techniques.

Part 4: 3D and Multi-View

Among the various audio-visual descriptors useful for image and video analysis and coding there are the descriptors related to 3D structure and multi-view. In this section of the book we cover this topic, considering both the issue of 3D stereo correspondences and 3DTV video coding.

The problem of 3D stereo correspondences is considered in "*A new evaluation criterion for point correspondences in stereo images*" by Aleksandar Stojanovic, and Michael Unger. In this chapter, the authors present a new criterion to evaluate point correspondences within a stereo setup. Many applications such as stereo matching, triangulation, lens distortion correction, and camera calibration require an evaluation criterion for point correspondences. The common criterion used is the epipolar distance. The uncertainty of the epipolar geometry provides additional information, and the proposed method uses this information for a new distance measure. The basic idea behind this criterion is to determine the most probable epipolar geometry that explains the point correspondence in the two views. This criterion considers the fact that the uncertainty increases for point correspondences induced by world points that are located at a different depth-level compared to those that were used for the fundamental matrix computation. Furthermore, the authors show that by using Lagrange multipliers, this constrained minimization problem can be reduced to solving a set of three linear equations with a computational complexity practically equal to the complexity of the epipolar distance.

A novel learning-based approach used to estimate local homography of points belonging to a given surface is proposed in "*Local homography estimation using keypoint descriptors*" by Alberto Del Bimbo, Fernando Franco, and Federico Pernici. In this chapter the authors present a new learning-based approach used to estimate local homography of points belonging to a given surface and show that it is more accurate than specific affine region detection methods. While other works attempt to do this task by using iterative algorithms developed for template matching, this method introduces a direct estimation of the transformation. In more details, it performs the following steps. First, a training set of features captures the geometry and appearance information about keypoints taken from multiple views of the surface. Then, incoming keypoints are matched against the training set in order to retrieve a cluster of features representing their identity. Finally the retrieved clusters are used to estimate the local homography of the regions around keypoints. Thanks to the high accuracy, outliers and bad estimates are filtered out by multiscale Summed Square Difference test.

The problem of 3DTV multiple description coding is addressed by Simone Milani and Giancarlo Calvagno in the book chapter titled "*A cognitive source coding scheme for multiple description 3DTV transmission*". In this framework, Multiple Description Coding has recently proved to be an effective solution for the robust transmission of 3D video sequences over unreliable channels. However, adapting the characteristics of the source coding strategy (Cognitive Source Coding) permits improving the quality of 3D visualization experienced by the end-user. This strategy has been successfully employed for standard video signals, but it can be applied to Multiple Description video coding for an effective transmission of 3D signals. The chapter presents a novel Cognitive Source Coding scheme that improves the performance of traditional Multiple Description Coding approaches by adaptively combining traditional predictive and Wyner-Ziv coders according to the characteristics of the video sequence and to the channel conditions. The approach is employed for video + depth 3D transmissions improving the average PSNR value up to 2.5 dB with respect to traditional MDC schemes.

Part 5: Multimedia Delivery

In the final section of the book we consider the important aspects related to the problem of multimedia documents delivery, focusing the attention on both images and video.

In "*An efficient prefetching strategy for remote browsing of JPEG 2000 image sequences*", Juan Pablo García Ortiz, Vicente González Ruiz, Inmaculada García, Daniel Müller, and George Dimitoglou propose an efficient prefetching strategy for interactive remote browsing of sequences of high resolution JPEG 2000 images. As a result of the inherent latency of client–server communication, the experiments of this study prove that a significant benefit can be achieved, in terms of both quality and responsiveness, by anticipating certain data from the rest of the sequence while an image is being explored. In this work a model based on the quality progression of the image is proposed in order to estimate which percentage of the bandwidth will be dedicated to prefetching. This solution can be easily implemented on top of any existing remote browsing architecture.

Matteo Naccari and Fernando Pereira in "*Comparing spatial masking modelling in just noticeable distortion controlled H.264/AVC video coding*" study the integration of a just noticeable distortion model in the H.264/AVC standard video codec to improve the final rate-distortion performance. Three masking aspects related to lossy transform coding and natural video contents are considered: frequency band decomposition, luminance component variations and pattern masking. For the latter aspect, three alternative models are considered, namely the Foley-Boynton, Foley-Boynton adaptive, and Wei-Ngan models. Their performance, measured for high definition video contents, and reported in terms of bitrate improvement and objective quality loss, reveals that the Foley-Boynton and its adaptive version provide the best performance with up to 35.6 % bitrate reduction at the cost of at most 1.4 % objective quality loss.

In traditional motion compensated predictive video coding, both the motion vector and the prediction residue are encoded and stored or sent for every predicted block. The motion vector brings displacement information with respect to a reference frame while the residue represents what we really consider to be the innovation of the current block with respect to that reference frame. This encoding scheme has proved to be extremely effective in terms of rate distortion performance. Nevertheless, one may argue that full description of motion and residue could be avoided if the decoder could be made able to exploit a proper a priori model for the signal to be reconstructed. In particular, it was recently shown that a smart enough decoder could exploit such an a priori model to partially infer motion information for a single block given only neighboring blocks and the innovation of that block. The last contribution, given by Claudia Tonoli and Marco Dalai presents an improvement over the single-block method. In the book chapter "*Coherent video reconstruction with motion estimation at the decoder*" the authors show that higher performance can be achieved by simultaneously reconstructing a frame region composed of several blocks, rather than reconstructing those blocks separately. A trellis-based algorithm is developed in order to make a global decision on many motion vectors at a time instead of many single separate decisions on different vectors.

Brescia, Italy
London, UK

Nicola Adami
Andrea Cavallaro
Riccardo Leonardi
Pierangelo Migliorati

Contributors

Michel Barlaud CNRS/University of Nice-Sophia Antipolis, Sophia Antipolis, France, e-mail: barlaud@i3s.unice.fr

Jenny Benois-Pineau Laboratoire Bordelais de Recherche en Informatique (LaBRI), CNRS (UMR 5800), Université Bordeaux 1, 351, cours de la Libération, 33405, Talence cedex, France, e-mail: jenny.benois@labri.fr

Alberto Del Bimbo Media Integration and Communication Center (MICC), University of Florence, Italy

Miguel Bugalho INESC-ID Lisboa, Rua Alves Redol 9, Lisboa 1000-029, Portugal; IST/UTL, Rua Alves Redol 9, Lisboa 1000-029, Portugal, e-mail: mmfb@l2f.inesc-id.pt

Giancarlo Calvagno Department of Information Engineering, University of Padova, via Gradenigo 6/B, 35131 Padova, Italy, e-mail: calvagno@dei.unipd.it

Nicola Conci Multimedia Signal Processing and Understanding Lab, DISI—University of Trento, Via Sommarive 14, 38123 Trento, Italy

Marco Dalai Department of Information Engineering, University of Brescia, Via Branze 38, 25123 Brescia, Italy, e-mail: marco.dalai@ing.unibs.it

Mattia Daldoss Multimedia Signal Processing and Understanding Lab, DISI—University of Trento, Via Sommarive 14, 38123 Trento, Italy

George Dimitoglou Department of Computer Science, Hood College, Frederick, MD 21701, USA

Anastasios Dimou Centre for Research and Technology Hellas, Informatics and Telematics Institute, 6th Km Charilaou-Thermi Road, 57001 Thermi, Greece, e-mail: dimou@iti.gr

Mario Döller Chair of Distributed Information Systems, University of Passau, Passau, Germany

Marko Esche Communication Systems Group, Technische Universität Berlin, Sekr. EN1, Einsteinufer 17, 10587 Berlin, Germany

Fernando Franco Media Integration and Communication Center (MICC), University of Florence, Italy

Juan Pablo García Ortiz Computer Architecture and Electronics Department, University of Almería, 04120 Almería, Spain

Antoni Gasull Technical University of Catalonia, Barcelona, Spain, e-mail: antoni.gasull@upc.edu

Seunghan Han Institute for Integrated Systems, Technische Universität München, Arcisstrasse 21, Munich, Germany; Siemens AG, Corporate Technology, Otto-Hahn-Ring 6, Munich, Germany, e-mail: hanseunghan@gmail.com

Andreas Hutter Siemens AG, Corporate Technology, Otto-Hahn-Ring 6, Munich, Germany; Corporate Technology, Siemens AG, 81739 Munich, Germany, e-mail: andreas.hutter@siemens.com

Mustafa Karaman Communication Systems Group, Technische Universität Berlin, Sekr. EN1, Einsteinufer 17, 10587 Berlin, Germany

André Kaup Chair of Multimedia Communications and Signal Processing, University of Erlangen-Nuremberg, Cauerstr. 7, 91058 Erlangen, Germany

Ioannis Kompatsiaris Centre for Research and Technology Hellas, Informatics and Telematics Institute, 6th Km Charilaou-Thermi Road, 57001 Thermi, Greece, e-mail: ikom@iti.gr

Bonjung Koo Siemens AG, Corporate Technology, Otto-Hahn-Ring 6, Munich, Germany; Department of Computer Science, Sogang University, Shinsu dong 1, Seoul, Korea, e-mail: kbj9090@gmail.com

Harald Kosch Chair of Distributed Information Systems, University of Passau, Passau, Germany

Miriam Leon Technical University of Catalonia, Barcelona, Spain, e-mail: mleon@tsc.upc.edu

Ferran Marques Technical University of Catalonia, Barcelona, Spain, e-mail: ferran.marques@upc.edu

Hugo Meinedo INESC-ID Lisboa, Rua Alves Redol 9, Lisboa 1000-029, Portugal, e-mail: hugo.meinedo@l2f.inesc-id.pt

Vasileios Mezaris Centre for Research and Technology Hellas, Informatics and Telematics Institute, 6th Km Charilaou-Thermi Road, 57001 Thermi, Greece, e-mail: bmezaris@iti.gr

Simone Milani Department of Information Engineering, University of Padova, via Gradenigo 6/B, 35131 Padova, Italy, e-mail: simone.milani@dei.unipd.it

Daniel Müller European Space Agency ESTEC, Noordwijk, Netherlands

Matteo Naccari Instituto Superior Técnico—Instituto de Telecomunicações, 1049-001 Lispon, Portugal, e-mail: matteo.naccari@lx.it.pt

Francesco G. B.De Natale Multimedia Signal Processing and Understanding Lab, DISI—University of Trento, Via Sommarive 14, 38123 Trento, Italy

Henri Nicolas Laboratoire Bordelais de Recherche en Informatique (LaBRI), CNRS (UMR 5800), Université Bordeaux 1, 351, cours de la Libération, 33405, Talence cedex, France, e-mail: henri.nicolas@labri.fr

Frank Nielsen LIX/Ecole Polytechnique, Palaiseau, France, e-mail: nielsen@lix.polytechnique.fr

Richard Nock CEREGMIA/University of Antilles-Guyane, Martinique, France, e-mail: richard.nock@martinique.univ-ag.fr

Juan Pablo García Ortiz Computer Architecture and Electronics Department, University of Almería, 04120 Almería, Spain

Fernando Pereira Instituto Superior Técnico—Instituto de Telecomunicações, 1049-001 Lispon, Portugal, e-mail: fernando.pereira@lx.it.pt

Federico Pernici Media Integration and Communication Center (MICC), University of Florence, Italy, e-mail: pernisi@dsi.unifi.it

Nicola Piotto Multimedia Signal Processing and Understanding Lab, DISI—University of Trento, Via Sommarive 14, 38123 Trento, Italy

Paolo Piro CNRS/University of Nice-Sophia Antipolis, Sophia Antipolis, France, e-mail: piro@i3s.unice.fr

Katharina Quast Chair of Multimedia Communications and Signal Processing, University of Erlangen-Nuremberg, Cauerstr. 7, 91058 Erlangen, Germany, e-mail: quast@lnt.de

Thomas Riegel Corporate Technology, Siemens AG, 81739 Munich, Germany

Vicente González Ruiz Computer Architecture and Electronics Department, University of Almería, 04120 Almería, Spain, e-mail: vicente.gonzalez.ruiz@gmail.com

Panagiotis Sidiropoulos Centre for Research and Technology Hellas, Informatics and Telematics Institute, 6th Km Charilaou-Thermi Road, 57001 Thermi, Greece; Faculty of Engineering and Physical Sciences, Center for Vision, Speech and Signal Processing, University of Surrey, Guildford, Surrey GU2 5XH, UK, e-mail: psid@iti.gr

Thomas Sikora Communication Systems Group, Technische Universität Berlin, Sekr. EN1, Einsteinufer 17, 10587 Berlin, Germany

Walter Stechele Institute for Integrated Systems, Technische Universität München, Arcisstrasse 21, Munich, Germany, e-mail: Walter.Stechele@tum.de

Florian Stegmaier Chair of Distributed Information Systems, University of Passau, Passau, Germany, e-mail: stegmai@dimis.fim.uni-passau.de

Aleksandar Stojanovic DEA—Facoltà di Ingegneria, University of Brescia, Studio N. 5, Via Branze 38, 25123 Brescia, Italy, e-mail: stojanovic@ient.rwth-aachen.de

Claudia Tonoli Department of Information Engineering, University of Brescia, Via Branze 38, 25123 Brescia, Italy, e-mail: claudia.tonoli@ing.unibs.it

Isabel Trancoso INESC-ID Lisboa, Rua Alves Redol 9, Lisboa 1000-029, Portugal; IST/UTL, Rua Alves Redol 9, Lisboa 1000-029, Portugal, e-mail: Isabel.Trancoso@inesc-id.pt

Michael Unger RWTH Aachen University, Aachen, Germany

Veronica Vilaplana Technical University of Catalonia, Barcelona, Spain, e-mail: veronica.vilaplana@upc.edu

Yifan Zhou Laboratoire Bordelais de Recherche en Informatique (LaBRI), CNRS (UMR 5800), Université Bordeaux 1, 351, cours de la Libération, 33405, Talence cedex, France, e-mail: yifanzhou67@yahoo.fr

Part I
Multimedia Content Analysis

Chapter 1
On the Use of Audio Events for Improving Video Scene Segmentation

Panagiotis Sidiropoulos, Vasileios Mezaris, Ioannis Kompatsiaris, Hugo Meinedo, Miguel Bugalho and Isabel Trancoso

Abstract This work deals with the problem of automatic temporal segmentation of a video into elementary semantic units known as scenes. Its novelty lies in the use of high-level audio information, in the form of audio events, for the improvement of scene segmentation performance. More specifically, the proposed technique is built upon a recently proposed audio-visual scene segmentation approach that involves the construction of multiple scene transition graphs (STGs) that separately exploit information coming from different modalities. In the extension of the latter approach presented in this work, audio event detection results are introduced to the definition of an audio-based scene transition graph, while a visual-based scene transition graph is also defined independently. The results of these two types of STGs are subsequently combined. The results of the application of the proposed technique to broadcast videos demonstrate the usefulness of audio events for scene segmentation

---

P. Sidiropoulos · V. Mezaris (✉) · I. Kompatsiaris
Information Technologies Institute, Centre for Research and Technology Hellas,
6th Km Charilaou-Thermi Road, Thermi 57001, Greece
e-mail: bmezaris@iti.gr

I. Kompatsiaris
e-mail: ikom@iti.gr

P. Sidiropoulos
Faculty of Engineering and Physical Sciences, Center for Vision, Speech and Signal Processing,
University of Surrey, Guildford, Surrey GU2 5XH, UK
e-mail: psid@iti.gr

H. Meinedo · M. Bugalho · I. Trancoso
INESC-ID Lisboa, Rua Alves Redol 9, Lisboa 1000-029, Portugal
e-mail: hugo.meinedo@l2f.inesc-id.pt

M. Bugalho · I. Trancoso
IST/UTL, Rua Alves Redol 9, Lisboa 1000-029, Portugal
e-mail: mmfb@l2f.inesc-id.pt

I. Trancoso
e-mail: Isabel.Trancoso@inesc-id.pt

N. Adami et al. (eds.), *Analysis, Retrieval and Delivery of Multimedia Content*,
Lecture Notes in Electrical Engineering 158, DOI: 10.1007/978-1-4614-3831-1_1,

and highlight the importance of introducing additional high-level information to the scene segmentation algorithms.

Keywords Video analysis · Scene segmentation · Audio events · Scene transition graph

1.1 Introduction

Video temporal decomposition into elementary semantic units is an essential preprocessing task for a wide range of video manipulation applications, such as video indexing, non-linear browsing, classification, etc. Video decomposition techniques aim to partition a video sequence into segments, such as shots and scenes, according to semantic or structural criteria. Shots are elementary structural segments that are defined as sequences of images taken without interruption by a single camera [1]. On the other hand, scenes are often defined as Logical Story Units (LSU) [2], i.e., as a series of temporally contiguous shots characterized by overlapping links that connect shots with similar content. Figure 1.1 illustrates the relations between different kinds of temporal segments of a video.

Early approaches to scene segmentation focused on exploiting visual-only similarity among shots [2, 3], to group them into scenes. In [3], the Scene Transition Graph (STG) was originally presented. The Scene Transition Graph method exploits the visual similarity between key-frames of video shots to construct a connected graph, whose cut-edges constitute the set of scene boundaries. Another recent uni-modal scene segmentation technique [4] uses spectral clustering to conduct shot grouping, without taking into account temporal proximity. Subsequently, the clustering outcome is used for assigning class labels to the shots, and the similarity between label sequences is used for identifying the scene boundaries.

In the last years, several scene segmentation methods that exploit both the visual and auditory channel have been developed, including [5–8]. In [5], a fuzzy k-means algorithm is used for segmenting the auditory channel of a video into audio segments, each belonging to one of 5 classes (silence, speech, music etc.). Following the assumption that a scene change is associated with simultaneous change of visual and audio characteristics, scene breaks are identified when a visual shot boundary exists within an empirically-set time interval before or after an audio segment boundary. In [6], visual information usage is limited to the stage of video shot segmentation. Subsequently, several low-level audio descriptors (i.e., volume, sub-band energy, spectral and cepstral flux) are extracted for each shot. Finally, neighboring shots whose Euclidean distance in the low-level audio descriptor space exceeds a dynamic threshold are assigned to different scenes. In [7], audio and visual features are extracted for every visual shot and serve as input to a Support Vector Machines (SVM) classifier, which decides on the class membership (scene-change / non-scene-change) of every shot boundary. However, this requires the availability of sufficient training data. Although audio information has been shown in these and other pre-

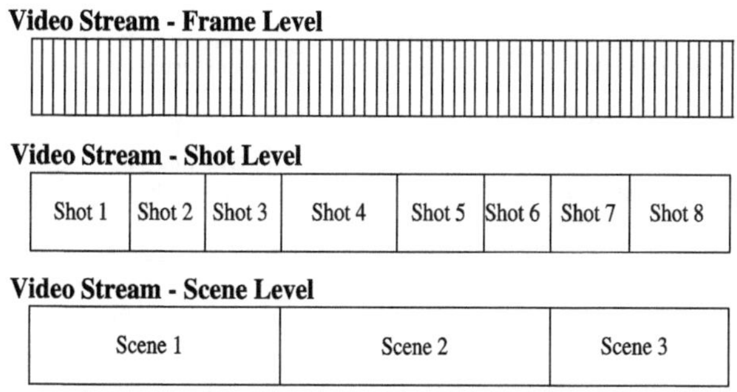

Fig. 1.1 Video stream decomposition to frames, shots and scenes

vious works to be beneficial for the task of scene segmentation, higher-level audio features such as speaker clustering or audio event detection results are not frequently exploited. In a recent work [8], the use of audio scene changes and automatic speech recognition (ASR) transcripts together with visual features is proposed; audio scene changes are detected using a multi-scale Kullback-Leibler distance and low-level audio features, while latent semantic analysis (LSA) is used for calculating the similarity between temporal fragments of ASR transcripts. In [9], the combined use of visual features and some high-level audio cues (namely, speaker clustering and audio background characterization results) for constructing scene transition graphs was proposed.

In this work, this definition of the scene as a Logical Story Unit is adopted and the method of [9] is extended in order to exploit richer high-level audio information. To this end, a large number of audio event detectors is employed, and their detection scores are used for representing each temporal segment of the audio-visual medium in an audio event space. This representation together with an appropriate distance measure is used, in combination with previously exploited high-level audio (e.g. speaker clustering results) and low-level visual cues, for constructing a combination of different scene transition graphs (Multi-Evidence STG—MESTG) that identifies the scene boundaries. The rest of the chapter is organized as follows: an overview of the proposed approach is presented in Sect. 1.2. Audio event definition and the use of audio events in representing video temporal segments are discussed in Sects. 1.3 and 1.4, while Sect. 1.5 presents the proposed MESTG approach. Experimental results are presented in Sect. 1.6 and conclusions are drawn in Sect. 1.7.

1.2 Overview of the Proposed Approach

Scene segmentation is typically performed by clustering contiguous video shots; the proposed MESTG approach is no exception to this rule. Thus, scene segmentation starts with the application of the method of [1] for generating a decomposition S of

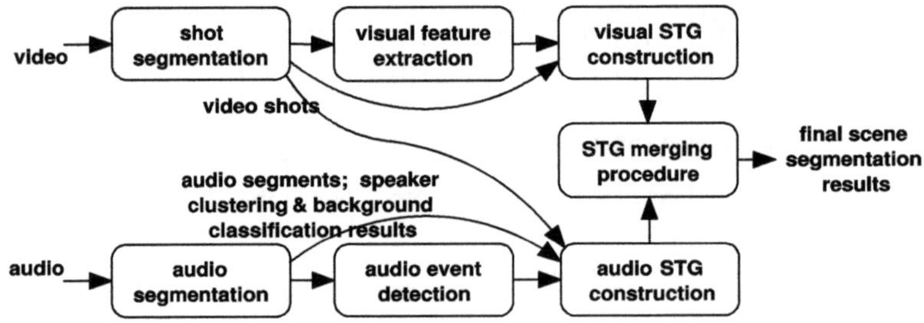

Fig. 1.2 Overview of the proposed scene segmentation scheme

the video to visual shots,

$$S = \{s_i\}_{i=1}^{I}. \tag{1.1}$$

Subsequently, as illustrated in Fig. 1.2, visual feature extraction is performed. Audio segmentation, which includes, among others, speaker clustering and background classification stages [10] [11], is also performed in parallel. This audio segmentation process results in the definition of a partitioning of the audio stream,

$$\mathcal{A} = \{\alpha_x\}_{x=1}^{X}, \alpha_x = [t_x^1, t_x^2], \tag{1.2}$$

where t_x^1 and t_x^2 are the start- and end-times of audio segment α_x. For each α_x, the speaker identity of it and its background class are also identified during audio segmentation; we use $\sigma(\alpha_x)$ to denote the speaker identity of α_x, if any, and $\beta(\alpha_x)$ to denote its background class. Audio event detection, as discussed in detail in the following section, is also performed. Using the resulting features, i.e.,

- HSV histograms of shot key-frames,
- Speaker clustering results,
- Audio background classification into one of three categories (noise, silence, music),
- Detection results (confidence values) for a multitude of audio events,

the proposed MESTG method proceeds with the definition of two types of scene transition graphs (audio STG, visual STG) and a procedure for subsequently merging their results.

1.3 Audio Events

For the purpose of scene segmentation, let us define an audio event as a semantically elementary piece of information that can be found in the audio stream of a video. Telephone ringing, dog barking, music, child voice, traffic noise, explosions are only at few of a wide range of possible audio events. As can be deduced from the audio event definition, more than one audio events may coexist in one temporal segment and may even temporally overlap with each other. For example, in a shot where a person stands by a street and talks, several speech- and traffic-related audio events are expected to coexist.

It is intuitionally expected that taking into account audio event detection results may contribute to improved video scene segmentation. This is based on the reasonable assumption that the presence of the same audio event in more than one adjacent or neighboring audio segments may be a good indication of their common scene membership. On the contrary, the presence of completely different audio events in adjacent temporal segments may be a good indication of their different scene membership, which reveals the presence of a scene boundary.

The first step in testing the validity of the above assumptions is the definition of a number of meaningful audio events and of appropriate methods for their detection. This work integrates two different sets of audio events. Different detection methodologies are used for each set.

1.3.1 Audio Segmentation

The first set includes the type of audio event that is dealt with by an audio segmentation (or diarization) module. Audio segmentation can mean many different things. In this chapter, we restrict its meaning to the type of segmentation that can be performed on the audio signal alone, without taking into account its linguistic contents. This type of segmentation can be done in several tasks. Acoustic Change Detection (ACD) is the task responsible for the detection of audio locations where speakers or background conditions have changed. Speech/Non-Speech (SNS) classification is responsible for determining if the audio contains speech or not (i.e., it results in a binary classification of the audio signal to either Speech or Non-Speech). Gender Detection (GD) distinguishes between male and female gender speakers (i.e., given a speech segment, it results in a binary classification of it to either Male or Female); however, an age-directed segmentation can be also useful as part of the gender detection task, for detecting children voices for instance. Background Conditions (BC) classification indicates whether the background audio signal (i.e, the audio signal, excluding any speech that may be part of it) is clean (: nothing is heard), musical (: music is heard), or noisy. Speaker Clustering (SC) identifies all the speech segments produced by the same speaker. Speaker Identification (SID) is the task of recognizing the identity of certain often recurring speakers, such as news anchors or very important personalities, by their voice, based on a classifier that is trained specifically for such speakers of interest (similarly in principle to how face recognition algorithms can be trained to identify specific people of interest, e.g. a particular political figure, by their faces). More recently, the term speaker diarization (SD) became synonymous to segmentation into speaker-homogeneous regions, answering the question "Who spoke when?". Altogether, 14 different events are automatically detected by this audio segmentation module (ex: Male Voice, Voice with Background Noise,

Music, etc.) [12]. The list is included in Table 1.1. Note, however, that this figure does not include the information provided by the Speaker Clustering component on the cluster identity, which is also exploited for scene segmentation in this work.

The audio segmentation components are mostly model-based, making extensive use of feed-forward fully connected Multi-Layer Perceptrons (MLPs) that are trained with the back-propagation algorithm. All the classifiers (realizing tasks SNS, GD, BC, and SID, as defined above) share a similar architecture: a MLP with 9 input context frames of 26 coefficients (12th order Perceptual Linear Prediction (PLP) plus energy and deltas), two hidden layers with 250 sigmoidal units each and the appropriate number of softmax output units (one for each class), which can be viewed as giving a probabilistic estimate of the input frame belonging to that class. Despite the Acoustic Change Detection and Speech/Non-speech blocks being conceptually different, they were implemented simultaneously in the SNS component, considering that a speaker turn is most often preceeded by a small non-speech segment. The output of the SNS MLP classifier is smoothed using a median filter, and processed by a finite-state machine, involving confidence and duration thresholds. When a speaker change is detected, the first t_{sum} frames of that segment are used to calculate gender, background conditions, and speaker identification classifications (e.g. anchors). Each classifier computes the decision with the highest average probability over all the t_{sum} frames. The Speaker Clustering component, which uses an online leader-follower strategy, tries to group all segments uttered by the same speaker. The first t_{sum} frames (at most) of a new segment are compared with all the same-gender clusters found so far. Two SC components are used in parallel (one for each gender). A new speech segment is merged with the cluster with the lowest distance, provided that it falls below a predefined threshold. The distance measure for merging clusters is a modified version of the Bayesian Information Criteria [11]. Our latest addition to the audio segmentation module is a telephone bandwidth detector. Given the lack of a large manually labeled corpus, a bootstrapping approach has been adopted in which a simple Linear Discriminant Analysis (LDA) classifier has been trained with a small amount of manually labeled data in order to generate automatic transcriptions for the posterior development of a binary MLP classifier. The adopted feature set consisted of 15 logarithmic filter bank energies extracted at a frame rate of 20 ms with a time shift of 10 ms, and corresponding deltas.

The background classifier was initially trained with only broadcast news data that had very limited examples of music and noisy backgrounds, and were inconsistently labeled in terms of these conditions. This motivated the development of alternative classifiers with extended training data reflecting a wide variety of conditions. The related detectors are: Music, Vocal Music, Non-Vocal Music and Speech (another speech detector, using multi-layer perceptrons, also exists, corresponding to the People Talking event).

The new Gaussian mixture models (GMMs) included 1,024 mixtures, and were trained using a different set of features (Brightness, Bandwidth, Zero Crossing Rate, Energy, Audio Spectrum Envelope and Audio Spectrum Centroid), extracted from 16 kHz audio, with 500 ms windows and 10 ms step. Silences were removed from

Table 1.1 List of the 14 audio-segmentation related events

Child Voice	Female Voice	Male Voice
Speech	Voice with Background Noise	Voice with Background Music
Music	Non-Vocal Music	Vocal Music
Clean Background	Noise Background	Music Background
Telephone Band	People Talking	

the audio. Four models were trained: World, Speech, Non-Vocal Music, and Vocal Music.

Each of the GMM models was used to retrieve log likelihood values for each frame. Frame confidence values were calculated by dividing the log likelihood values for each model by the sum of all log likelihood values for all four models. The Vocal, Non-Vocal and Speech models were used for the Vocal Music, Non-Vocal Music and Speech event detectors. The Music detector is the sum of the confidence values for the Vocal and Non-Vocal models. Segment confidence values were obtained by averaging the frame confidence values.

1.3.2 Finer Discrimination of Noisy Events

The second set of events targets a finer discrimination of noise-like sounds, such as Dog Barking, Siren, Crowd Applause, Explosion, etc. [13]. The greatest difficulty in building automatic detectors for this type of event is the lack of corpora manually labeled in terms of these events. This motivated the adoption of a very large sound effect corpus for training, given that it is intrinsically labeled, as each file typically contains a single type of sound. The corpus includes approximately 18,700 files with an estimated total duration of 289.6 h, and was provided by one of the partners in the VIDIVIDEO project (B&G).[1] The list of 61 events for which this corpus provided enough training material is shown in Table 1.2.

Most of the training files have a sampling rate of 44.1 kHz. However, many were recorded with a low bandwidth (<10 kHz). This motivated a uniform downsampling to 16 kHz. This corpus was used to train one-against-all detectors for each concept by building concept-specific and world models. Our initial set of detectors was SVM-based, and the experiments were made using the LIBSVM toolkit [14]. Preliminary experiments compared the performance of a limited set of features: Perceptual Linear Prediction (PLP) or Mel-Frequency Cepstral Coefficients (MFCC) coefficients (19 + energy + deltas), Zero Crossing Rate (ZCR), brightness, and bandwidth. The latter are, respectively, the first and second order statistics of the spectrogram, and they roughly measure the timbre quality. The world model was build using between 92 and 96 files, of which an average of 31 were used as the development set. As a

[1] Netherlands Institute for Sound and Vision, http://www.instituut.beeldengeluid.nl/

Table 1.2 List of 61 additional audio events corresponding to noise-like sounds

Airplane Engine Jet	Airplane Engine Propeller	Animal Hiss
Baby Whining or Crying	Bear	Bell Electric
Bell Mechanic	Big Cat	Birds
Bite Chew Eat	Bus	Buzzer
Car	Cat Meowing	Chicken Clucking
Child Laughing	Cow	Crowd Applause
Digital Beep	Dog Barking	Dolphin
Donkey	Door Open or Close	Drink
Elephant or Trumpet	Electricity	Explosion
Fire	Fireworks	Frog
Glass	Gun Shot Heavy	Gun Shot Light
Hammering	Helicopter	Horn Vehicle
Horse Walking	Insect Buzz	Insect Chirp
Moose or Elk or Deer	Morse Code	Motorcycle
Paper	Pig	Rattlesnake
Saw Electric	Saw Manual	Sheep
Siren	Telephone Ringing Bell	Telephone Ringing Digital
Thunder	Traffic	Train
Typing	Walk or Run or Climb Stairs (Hard)	Walk or Run or Climb Stairs (Soft)
Water	Whistle	Wind
Wolf or Coyote or Dog Howling		

starting point, analysis windows of 0.5 s with 0.25 s overlap were adopted. Three different kernels were considered for the SVM (linear, polynomial and radial basis function (RBF)). Overall, the best results were obtained with the latter kernel. The difference between the performance of MFCC and PLP coefficients was not significant.

As a result of the event detection process discussed in this and the previous section, a total of 75 audio events are defined and, based on the output of the corresponding detectors, a vector EV,

$$EV = [ev(1), ev(2), \ldots, ev(J)], \quad J = 75, \tag{1.3}$$

of confidence values is extracted and stored for each audio segment.

1.3.3 Audio Event Detection Performance

The Audio Segmentation components were tuned to the Broadcast News (BN) domain, which justifies the evaluation of their performance in a test set of six 1-h long

BN shows. The classification error rate of the SNS and GD blocks are comparable to the state of the art: 4.7 and 2.4 %, respectively.[2] As explained, the BC results could not be considered reliable as the manual labels unfortunately lacked consistency.

The speaker clustering performance for news anchors shows very good results due to the SID models (4.1 % Diarization Error Rate (DER)). For the other speakers the results are not so good (26.0 % DER). In part, these results can be attributed to the long duration of the BN shows, which have an average of 64 different speakers per news show, and also to the very large percentage of speech with loud background noise, mainly from street interviews.

The telephone bandwidth classifier was not evaluated in this data set, which did not include telephone data labels. The rate of correctly classified frames in the validation data set, obtained by the LDA classifier, was 99.8 %. In other BN test sets, the rate achieved by the MLP was lower, which we also attributed to the high variability of the training data.

For the audio events of the second set, the performance was first evaluated in terms of F-measure, in a development set of sound effects. The results were generally very good (above 0.8). The worst results were obtained with Door, Fireworks, Hammering, and Saw Manual. The performance with real-life data (movies, documentaries, talk shows and broadcast news), however, is much more challenging than the classification of isolated events. The worse performance can often be due to the fact that audio events almost never occur separately, being corrupted by music, speech, background noise and/or other audio events.

1.4 Audio Event-Based Segment Representation and Similarity Evaluation

For enabling the effective representation of temporal segments in the audio event space, and the evaluation of segment dissimilarity on the basis of audio events, two tasks are necessary: the normalization of the extracted audio event vectors, and the definition of an appropriate event vector distance measure.

Audio event vector normalization is motivated by the diversity of the distributions of confidence values among different event detectors for a given video. This is in part due to the differences in the actual frequency of appearance of different events within the video. For example, in a video with a female narrator speaking throughout the entire video and a thunder-like sound being heard in just a couple of shots, it is expected that the "female voice" audio event will receive very high confidence values in many shots, while the "thunder" audio event is likely to receive high or moderate confidence values in just the shots where the thunder-like sound is heard, and even lower values in all others. However, the high or moderate confidence values that the latter audio event receives should be considered as a strong indication in favor of

[2] A recent version of the GD component achieved the first place in the Interspeech 2010 Paralinguistic Challenge in the category of Male/Female/Child classification [15].

those shots' common scene membership. In order for them to receive the due attention during scene segmentation, the normalization of confidence values depending on their distribution for each audio event is proposed, and a very simple (most likely non-optimal) normalization approach is adopted in this work. Specifically, if $ev(j)$ is the initial confidence value of the jth audio event in a temporal segment, and $maxev_j$ is the maximum value of the jth audio event in all the temporal segments of the video, then the normalized confidence value $\tilde{ev}(j)$ is:

$$\tilde{ev}(j) = \frac{ev(j)}{maxev_j}. \tag{1.4}$$

Following event vector normalization, the definition of a shot dissimilarity measure is based on the assumption that not only the difference of audio event confidence values between two segments, but also the absolute confidence values themselves, are important. Indeed, if for a given audio event two segments present similarly low confidence values, the only deduction that can be made is that this audio event is most probably not present in both segments; no conclusion can be drawn on the semantic similarity of these two segments. On the contrary, if two segments present similarly high confidence values, then it can be inferred that the same audio event is present in both segments, and this concurrence reveals a significant semantic similarity. The commonly used L1 distance or other Minkowski distances would not satisfy the above requirements, since they depend only on the difference of the confidence values. Instead of them, a variation of the Chi-test distance is employed in this work. If $\tilde{EV}_1$, $\tilde{EV}_2$ are two normalized audio event vectors, then their distance D is defined as:

$$D(\tilde{EV}_1, \tilde{EV}_2) = \sqrt{\sum_{j=1}^{J} \frac{(\tilde{ev}_1(j) - \tilde{ev}_2(j))^2}{\tilde{ev}_1(j) + \tilde{ev}_2(j)}}. \tag{1.5}$$

It can be seen that this dissimilarity measure does not depend only on the difference of the audio event vectors, satisfying the previously discussed dissimilarity measure requirements.

1.5 Multi-Evidence Scene Transition Graph Method

1.5.1 Audio STG Definition

The definition of the ASTG is based on the following assumptions:

- Scene boundaries are a subset of the visual shot boundaries of the video (i.e., a visual shot cannot belong to more than one scenes).
- Each audio segment cannot belong to more than one scenes. The same holds for a set of temporally consecutive audio segments that share the same $\sigma(\cdot)$, $\beta(\cdot)$

values and exhibit similar audio events. Two audio segments are said to exhibit similar audio events if the distance between their audio event vectors, as defined in Sect. 1.4, is lower than an empirical threshold.

- Audio event similarity and the distribution of speaker identities across two shots (or two larger temporally contiguous video segments) can serve as measures of audio similarity.

Based on these assumptions, an ASTG is constructed as follows (Fig. 1.3):

- Step 1. The similarity of temporally adjacent audio segments α_x, α_{x+1} is examined, starting from α_1. Denoting $\tilde{EV}_x$, $\tilde{EV}_{x+1}$ the audio event vectors of α_x, α_{x+1} respectively, the two audio segments are merged if $\sigma(\alpha_x) = \sigma(\alpha_{x+1})$, $\beta(\alpha_x) = \beta(\alpha_{x+1})$, and $D(\tilde{EV}_x, \tilde{EV}_{x+1}) < T_{ev}$, where T_{ev} is an empirically defined threshold. For simplicity, the audio segments resulting from this merging step and used in the next step continue to be denoted α_x.
- Step 2. Merging of visual shots is performed: for every α_x, the visual shots that temporally overlap with it by at least T_a ms are merged to a video unit.
- Step 3. The video units formed in Step 2 are clustered according to the dissimilarity $\Delta(\cdot)$ of their speaker identity distributions and the distance $D(\cdot)$ of their audio event vectors. The two dissimilarity measures are linearly combined to produce a one-dimensional distance measure. Assignment of two video units to the same cluster requires both this distance measure and the temporal distance between them to be lower than certain thresholds.
- Step 4. A connected graph is formed, in which the nodes represent the clusters of video units and a directed edge is drawn from a node to another if there is a shot included the first node that immediately precedes any shot included in the second node [3, 9]. The collection of cut-edges, i.e., the edges that, if removed, result in two disconnected graphs, constitutes the set of estimated video scene boundaries.

It should be noted that the speaker identity distribution of a video unit is:

$$H_x = [h_1, h_2, \ldots, h_G], \tag{1.6}$$

where G is the total number of speakers in the video, according to the speaker clustering results, and h_g, $g = 1, \ldots, G$, is defined as the time that speaker g is active in the video unit divided by the total duration of the same video unit. The $L1$ metric is used as similarity function $\Delta(H_x, H_y)$.

1.5.2 Visual STG Definition

Similarly to ASTG, a scene transition graph based on visual information (VSTG) is defined. The VSTG comprises nodes, which contain a number of visually similar and temporally neighboring shots, and edges which represent the time evolution of the story. Visual similarity of shots is evaluated by calculating the Euclidean

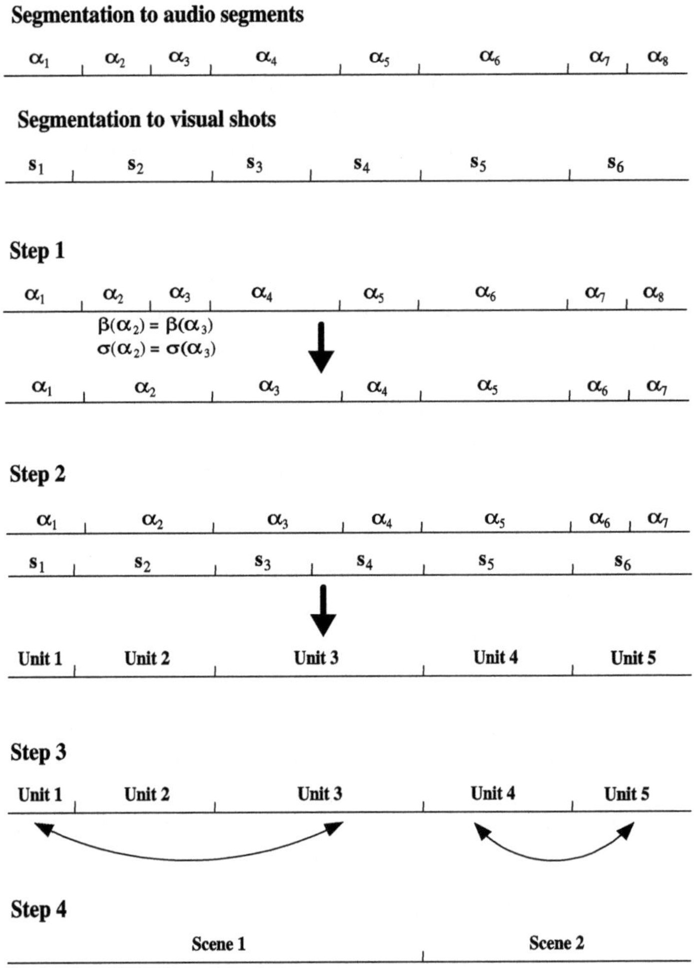

Fig. 1.3 An example of ASTG construction according to the algorithm of Sect. 1.5.1. The video stream is initially decomposed into 8 audio segments (a_1 to a_8) and 6 video shots (s_1 to s_6). Firstly, the audio segments that are adjacent and present same background class, speaker identity and also similar audio events are merged (a_2 and a_3). Subsequently, in Step 2 shots s_3 and s_4, which overlap with audio segment a_3 by more than a threshold, are merged into a video unit. On the contrary, shots s_1 and s_2 are not merged, since the overlapping of s_2 with audio segment a_1 is minimal. In the third step, speaker identity distributions and audio event vectors are estimated for each video unit and their dissimilarity is used to determine which video units should be assigned to the same cluster (Unit 1 and Unit 3 are assigned to the same cluster; Unit 4 and Unit 5 are also assigned to a single cluster). Finally, the scene transition graph is constructed and as a result, in this example, the video units are joined to form 2 scenes

distance of HSV-histogram vectors of shot key-frames. More details on the visual scene transition graph can be found in [3].

1.5.3 Visual and Audio Scene Transition Graph Merging

In [9] we introduced a probabilistic scene transition graph merging approach that combines the visual and audio STGs and simultaneously reduces the dependency of the proposed approach on STG construction parameters. Similarly to this approach, in this work multiple VSTGs are created, each using a different randomly selected set of parameter values. Then, the fraction p_i^v of VSTGs that identify the boundary between shots s_i and s_{i+1} as a scene boundary (i.e., the number of such VSTGs, divided by the total number of generated VSTGs) is calculated and used as a measure of our confidence on this being a scene boundary, based on visual information. The same procedure is followed for audio information using multiple ASTGs, resulting in confidence values p_i^a. Subsequently, these confidence values are linearly combined to result in an audio-visual confidence value p_i:

$$p_i = V \cdot p_i^v + U \cdot p_i^a. \tag{1.7}$$

Finally, all shot boundaries for which p_i exceeds a threshold form the set of scene boundaries estimated by the proposed MESTG approach. In the above formula, U and V are global parameters that control the relative weight of the ASTGs and VSTGs in the audio-visual scene boundary estimation.

1.6 Experimental Results

For experimentation, a test-set of 7 documentary films (229 min in total) from the collection of B&G was used. Application of the shot segmentation algorithm of [1] to this test-set and manual grouping of the shots to scenes resulted in 237 ground truth scenes. For evaluating the results of the proposed and other scene segmentation techniques, the Coverage and Overflow measures, proposed in [16] for scene segmentation evaluation, were employed. Coverage measures to what extent frames belonging to the same scene are correctly grouped together, while Overflow evaluates the quantity of frames that, although not belonging to the same scene, are erroneously grouped together. More detailed definitions of these two measures can be found in [16]. The optimal values for Coverage and Overflow are 100 and 0 % respectively. The F-score is defined in this work as the harmonic mean of C and $1 - O$, to combine Coverage and Overflow in a single measure,

$$F = \frac{2C(1-O)}{C + (1-O)}, \tag{1.8}$$

where $1 - O$ is used in the above definition instead of O to account for 0 being the optimal value of the latter, instead of 1.

Table 1.3 Performance evaluation of MESTG and comparison with literature works

Method	VSTG [3]	[5]	[4]	AVSTG [9]	MESTG
Coverage (%)	79.18	77.93	70.13	83.86	**85.75**
Overflow (%)	17.81	13.88	21.93	11.05	**10.71**
F-Score	80.66	81.82	73.89	86.33	**87.48**

Table 1.4 Performance evaluation of 4 different audio STG variations in the documentary database

Method	SP1	SP2	SPAE1	SPAE2
Coverage (%)	58.7	67.14	58.86	**69.29**
Overflow (%)	**20.32**	26.67	28.77	31.72
F-Score	67.6	**70.1**	64.46	68.78
Coverage (%)	78.5	83.86	84.43	**85.75**
Overflow (%)	10.73	11.05	11.53	**10.71**
F-Score	83.54	86.33	86.41	**87.48**

The first part of the table reports the performance of each variation when used by itself for scene segmentation. The second part reports the overall performance when each variation is combined with the visual STG as described in Sect. 1.5.3

Using the above test-set and measures, the proposed approach (MESTG) was compared with the audio-visual scene segmentation technique (AVSTG) of [9], the methods of [4, 5], and the visual scene transition graph (VSTG). For constructing the latter, the required parameter values were chosen by experimentation, as in [3]. For the MESTG and AVSTG approaches, the probabilistic merging procedure discussed in Sect. 1.5.3 was followed, involving the creation of 1,000 ASTGs and 1,000 VSTGs with different parameters for estimating the required probability values. Weights V, U of (1.7) were tuned with the use of least squares estimation and one video manually segmented into scenes; the resulting values were 0.482 and 0.518 respectively. The results of experimentation are shown in Table 1.3, where it can be seen that the use of audio events in MESTG leads to an increase of Coverage by 1.89 % and a decrease of Overflow by 0.34 %, compared to the AVSTG. The MESTG approach also significantly outperforms the methods of [3–5].

Furthermore, we have compared four different alternatives for constructing the audio scene transition graph. The first one (SP1) uses only the speaker identity distribution, while omitting Steps 1 and 2 of the ASTG construction algorithm of Sect. 1.5.1. The other 3 variations use the proposed ASTG construction algorithm and differentiate only in terms of the considered audio descriptors. Specifically, SP2 makes use only of speaker identity distribution (1.6), whereas SPAE1 additionally employs the 14 audio events of Table 1.1. Finally, in SPAE2 the ASTG is built as proposed in this work, i.e, it exploits the speaker identity distribution, the 14 audio events of Table 1.1 and the 61 audio events of Table 1.2.

In the experimentation we examined the results of these variations both when they are used by themselves for scene segmentation and when each of them is com-

Fig. 1.4 A scene segmentation example. In each row, video shots are represented by one keyframe. According to the ground truth segmentation, all depicted shots belong to a single scene, related to field sprays in which shots of airplanes spraying interchange with farmers talking. It can be seen that the VSTG alone erroneously detects 3 scene boundaries, i.e., a scene boundary is declared in all shot boundary positions where the visual signal changes significantly, providing that there is no repetitive pattern (e.g. the same person re-appearing, as is the case with the second of the two farmers shown above). AVSTG cannot fully remedy this over-segmentation, whereas MESTG manages to assign all 8 shots to the same scene, making use of the common airplane sound that is found in all shots in which a speaker is not included

bined with the visual STG, using the merging approach of Sect. 1.5.3. It should be noted that the combination of SP2 and the visual STG leads to the technique that is proposed in [9] (AVSTG), while the combination of SPAE2 and the visual STG results in the MESTG, presented in this work. The results of experimentation are shown in Table 1.4. It can be seen that none of the audio segmentation techniques can provide adequate scene segmentation accuracy when used in isolation. However, when combined with the visual STG, the additional improvement that each portion of the audio information contributes to can be seen by comparing the results of the last row of Table 1.4. Specifically, the proposed approach is shown to outperform the other 3 variations by at least 1.07 % when used along with the VSTG. Finally, as it is shown in Table 1.4, omitting Steps 1 and 2 of the ASTG construction algorithm reduces the system performance by 2.79 %.

In Figs. 1.4 and 1.5 two examples of the outcome of MESTG, AVSTG and VSTG are shown. In contrary to MESTG, both the VSTG and AVSTG approaches fail to cluster all shots into a single scene in these examples.

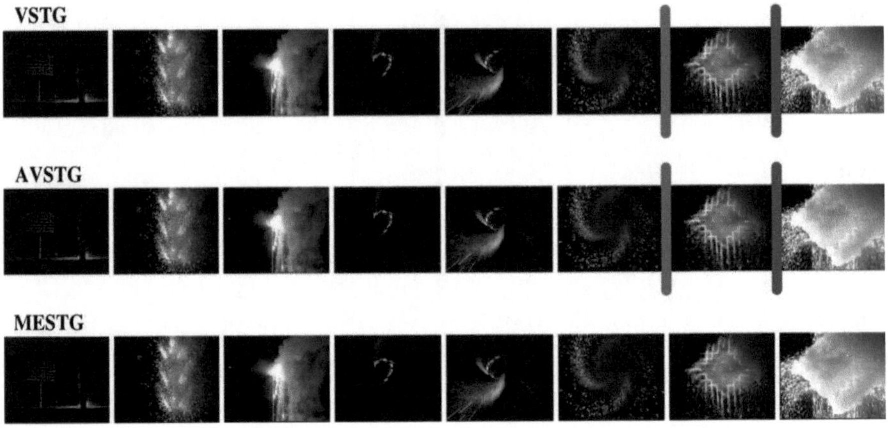

Fig. 1.5 A scene segmentation example. In each row, video shots are represented by one keyframe. These correspond to part of a single scene, formed by shots from a firework contest. No speech is contained in this part of the video; the audio content is limited to the sounds caused by the fireworks. As can be seen, both VSTG and AVSTG fail to recognize that the 7th and 8th shot also belong to the same scene with the rest of the shots, due to the fact that these are neither very similar in terms of appearance nor can be linked to the same speakers, in the absence of speech. On the contrary, MESTG manages to cluster all shots into a single scene, again demonstrating the significance of non-speech-related audio events

1.7 Conclusions

In this work the use of high-level audio events for the improvement of scene segmentation performance was examined, and a multi-modal scene segmentation technique exploiting audio events and other audio-visual information was proposed. The proposed technique was shown to outperform previous approaches that did not exploit high-level audio events. Future extensions of this work include experimentation with additional measures for evaluating similarity in the audio event space, and the use of additional audio events, as well as other high-level audio-visual information, for further improving the accuracy of the results.

Acknowledgments This work was supported by the European Commission under contracts FP6-045547 VIDI-Video and FP7-248984 GLOCAL.

References

1. Tsamoura E, Mezaris V, Kompatsiaris I (2008) Gradual transition detection using color coherence and other criteria in a video shot meta-segmentation framework. In: Proceedings of IEEE international conference on image processing, workshop on multimedia information retrieval (ICIP-MIR 2008), pp 45–48

2. Hanjalic A, Lagendijk RL, Biemond J (1999) Automated high-level movie segmentation for advanced video-retrieval systems. IEEE Trans Circ Syst Video Technol 9(4):580–588
3. Yeung M, Yeo BL, Liu B (1998) Segmentation of video by clustering and graph analysis. Comput Vis Image Understand 71(1):94–109
4. Chasanis V, Likas A, Galatsanos N (2009) Scene detection in videos using shot clustering and sequence alignment. IEEE Trans Multimed 11(1):89–100
5. Nitanda N, Haseyama M, Kitajima H (2005) Audio signal segmentation and classification for scene-cut detection. In: Proc IEEE Int Symp Circ Syst 4:4030–4033
6. Chianese A, Moscato V, Penta A, Picariello A (2008) Scene detection using visual and audio attention. In: Proceedings of Ambi-Sys workshop on ambient media delivery and interactive television
7. Wilson K, Divakaran A (2009) Discriminative genre-independent audio-visual scene change detection. In: Proceedings of SPIE conference on multimedia content access: algorithms and systems III, vol 7255
8. Wang J, Duan L, Liu Q, Lu H, Jin J (2008) A multimodal scheme for program segmentation and representation in broadcast video streams. IEEE Trans Multimed 10(3):393–408
9. Sidiropoulos P, Mezaris V, Kompatsiaris I, Meinedo H, Trancoso I (2009) Multi-modal scene segmentation using scene transition graphs. In: Proceedings of ACM Multimedia, pp 665–668
10. Amaral R, Meinedo H, Caseiro D, Trancoso I, Neto J (2007) A prototype system for selective dissemination of broadcast news in European Portuguese. EURASIP J Adv Sig Proces 2007:1–11
11. Meinedo H (2008) Audio pre-processing and speech recognition for Broadcast News. PhD thesis, IST, Technical University of Lisbon
12. Trancoso I, Pellegrini T, Portelo J, Meinedo H, Bugalho M, Abad A, Neto J (2009) Audio contributions to semantic video search. In: Proceedings of IEEE international conference on multimedia and expo, pp 630–633
13. Bugalho M, Portelo J, Trancoso I, Pellegrini T, Abad A (2009) Detecting audio events for semantic video search. In: Proceedings of interspeech 2009
14. Chang CC, Lin CJ (2001) LIBSVM: a library for support vector machines. http://www.csie.ntu.edu.tw/~cjlin/libsvm
15. Meinedo H, Trancoso I (2010) Age and gender classification using fusion of acoustic and prosodic features. In: Proceedings of Interspeech 2010
16. Vendrig J, Worring M (2002) Systematic evaluation of logical story unit segmentation. IEEE Trans Multimed 4(4):492–499

Chapter 2
Region-Based Caption Text Extraction

Miriam Leon, Veronica Vilaplana, Antoni Gasull and Ferran Marques

Abstract This chapter presents a method for caption text detection. The proposed method will be included in a generic indexing system dealing with other semantic concepts which are to be automatically detected as well. To have a coherent detection system, the various object detection algorithms use a common image description, a hierarchical region-based image model. The proposed method takes advantage of texture and geometric features to detect the caption text. Texture features are estimated using wavelet analysis and mainly applied for *text candidate spotting*. In turn, *text characteristics verification* relies on geometric features, which are estimated exploiting the region-based image model. Analysis of the region hierarchy provides the final caption text objects. The final step of *consistency analysis for output* is performed by a binarization algorithm that robustly estimates the thresholds on the caption text area of support.

Keywords Text detection and localization · Binary Partition Tree

M. Leon (✉) · V. Vilaplana · A. Gasull · F. Marques
Technical University of Catalonia,
Barcelona, Spain
e-mail: mleon@tsc.upc.edu

V. Vilaplana
e-mail: veronica.vilaplana@upc.edu

A. Gasull
e-mail: antoni.gasull@upc.edu

F. Marques
e-mail: ferran.marques@upc.edu

N. Adami et al. (eds.), *Analysis, Retrieval and Delivery of Multimedia Content*,
Lecture Notes in Electrical Engineering 158, DOI: 10.1007/978-1-4614-3831-1_2,

2.1 Introduction

Semantic image indexing relies on the annotation of the presence in the scene of some a priori defined semantic concepts. At a first level of abstraction, semantic concepts are commonly associated with objects. However, object detection is a non-solved problem in a general framework and the extraction of the text in the scene can provide additional relevant information for semantic scene analysis [1]. This is specially true for caption text which is usually synchronized with the scene content. Caption text is artificially superimposed on the video at the time of editing and it usually underscores or summarizes the video content. This makes caption text particularly useful for building keyword indexes [2]. For example, when recognizing a given location (for example, a street), in addition to the information obtained by recognizing the buildings in the image, a caption text associating the scene with a given city may help to confirm the location.

As proposed in [3], text detection algorithms can be classified in two categories: those working on the compressed domain and those working on the spatial domain. Independently of their domain, algorithms can be divided into three phases: (i) *text candidate spotting*, where an attempt to separate text from background is done; (ii) *text characteristics verification*, where text candidate regions are grouped to discard those regions wrongly selected; and (iii) *consistency analysis for output*, where regions representing text are modified to obtain a more useful character representation as input for an OCR. In this chapter, we develop the three phases of the algorithm within the context of caption text.

The caption text detector presented in this work will be included in a more generic indexing system. Actually, the global application is that of off-line enrichment of the current annotation of very large video databases (for instance, the whole repository of TV broadcasters) as well as of creation and instantiation of new descriptors for future annotation of new semantic concepts (for example, searching in the database for a person who previously did not require being explicitly annotated).

Two of the requirements imposed by this application are (i) analysis of the video at the temporal resolution provided by the key frames that are currently stored and (ii) use of an image representation and description which compacts all the scene information in a small number of elements and, at the same time, is as generic as possible, so that the representation can be reused in different contexts (for example, to detect other objects) [4].

Given the first constraint, we concentrate on the problem of caption text extraction in still images. Caption text presents some features that are typically used by text extraction algorithms. The horizontal intensity variations produced by the text are exploited in techniques that analyze the image in the transform domain, either using the DCT [5] or the wavelet transform [6]. Also spatial domain techniques take advantage of this feature by proposing edge detectors to spot the areas with high probability of containing text [7]. Next, spatial cohesion features, such as size, fill factor, aspect ratio or horizontal alignment, are applied to check if text candidate regions are consistent with its neighborhood and to discard false positives [8].

Note that all these techniques are specific for text detection and commonly independent of the approaches dealing with the detection of other semantic concepts. In the case of detecting text in a global indexing system, it is interesting to have a common image representation and a common set of descriptors.

Regarding the image representation, region-based image representations provide a simplification of the image in terms of a reduced number of representative elements, which are the regions. In a region-based image representation, objects in the scene are obtained by the union of regions in an initial partition. To reduce the number of possible region unions, it is useful to create a hierarchy of regions representing the image at different resolution levels. The idea is to have not only a single partition but a universe of partitions representing the image at various resolutions. In this context, object detection algorithms (and specifically text detection algorithms) only need to analyze the image at those positions and scales that are proposed by the regions in the hierarchy [4].

In a previous work, the tree of maxima (and minima) [9] was proposed as hierarchical region-based image model for text detection [10]. Nevertheless, in order to reuse the representation to detect other objects, the Binary Partition Tree (BPT) [11] is used in this work since its suitability for generic object detection was illustrated in [4] and, posteriorly, demonstrated in [12] for the case of various semantic objects of different nature such as human faces, sky regions, traffic signals and car plates.

Given these requirements, we proposed in [13] a method for caption text extraction in still images using a hierarchical region-based image representation. Here, improvements for the first two phases (*text candidate spotting* and *text characteristics verification*) and a solution for the third phase (*consistency analysis for output*) are proposed.

The presentation of these concepts is structured as follows. Section 2.2 summarizes the main ideas behind the image model [11] and its use for object detection and, specifically, text detection [4]. In Sect. 2.3, the region-based caption text detection approach is detailed. This section is structured in three sections in which every phase of the text detector is described. Section 2.3.1 discusses the use of wavelet information to spot the text candidates in the image [6]. The use of the Haar transform in the color domain is proposed to extract text candidates with low contrast in the luminance component. In Sect. 2.3.2, geometrical descriptors are used to confirm the spotted candidates and discard false positives [8]. In that case, we take advantage of the region-based representation to estimate the geometrical descriptors [13] and of the hierarchical image description to obtain the best set of text caption representatives. In turn, Sect. 2.3.3 describes the proposal for the final *consistency analysis for output* step. It is performed by an adaptive binarization algorithm that robustly estimates the thresholds on the area of support of the caption text candidate and provides the final input to the OCR. Section 2.4 discusses the results obtained by this technique. Finally, conclusions are drawn in Sect. 2.5.

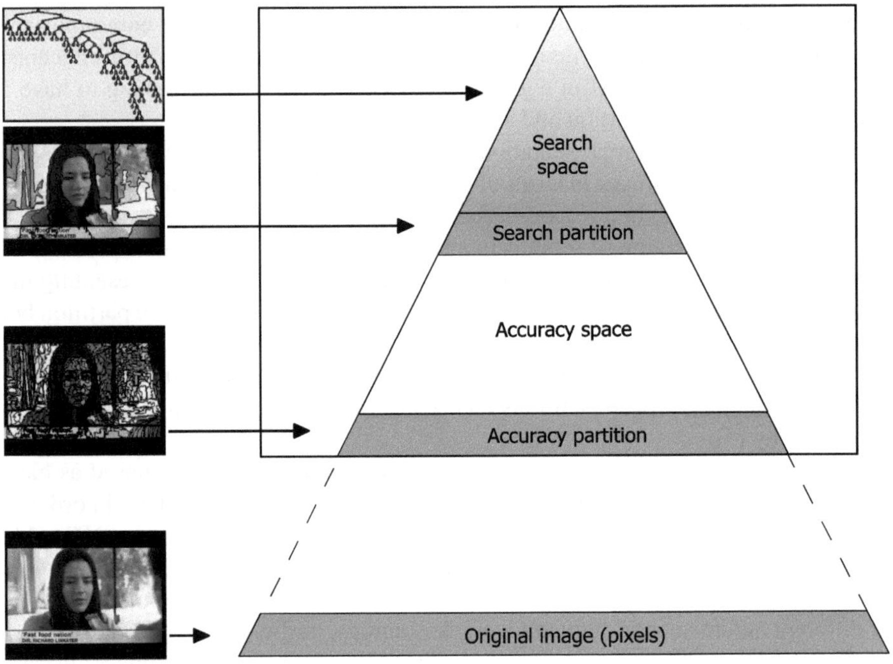

Fig. 2.1 Region-based hierarchical representation

2.2 Hierarchical Region-Based Image Model

The Binary Partition Tree (BPT) [11] reflects the similarity between neighboring regions. It proposes a hierarchy of regions created by a merging algorithm that can make use of any similarity measure. Starting from a given partition, the region merging algorithm proceeds iteratively by (1) computing a similarity measure for all pair of neighbor regions, (2) selecting the most similar pair of regions and merging them into a new region and (3) updating the neighborhood and the similarity measures. The algorithm iterates steps (2) and (3) until all regions are merged into a single region. The BPT stores the whole merging sequence from an initial partition to the one-single region representation. The leaves in the tree are the regions in the initial partition. A merging is represented by creating a parent node (the new region resulting from the merging) and linking it to its two children nodes (the pair of regions that are merged).

The BPT represents a set of regions at different scales of resolution and its nodes provide good estimates of the objects in the scene. Using the BPT representation in object detection, the image has to be analyzed only at the positions and scales that are proposed by the BPT nodes. Therefore, the BPT can be considered as a means of reducing the search space in object detection tasks.

The initial partition can be made of individual pixels or flat zones, which produce a very large BPT. In object detection applications, the use as initial partition of a very accurate partition with a fairly high number of regions is more appropriate [4]. Since this partition is used to ensure an accurate object representation, it is called the *accuracy partition* (see Fig. 2.1). Moreover, in the context of object detection, it is useless to analyze very small regions because they cannot represent meaningful objects. As a result, two zones are differentiated in the BPT: the accuracy space providing preciseness to the description (lower scales) and the search space for the object detection task (higher scales). A way to define these two zones is to specify a point of the merging sequence starting from which the regions that are created are considered as belonging to the search space. The partition that is obtained at this point of the merging process is called the *search partition* (see Fig. 2.1).

In the case of caption text detection, text bars are assumed to be the objects to be detected, and they are extracted by the analysis of the search space. In turn, in the case of scene text detection, characters are not always found in the search partition. This detection requires a more accurate image representation and it will be performed analyzing nodes in the accuracy space. Scene text detection will not be detailed since it is not under the scope of the work presented in this chapter.

2.3 Caption Text Detection Approach

Caption text can be described as text added inside a rectangular bar, horizontally aligned, which contrasts strongly with the bar background and has textured aspect. These features are commonly translated into two types of descriptors: texture and geometric descriptors which are typically used for *text candidate spotting* and *text characteristic verification*, respectively.

Textured areas can be detected using wavelet analysis. However, this approach produces many false positives (that have to be filtered out using geometric descriptors) and some misses in low contrast areas. On the other hand, given the generic framework of our application, the BPT has been created combining color homogeneity and contour complexity criteria [4]. Due to their homogeneous background and regular shape, caption text objects are likely to appear as single nodes in the BPT. Hence, we propose to combine the two approaches.

In a first stage, texture is estimated over the whole image by means of a multiresolution analysis using a Haar wavelet decomposition. Texture information is used to select highly textured regions (candidate regions) in the BPT. Candidate regions are anchor points for caption text detection. In order to correctly estimate useful descriptors to evaluate candidate regions, their area of support is conformed to the object shape characteristics. Region evaluation is carried out combining region-based texture information and geometric features. Among those nodes in the BPT that pass this text characteristic verification stage, final caption text nodes are selected analyzing the BPT structure.

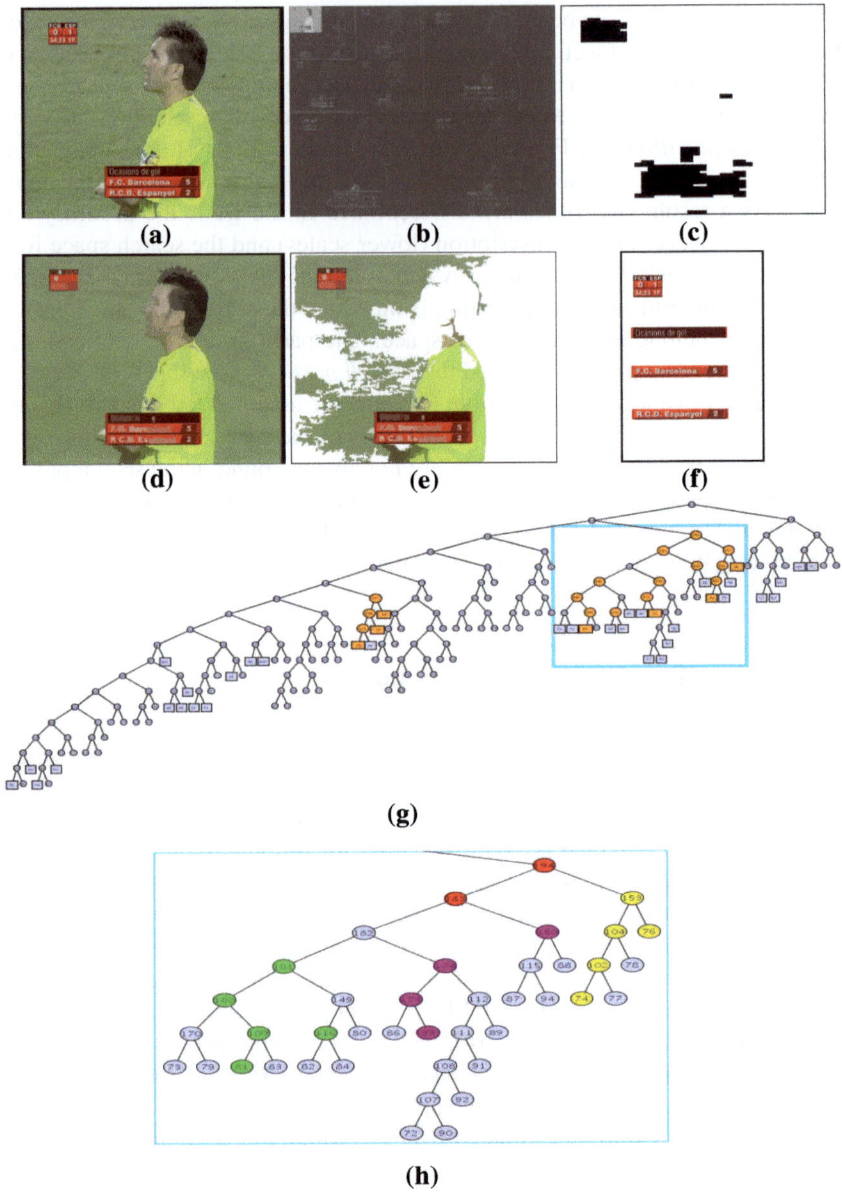

Fig. 2.2 Example of caption text detection. **a** Original image, **b** wavelet transform, **c** text candidate pixels, **d** search partition, **e** text candidate regions, **f** set of final selected regions, **g** BPT showing the selected leaves (*squared nodes*) and the candidate nodes (*orange nodes*), and **h** Detail of the BPT (*rectangle* in **g**) showing the final selected nodes for each text bar (*green*, *lilac and yellow nodes*) and the discarded nodes (*red nodes*)

2.3.1 Text Candidate Spotting

As proposed in [6], texture descriptors such as DWT coefficients give enough information to determine where textured areas can be found in an image. In [13] we proposed to use the power of the LH and HL subbands in a Haar transform (Fig. 2.2b) analyzed over a sliding window of fixed size H × W (W > H to consider horizontal text alignment):

$$P_{LH}^{l}(m,n) = \frac{1}{WH}\sum_{i=1}^{W}\sum_{j=1}^{H} LH^{l}(m+i,n+j)^2, \tag{2.1}$$

where l denotes the decomposition level and an analogous expression is used for P_{HL}^{l}. The window is moved over subbands of the transformed image with an overlapping of half the window size in both directions. Both subbands are analyzed because DWT power in windows containing text should present high values ($>T_1$) in at least one of the two subbands and relevant enough values ($>T_2$) in the other subband. This way, all pixels in a window are classified as text candidates if the power in the window satisfies the following condition:

$$\Big((P_{LH}^{l} > T_1) \wedge (P_{HL}^{l} > T_2)\Big) \vee \Big((P_{LH}^{l} > T_2) \wedge (P_{HL}^{l} > T_1)\Big), \tag{2.2}$$

where T_1 and T_2 are two thresholds, T_1 being more restrictive than T_2 ($T_1 > T_2$).

This wavelet analysis may produce misses in low contrasted areas. In the case of caption text, such misses are commonly related to text over a background with a similar luminance value but whose chrominance values are different enough to be distinguished by the human visual system. Taking into account this observation, the previous technique has been separately applied to the three YCbCr image components.

The final mask marking all the text candidates is obtained by performing the union of the (upsampled) masks at each decomposition level (Fig. 2.2c) and at each image component. For the results presented in Sect. 2.4, the size of the sliding window is 6×18, $l = 2$, $T_{1Y} = 1200$ and $T_{2Y} = 400$ for luminance, and $T_{1CbCr} = 18$ and $T_{2CbCr} = 10$ for chrominance.

Finally, regions in the search partition (Fig. 2.2d) are selected if they contain any text candidate pixel. Moreover, texture-based selection is propagated through the BPT so that all ancestors of the candidate regions are selected as well (Fig. 2.2g). This is a very conservative policy but, at this stage, it is important not to miss any possible region containing text (Fig. 2.2e).

2.3.2 Text Characteristic Verification

For every selected node, descriptors are estimated to verify if the region represents a caption text object. Initially, a region-based texture descriptor is computed as in Eq. (2.1) but now the sum is performed over interior pixels to avoid the influence of wavelet coefficients due to the gradient in the region boundary. This descriptor is mainly used to filter out regions that have been selected due to the presence in the mask (see Fig. 2.2c) of a few wrong candidate pixels in the surroundings of textured areas.

To complete the verification process, geometric descriptors are calculated for the remaining candidate nodes. Before computing these descriptors, the area of support of candidate nodes is modified by a hole filling process and an opening with a small structuring element (typically, 9×9).

This stage is needed to eliminate small leaks that the segmentation process may introduce due to the interlacing or to color degradation between regions. Such leaks result in very noisy contours that bias the geometric descriptor estimation. Finally, since the opening may split the region into several components, the largest connected component is selected as the area of support for computing geometric descriptors. Descriptors and the thresholds that nodes should accomplish (following a restrictive policy) are listed in the sequel. Values in brackets indicate the thresholds used for the experiments presented in Sect. 2.4 for standard PAL format 720×576 images.

- *Rectangularity* (*R*): R must be in the range [0, 1]. The calculation of the rectangularity is done using the Discrepancy method [14]. A rectangle is fitted to the region, and the discrepancies between the rectangle and region are measured. R must be greater than T_R; the nearer to 1, the more similar to a rectangle ($T_R = 0.85$).
- *Aspect ratio*: ($AR = Width_{BB}/Height_{BB}$) must be in the range $[T_{AR_1}, T_{AR_2}]$, the upper limit is not strictly necessary but is useful to discard line-like nodes ($T_{AR_1} = 1.33$, $T_{AR_2} = 20$). Given the regular shape (close to rectangular) of caption text objects, the AR is calculated with the bounding box (BB) of the node area of support.
- *Height*: must be in the range $[T_{H_1}, T_{H_2}]$ ($T_{H1} = 13$ pixels for character visibility and $T_{H_2} = 144$, a quarter of PAL format height).
- *Area*: must be in the range $[T_{A_1}, T_{A_2}]$ ($T_{A_1} = 225$, the area of a node with minimum height and minimum aspect ratio, and $T_{A_2} = 138.240$, a third of the PAL format image area).
- *Compactness*: ($CC = Perimeter^2/Area$) must be smaller than T_{CC}, to avoid nodes with long, thin elongations commonly due to interlacing ($T_{CC} = 800$).

The result of applying these descriptors and thresholds to the image shown in Fig. 2.2a, is presented in Fig. 2.2g, where the verified nodes are marked in orange.

At this stage, verified nodes may present two problems. First, as shown in Fig. 2.2h, several verified nodes may be in the same subtree; that is, several (complete or partial) instances of the same caption text object may be represented in a subtree. Second, if the image contains a collection of caption text bars laying close enough, they may be merged into a single node; that is, a single subtree may represent several caption text objects that, due to their proximity, can be understood as a single one.

The first problem leads to the presence of unnecessary verified nodes, actually representing the same caption text object, that are to be processed in the *consistency analysis for output* step. In that case, the best node in the subtree has to be selected. The straightforward solution of selecting the highest node in the subtree may lead to non-optimal solutions, as discussed in [13]. In that work, a confidence value was estimated for each node, and nodes in the subtree with the highest confidence value were finally preserved.

Nevertheless, a second problem has been detected due to the presence of several caption text objects in the image merged as a single node in the tree, that pass the verification stage. Such configurations are very common, for instance, in sport events, where the data of several participants are jointly presented. In that case, the problem can be more severe due to possible differences in the colors of fonts and backgrounds used in the neighbor caption text bars. If all the bars are selected as a single object, these differences result in a decrease in the performance of the subsequent *consistency analysis for output* step. This step relies on a binarization of the validated caption text bar area of support; if the two classes (character and background) are not homogeneous in color, the binarization may fail.

Having in mind these two problems, we propose here a new strategy to jointly handle both situations in a more robust manner. Subtrees are traversed in postorder. For each subtree, a list of possible caption text objects is created. Verified nodes in the subtree are compared with the previous caption text objects already stored in the list. If the geometrical features of the verified node under analysis allow us to assume that this node belongs to a caption text object already in the list, the verified node under analysis is assigned to this caption text object and the description of this caption text object is updated. If the verified node under analysis cannot be assigned to any already existing caption text object, it is added to the list as a new caption text object.

All these comparisons are performed using only simple geometrical descriptors previously extracted from the tree nodes. In particular, the features that are compared between a verified node under analysis and an already existing caption text object are the coordinates of its center of mass as well as the height and the width of the modified node bounding box. Combining these three elements, the following situations can be detected:

1. The node completes an already existing caption text object: This is the case of a caption text object that is mostly represented by a single node in the BPT but some parts of it (for instance, its interior) are missing. In that case, neither the y-coordinate of the center of mass nor the height or the width of the BB present a substantial change. The node is assigned to this caption text object and the object description is updated.
2. The node horizontally extends an already existing caption text object: This is the case of a caption text object that has been split in the BPT into two horizontal-neighbor regions. The y-coordinate of the center of mass and the height of

the bounding box do not present a substantial change, whereas the width of the bonding box increases. The node is assigned to this caption text object and the object description is updated.

For other situations, the overlap between the node under analysis and the extension of the area of support of the caption text object is analyzed. If they overlap, the node is assumed to be part of the caption text object and its description is updated. If they do not overlap, a new caption text object is defined.

In the example of Fig. 2.2a the largest text box is detected as three separated text bar objects. Figure 2.2h shows a subtree whose root node represents this text box. The search algorithm detects the nodes with the same color as nodes which are part of the same text bar object, obtaining each text bar independently (see Fig. 2.2f).

2.3.3 Consistency Analysis for Output

For every caption text object, a binarization step is carried out. Given the specific characteristics of caption text bars, the binarization is performed by analyzing a few lines in the image. N (typically N = 3) equidistant horizontal line segments are selected within the area of support of the caption text object. The mean and the variance of the pixels in each line segment are computed. Line segments with high variance are assumed to be formed by text and background and are used to characterize the probability density function of the text, which is assumed to be Gaussian.

In turn, low variance line segments are supposed to represent the background and can be used to characterize its probability density function that is assumed to be Gaussian as well. Then, binarization is performed by a Maximum Likelihood approach. An example illustrating this process is presented in Fig. 2.3. As it can be seen (and it will be further discussed in next section), this approach leads to good results. Other binarization approaches have been also tested leading to lower performance.

The output of the binarization method is directly used as input for the OCR system. In this work, we have used the opensource **tesseract-ocr** system[1] which can be trained for a specific language and vocabulary.

As previously commented, the binarization approach assumes that background and text can be statistically modeled by Gaussian probability density functions. This assumption does not stand when, for instance, the various words in a given caption text bar are not homogeneous in color. This situation typically leads to a wrong binarization of some of the words. In order to solve this problem, words within the same caption text bar are segmented and a word-by-word binarization is implemented. The segmentation is carried out by applying first an edge detector (in our case, the Canny edge detector [15] but any other similar system could be used) to the caption text bar

---

[1] http://www.code.google.com/p/tesseract-ocr/

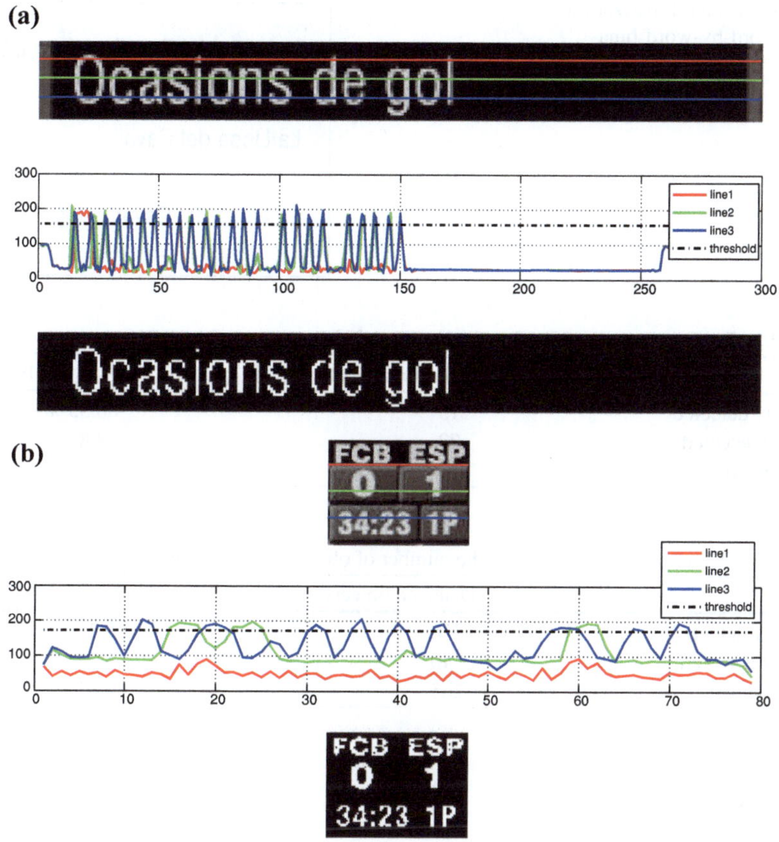

Fig. 2.3 Illustration of the caption textbinarization process for N = 3

area and second by performing a dilation of the detected contours using a rectangular structuring element. Every connected component is assumed to be a separate word and the previous binarization approach is applied.

An example of the usefulness of this word-by-word binarization process is illustrated in Fig. 2.4. The global binarization process fails due to the differences between the text representation in the first and third elements with respect to the second one. The result of the global binarization is illustrated in the second row of Fig. 2.4, where the second text element has been included in the background. The third row of Fig. 2.4 shows the correct result obtained when a word-by-word binarization is used. This example is further illustrated in Fig. 2.6b.

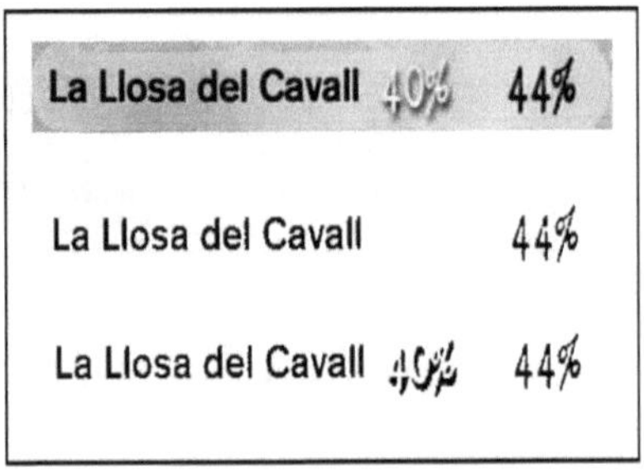

Fig. 2.4 Text bar binarization versus word-by-word binarization

Table 2.1 Detection results related to the number of objects in the first database

	Detected objects	% over 249 objects
Correctly detected	215	86.35 %
Partially detected	22	8.83 %
False negative	12	4.82 %

Table 2.2 Detection results related to the number of objects in the second database

	Detected objects	% over 2063 objects
Correctly detected	1758	85.21 %
Partially detected	40	1.93 %
False negative	265	12.84 %

2.4 Results

The technique has been tested in two corpus, one formed by news and sport event videos, and the other one by sport event videos.[2] In the first corpus, there is a total of 249 caption text objects extracted from a set of 150 challenging images with text of different size and color, and complex background textures. Results, classified as correctly detected, partially detected, false positives and false negatives, are summarized in Tables 2.1 and 2.3, and illustrated in Figs. 2.5 and 2.6.

If these values are expressed in terms of recall and precision, partially detected objects (PDO) can be considered as false negative or as detected objects since they represent good anchor points for the following step (see Table 2.3). The number of false positives is 24. Results do not differ significantly from [13] but text bars are detected separately instead of together in a single text box.

In the second corpus there are 2063 caption text objects extracted from 649 key frames. The most remarkable result is that the number of false positives is very high due to the presence in the images of advertising panels and spectators, whereas the

[2] All images used in this chapter belong to TVC, Televisió de Catalunya, and are copyright protected. These key-frames have been provided by TVC with the only goal of research under the framework of the i3media project.

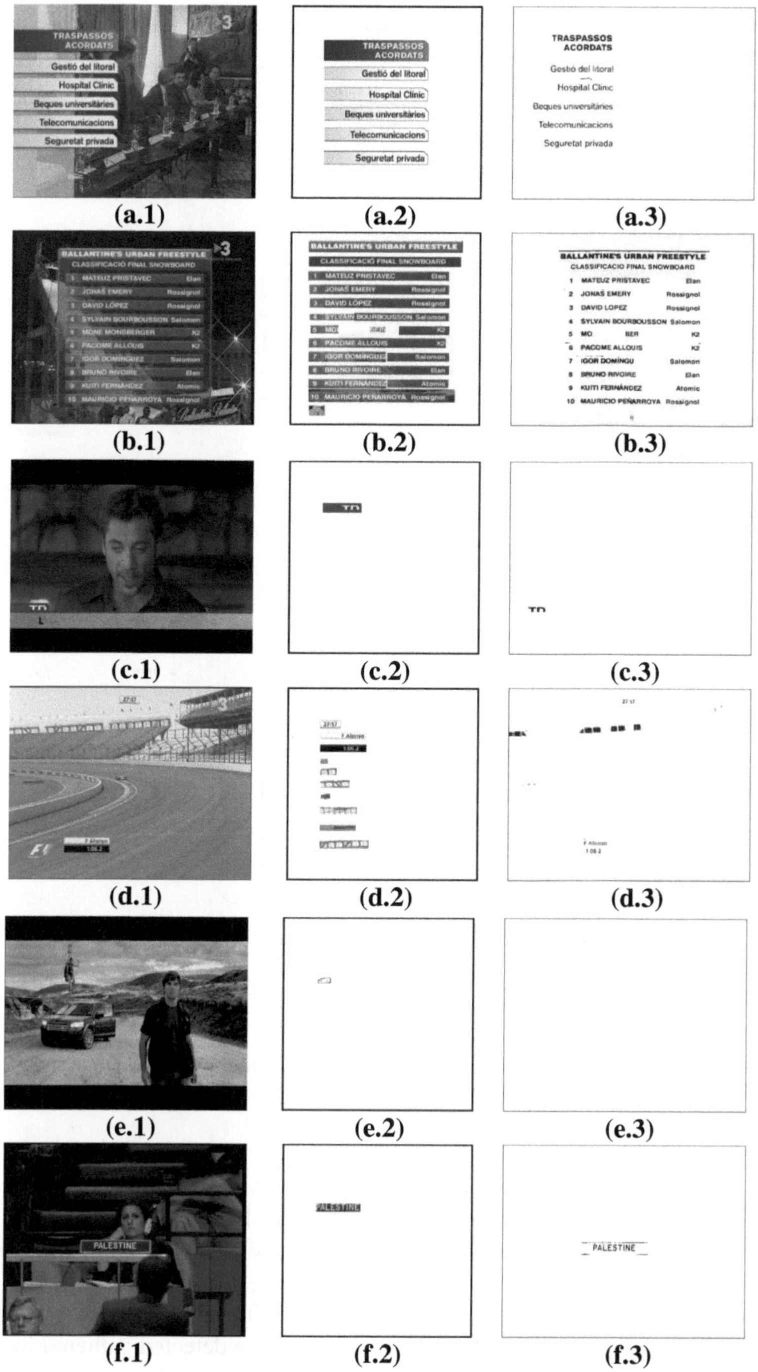

Fig. 2.5 Illustration of the caption text detection process. *First column* Original image; *Second column* List of the final selected regions; *Third column* Binarization result

Table 2.3 Detection results presented as precision and recall for the first and second databases, respectively

PDO	As outlier	As correct	As outlier	As correct
Recall	0.863	0.950	0.8521	0.871
Precision	0.885	0.894	0.692	0.697

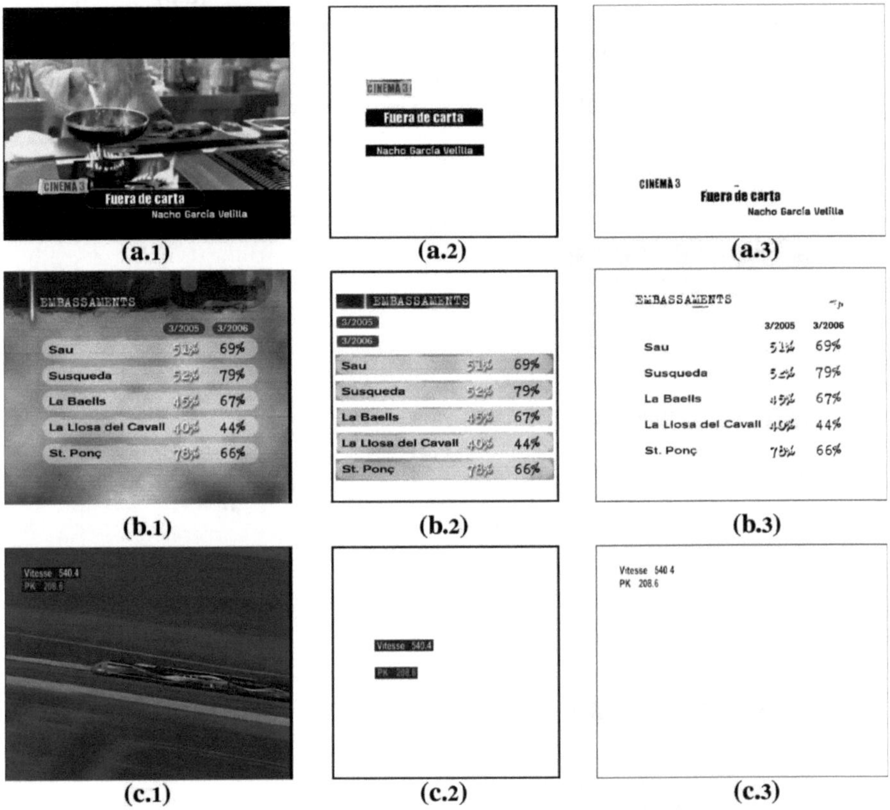

Fig. 2.6 Illustration of the caption text detection process. *First column* Original image; *Second column* List of the final selected regions; *Third column* Binarization result

number of detected text bar is satisfying (see Tables 2.2 and 2.3). Nevertheless, some of these elements are discarded in the third step (see Fig. 2.5d, e).

Figures 2.5 and 2.6 illustrate these results with some images that exemplify the performance and limitations of the algorithm. For every example, we present the different caption text bars that have been detected. To allow analyzing which bars have been detected isolately and which ones have been detected gathered in a single component, when necessary the detected text bars are separately presented regardless of their original position in the image.

Figure 2.5a presents an example of non-perfectly rectangular caption text objects. It is common that caption text objects present some modifications to make information more attractive for the viewer. As it can be observed, the assumed variability on the shape model allows the correct detection of such caption text objects.

Figure 2.5b is an example that illustrates false negatives and partial detections. The similarity between caption text background and objects around mislead the segmentation process and, in some cases, the caption text object is not correctly represented in the BPT. Moreover, we can illustrate as well an example of partial detection: caption text object marked with a "7" has been reported as partial detection since it has not been fully extracted as a single node.

Figure 2.5c is another example of false negative. In this case, although the bar is opaque and rectangular, the amount of text present in the lower caption bar (only a letter "L") does not produce an enough textured region to be detected in the *text candidate spotting* phase (See Sect. 2.3.1). Actually, this false negative is mostly due to a wrong selection of the key frame and in subsequent key frames the whole text in the caption bar appears and is completely detected.

Figure 2.5d, e shows representative examples of typical outliers. These structures correspond to highly textured, rectangular nodes in the BPT which are mistaken by caption text blocks. Nevertheless, they are removed in the following phase of *consistency analysis for output.* Figure 2.5d shows outliers, which are commonly present in sports sequences due to the presence in the image of stands or spectators.

Figure 2.5f shows an example of the behaviour of the algorithm in the presence of scene text. This type of text may present similar characteristics to the caption text (it may be placed in a close-to-rectangular bar and be highly contrasted to its background) and therefore it can be detected as such. Note that in the precision results provided in Table 2.3, 25 % of the detections classified as False Positive are related to scene text.

Figure 2.6a illustrates the robustness of the proposed algorithm to variations in the font type. Note that the algorithm exploits the texture appearance of the text (which is mostly common to all types of fonts) and, therefore, it presents similar performance independently of the font.

Figure 2.6b presents an example of the usefulness of the division of a caption text object into words and their subsequent separated binarization. In this example, the various text elements within each caption object do not share the same color features and, therefore, the global binarization, that assumes a Gaussian probability density function for all the text in the caption object, is wrong. The word-by-word binarization process allows us to correctly binaryze the various text elements.

Finally, Fig. 2.6c shows an image where the use of color information (see Sect. 2.3.1) provides good results. Letters in fluorescent green would be discarded in the *text candidate spotting* phase due to low contrast if only luminance information had been used.

2.5 Conclusions and Further Work

We have presented a new technique for caption text detection. This technique will be included in a global indexing system and, therefore, works on a common hierarchical region-based image representation. The technique combines texture information (through Haar wavelet decomposition) and geometric information (through the analysis of the regions proposed by the hierarchical image model) to robustly extract caption text objects in the scene.

Future work will focus on the creation of new text descriptors and on the analysis of the temporal redundancy of text. The former aims to improve the detection of text in textured areas. The latter, aims to take advantage of the fact that text has to appear at least 2 seconds on the screen so that the viewer can understand the information.

Acknowledgments This work was partially founded by the Catalan Broadcasting Corporation (CCMA) and Mediapro through the Spanish project CENIT-2007-1012 i3media and TEC2007-66858/TCM PROVEC of the Spanish Government.

References

1. Assfalg J, Bertini M, Colombo C, Del Bimbo C (2001) Extracting semantic information from news and sport video. In: Proceedings of the 2nd ISPA, pp 4–11
2. Crandall D, Antani S, Kasturi R (2002) Extraction of special effects caption text events from digital video. Int J Doc Anal Recog 2:138–157
3. Jung K, Kim K, Jain AK (2004) Text information extraction in images and video:a survey. Pattern Recog 37:977–997
4. Vilaplana V, Marqués F, Salembier P (2008) Binary partition trees for object detection. IEEE Trans Image Process 17(11):2201–2216
5. Zhong Y, Zhang H, Jain AK (2000) Automatic caption localization in compressed video. IEEE Trans PAMI 22(4):385–393
6. Li H, Doermann D, Kia O (2000) Automatic text detection and tracking in digital video. IEEE Trans Image Process 9(1):147–155
7. Tekinalp S, Alatan AA (2003) Utilization of texture, contrast and color homogeneity for detecting and recognizing text from video frames. In: IEEE ICIP 2003, Barcelona, Spain
8. Retornaz T, Marcotegui B (2007) Scene text localization based on the ultimate opening. Proc ISMM 1:177–188
9. Salembier P, Oliveras A, Garrido L (1998) Anti-extensive connected operators for image and sequence processing. IEEE Trans Image Process 7(4):555–570
10. Leon M, Mallo S, Gasull A (2005) A tree structured-based caption text detection approach. In: Proceedings of 5th IASTED VIIP, pp 220–225
11. Salembier P, Garrido L (2000) Binary partition tree as an efficient representation for image processing, segmentation and information retrieval. IEEE Trans Image Process 9(4):561–576
12. Vilaplana V, Marques F, Leon M, Gasull A (2010) Object detection and segmentation on a hierarchical region-based image representation. In: Proceedings of the ICIP-10, IEEE international conference on image processing, pp 3393–3396, Hong Kong, China
13. Leon M, Vilaplana V, Gasull A, Marques F (2009) Caption text extraction for indexing purposes using a hierarchical region-based image model. In: IEEE ICIP 2009, El Cairo, Egypt
14. Rosin PL (1999) Measuring rectangularity. Mach. Vis. Appl. 11(4):191–196
15. Canny J (1986) A computational approach to edge detection. IEEE Trans Pattern Anal Mach Intell 8(6):679–698

Chapter 3
k-NN Boosting Prototype Learning for Object Classification

Paolo Piro, Michel Barlaud, Richard Nock and Frank Nielsen

Abstract Image classification is a challenging task in computer vision. For example fully understanding real-world images may involve both scene and object recognition. Many approaches have been proposed to extract meaningful descriptors from images and classifying them in a supervised learning framework. In this chapter, we revisit the classic k-nearest neighbors (k-NN) classification rule, which has shown to be very effective when dealing with local image descriptors. However, k-NN still features some major drawbacks, mainly due to the uniform voting among the nearest prototypes in the feature space. In this chapter, we propose a generalization of the classic k-NN rule in a supervised learning (boosting) framework. Namely, we redefine the voting rule as a strong classifier that linearly combines predictions from the k closest prototypes. In order to induce this classifier, we propose a novel learning algorithm, MLNN (Multiclass Leveraged Nearest Neighbors), which gives a simple procedure for performing prototype selection very efficiently. We tested our method first on object classification using 12 categories of objects, then on scene recognition as well, using 15 real-world categories. Experiments show significant improvement over classic k-NN in terms of classification performances.

Keywords Boosting · k-NN classification · Object recognition · Scene categorization

---

P. Piro (✉) · M. Barlaud
CNRS/University of Nice-Sophia Antipolis, Sophia Antipolis, France
e-mail: paolo.piro@gmail.com

M. Barlaud
e-mail: barlaud@i3s.unice.fr

R. Nock
CEREGMIA/University of Antilles-Guyane, Martinique, France
e-mail: richard.nock@martinique.univ-ag.fr

F. Nielsen
LIX/Ecole Polytechnique, Palaiseau, France
e-mail: nielsen@lix.polytechnique.fr

N. Adami et al. (eds.), *Analysis, Retrieval and Delivery of Multimedia Content*,
Lecture Notes in Electrical Engineering 158, DOI: 10.1007/978-1-4614-3831-1_3,

3.1 Introduction

In this chapter, we address the two main tasks involved in multi-class real-world image classification, i.e. object recognition and scene categorization. The first consists in automatically classifying an unlabeled region extracted from an image (e.g. by segmentation) according to a set of predefined objects. The latter task consists in labeling the overall image according to a set of real-world scenes. Such tasks are very challenging, and are attracting more and more research effort from the computer vision community, as prompted by the plethora of classification approaches proposed for PASCAL 2009 competition.[1] A wide range of image descriptors has been investigated for object categorization purposes, which generally rely on detecting relevant local characteristics of objects (e.g., local shape and appearance). The best known examples of such descriptors are SIFT [13], which are commonly extracted for most state-of-the-art image representations, like into Bags-of-Features (BoF) [21] and Fisher vectors [15].

Despite lots of works, much remains to be done to challenge human level performances. In fact, images carry only part of the information that is used by humans to recognize scenes or objects, and parts of the information available from images may be highly misleading: e.g. real object categories may exhibit high intra-class variability (i.e. visually different objects may belong to the same category) and low inter-class variability (i.e. distinct categories may contain visually similar objects). The same holds for natural scene categories.

Voting classification techniques, like k-nearest neighbors (k-NN), have been shown to be very effective when dealing with local image descriptors. However, they may suffer from high sensitivity to "noisy" prototypes, thus requiring suitable learning procedures for rejecting unreliable matches. Moreover, it is a critical challenge to reduce the computational cost of descriptor matching without impairing classification performances. In order to cope with these issues, the literature has favored two main approaches so far: improve categorization by means of local classifiers [6, 9, 23], or filter out ill-defined examples [3].

In this chapter, we propose a novel solution: a new provable boosting algorithm for k-nearest neighbor (k-NN) rules in a multiclass framework. Our algorithm, called MLNN (Multiclass Leveraged Nearest Neighbors), induces a multiclass leveraged k-nearest neighbor rule that generalizes the uniform k-NN rule, using directly the examples as weak hypotheses. Compared to other local learning methods for k-NN classification [23], MLNN also speeds up query processing: instead of learning a local classifier for each query, MLNN performs learning upwards, once and for all, and does not need to be run again or updated depending on queries. Finally, the most significant advantage of MLNN lies in its ability to find out the most relevant prototypes for categorization, thus enabling to filter out the remaining examples.

In the following section we present MLNN, along with the statement of its theoretical properties (Sect. 3.2). Then, we present and discuss experimental results of

[1] http://pascallin.ecs.soton.ac.uk/challenges/VOC/voc2009/

both object recognition using SIFT descriptors (Sect. 3.3.1) and scene categorization using Bag-of-Features histograms (Sect. 3.3.2).

3.2 Method

3.2.1 Problem Statement and Notations

Instead of splitting the multiclass classification problem in as many *one-versus-all* (two-class) problems—a frequent approach in boosting [19]—we directly tackle the *multiclass* problem, following [22]. For a given query, we compute its *classification score* for all categories (or classes, or labels). Then, we select the label with the maximum score. We suppose given a set $\mathcal{S}$ of m annotated descriptors arising from images (or image regions). Each image descriptor provides a training *example* $(\boldsymbol{x}, \boldsymbol{y})$, where $\boldsymbol{x}$ is the image feature vector and $\boldsymbol{y}$ the *class vector* that specifies the category membership of the descriptor. In particular, the sign of component y_c gives the positive/negative membership of the example to class c ($c = 1, 2, \ldots, C$). Inspired by the multiclass boosting analysis of [22], we constrain the class vector to be *symmetric*, i.e. $\sum_{c=1}^{C} y_c = 0$, by setting: $y_{\tilde{c}} = 1$, $y_{c \neq \tilde{c}} = -\frac{1}{C-1}$, where $\tilde{c}$ is the true image category.

3.2.2 (Leveraged) Nearest Neighbors

The regular k-NN rule is based on majority vote among the k nearest neighbors in set $\mathcal{S}$, to decide the class of query $\boldsymbol{x}$. It can be defined as the following multiclass classifier $\boldsymbol{h} = \{h_c,\ c = 1, 2, \ldots, C\}$:

$$h_c(\boldsymbol{x}) = \frac{1}{k} \sum_{i \sim_k x} [y_{ic} > 0] \ , \tag{3.1}$$

where $h_c \in [0, 1]$ is the classification score for class c, $i \sim_k x$ denotes an example $(\boldsymbol{x}_i, \boldsymbol{y}_i)$ belonging to the k nearest neighbors of $\boldsymbol{x}$ and square brackets denote the indicator function.

In this chapter, we propose to generalize (3.1) to the following *leveraged* k-NN rule $\boldsymbol{h}^\ell = \{h_c^\ell,\ c = 1, 2, \ldots, C\}$:

$$h_c^\ell(\boldsymbol{x}) = \sum_{j \sim_k x} \alpha_j y_{jc} \in \mathbb{R} \ , \tag{3.2}$$

where the uniform voting of (3.1) is replaced by a weighted voting with weighting coefficients α_j. Furthermore, in (3.2) the k nearest neighbors are searched either in

$\mathscr{S}$, or in a sparse subset $\mathscr{P} \subseteq \mathscr{S}$ obtained after a *prototype selection* step, achieved before any query is presented. Each example in $\mathscr{P}$ is a relevant category *prototype*. Prototype selection is achieved by using the leveraging coefficients α_j (3.2) computed at training time, which are expected to represent the "confidence" of prototypes for classifying new data.

In the following sections we describe the boosting-like procedure we propose to compute the α_j's. In particular, we propose to minimize a particular upperbound of the risk functional on training data, thus exploiting a very important trick that has been at the center of major advances in classification over the last ten years.

3.2.3 Multiclass Surrogate Risk Minimization

In order to fit our classification rule (3.2) onto training set $\mathscr{S}$, we focus on the minimization of a multiclass exponential (surrogate[2]) risk:

$$\varepsilon^{\exp}\left(\boldsymbol{h}^{\ell}, \mathscr{S}\right) \doteq \frac{1}{m}\sum_{i=1}^{m} \exp\left(-\frac{1}{C}\sum_{c=1}^{C} y_{ic} h_c^{\ell}(\boldsymbol{x}_i)\right) . \tag{3.3}$$

This function is an upper bound of the *empirical risk*:

$$\varepsilon^{0/1}\left(\boldsymbol{h}^{\ell}, \mathscr{S}\right) \doteq \frac{1}{mC}\sum_{i=1}^{m}\sum_{c=1}^{C} \left[y_{ic} h_c^{\ell}(\boldsymbol{x}_i) < 0 \right] , \tag{3.4}$$

which is not differentiable and often computationally hard to directly minimize [14]. Remark that both risks (3.3, 3.4) depend on quantity $y_{ic} h_c^{\ell}(\boldsymbol{x}_i)$, the *edge* of classifier $\boldsymbol{h}^{\ell}$ on example $(\boldsymbol{x}_i, \boldsymbol{y}_i)$ for class c. This edge is positive iff the category membership predicted by the classifier agrees with the true membership of the example. Plugging definition (3.2) into surrogate risk (3.3) gives:

$$\varepsilon^{\exp}\left(\boldsymbol{h}^{\ell}, \mathscr{S}\right) \doteq \frac{1}{m}\sum_{i=1}^{m} \exp\left(-\sum_{j=1}^{m} \alpha_j \mathrm{r}_{ij}\right) , \tag{3.5}$$

which highlights an essential ingredient of our algorithm, i.e. the *multiclass* k-NN *edge matrix* $[\mathrm{r}_{ij}]_{m \times m}$, whose entry r_{ij} is different from zero iff example j is a neighbor of i, whereas the positive (negative) sign of r_{ij} specifies the membership of the two examples to the same (not the same) class. (See definition (7) in Algorithm 1.) Finally, after computing the edge matrix, which is a constant term in (3.5), the unknown

---

[2] We call *surrogate* a function that upperbounds the risk functional we should minimize, and thus can be used as a primer for its minimization.

leveraging coefficients α_j can be fitted by running the algorithm described in the following section, which iteratively minimizes the surrogate risk.

3.2.4 MLNN: *Multiclass Leveraged k-NN Rule*

Algorithm Pseudocode of MLNN is shown in Algorithm 1. Like common boosting algorithms, MLNN operates on a set of weights w_i $(i = 1, 2, \ldots, m)$ defined over training data. These weights are repeatedly updated based on δ_j, which is a local measure of the class density around a given example j. Namely, at each iteration t of the algorithm, a *weak index chooser* oracle $\textsc{Wic}(\{1, 2, \ldots, m\}, t)$ determines index $j \in \{1, 2, \ldots, m\}$ of the example to leverage (step I.0). Various choices are possible for this oracle. The simplest is perhaps to compute Eq. (3.10, 3.11) for all the training examples, then to pick j maximizing δ_j:

$$j \leftarrow \textsc{Wic}(\{1, 2, \ldots, m\}, t) \;:\; \delta_j = \max_{j \in \{1,2,\ldots,m\}} \delta_j^t \;. \tag{3.6}$$

Furthermore, notice that, when whichever w_j^+ or w_j^- is zero, δ_j in (3.11) is not finite. We propose a simple strategy to eliminate this drawback, inspired by [19], i.e. to add $1/m$ to both the numerator and the denominator of the fraction in the log term of (3.11). This smoothes out δ_j, guaranteeing its finiteness without impairing convergence of MLNN. This oracle allows an example to be chosen more than once, thus letting its leveraging coefficient α_j be updated several times (step I.3). It is known that, to be statistically consistent some boosting algorithms require to be run for $T \ll m$ rounds [1]. Cast in the setting of MLNN, this constraint precisely supports prototype selection, as T is an upperbound for the number of examples with non-zero leveraging coefficients.

Complexity MLNN shares the property with boosting algorithms of being resources-friendly: since computing the leveraging coefficients scales linearly with the number of neighbors, its time complexity bottleneck does not rely on boosting, but on the complexity of nearest neighbor search. Furthermore, its space complexity is also reduced: since weak hypotheses are examples, example j can be a classifier only for its *reciprocal* nearest neighbors—those examples for which j itself is a neighbor—,corresponding to non-zero entries in column j of edge matrix (7). This matrix is thus extremely sparse for reasonable values of k. As a consequence, update rule (3.12) is to be computed on a small number of examples.

Convergence Using known arguments of the boosting theory [14], we proved the convergence of MLNN to the minimum of the surrogate risk, along with a convergence rate, which is based on the following *weak index assumption* (**WIA**):

WIA: let $p_j \doteq w_j^+/(w_j^+ + w_j^-)$. There exist some $\gamma > 0$ and $\eta > 0$ such that the following two inequality holds for index j returned by $\textsc{Wic}(\{1, 2, \ldots, m\}, t)$:

Algorithm 1: MULTICLASS LEVERAGED k- NN MLNN($\mathscr{S}$)

Input: $\mathscr{S} = \{(\boldsymbol{x}_i, \boldsymbol{y}_i), \quad i = 1, 2, \ldots, m, \quad \boldsymbol{y}_i \in \{-\frac{1}{C-1}, 1\}^C\}$

Let $$\mathrm{r}_{ij} \doteq \begin{cases} \frac{1}{C}\sum_{c=1}^{C} y_{ic}y_{jc} & \text{if } j \sim_k i \\ 0 & \text{otherwise} \end{cases} . \quad (9)$$

Let $\alpha_j \leftarrow 0, \quad \forall j = 1, 2, \ldots, m$.

Let $w_i \leftarrow 1/m, \quad \forall i = 1, 2, \ldots, m$.

for $t = 1, 2, \ldots, T$ **do**

[I.0] Weak index chooser oracle:

Let $j \leftarrow$ WIC($\{1, 2, \ldots, m\}, t$);

[I.1] Let

$$w_j^+ = \sum_{i:\, \mathrm{r}_{ij}>0} w_i, \quad w_j^- = \sum_{i:\, \mathrm{r}_{ij}<0} w_i , \tag{3.10}$$

$$\delta_j \leftarrow \frac{(C-1)^2}{C}\log\left(\frac{(C-1)w_j^+}{w_j^-}\right); \tag{3.11}$$

[I.2] Let

$$w_i \leftarrow w_i \exp(-\delta_j \mathrm{r}_{ij}), \quad \forall i : j \sim_k i ; \tag{3.12}$$

[I.3] Let $\alpha_j \leftarrow \alpha_j + \delta_j$.

end

Output: $h_c^\ell(\boldsymbol{x}) = \sum_{j \sim_k x} \alpha_j y_{jc}, \quad \forall c = 1, 2, \ldots, C$.

$$|p_j - \frac{1}{C}| \geq \gamma , \tag{3.7}$$

$$(w_j^+ + w_j^-)/||\boldsymbol{w}||_1 \geq \eta . \tag{3.8}$$

We summarize this fundamental convergence property in the following theorem:

Theorem 3.1 *If the* **WIA** *holds for* $\tau \leq T$ *steps, then* MLNN *converges with* τ *to* $\boldsymbol{h}^\ell$ *realizing the* **global** *minimum of the surrogate risk* (3.3), *and* $\varepsilon^{0/1}(\boldsymbol{h}^\ell, \mathscr{S}) \leq \exp(-\frac{C}{C-1}\eta\gamma^2\tau)$.

Inequality (3.7) is the usual weak learning assumption, used to analyze classical boosting algorithms [7, 19], when considering examples as weak classifiers. A *weak coverage assumption* (3.8) is needed as well, because insufficient *coverage* of the reciprocal neighbors could easily wipe out the surrogate risk reduction due to a large γ in (3.7). For a deeper insight into the properties of our k-NN boosting method, see [17].

Fig. 3.1 Twelve categories from the Caltech-101 database

3.3 Experiments

3.3.1 Object Recognition

In this section we present experimental results of MLNN vs plain k-NN on a database of real objects. The reason to use, and boost, plain k-NN instead of particular sophisticated approaches on nearest neighbors (e.g. spatial pyramids [12] or Bag-of-Features [21]) was more than merely remaining as general as possible. We carried out experiments in order to investigate the improvements brought by boosting on nearest neighbor voting. This necessitates to remove all unnecessary adjustments which could potentially interfere. Namely, we used 12 categories from the well-known Caltech-101 database for object classification: *accordion, airplanes, car side, cellphone, cup, ewer, ferry, grand piano, laptop, motorbikes, watch, Windsor chair* (Fig. 3.1). This database contains a large variety of objects, and also exhibits high intra-class variability, i.e. visually different objects may be in the same category.

3.3.1.1 Training

We used 40 training images per category, and extracted dense SIFT descriptors [13] from image regions corresponding to objects. For this purpose, we used the ground-truth object masks provided with the database. We computed dense descriptors of 16×16 patches over a grid with spacing of 8 pixels, as proposed in [12]. We used the descriptors of all training objects for learning prototypes, i.e. a subset of relevant object descriptors with their leveraging coefficients. Namely, we retained only examples with positive α_j as prototypes for classifying test images. In Fig. 3.2 we show how values of prototype leveraging coefficients are distributed in each category. The best represented categories are those maximizing the integral of such

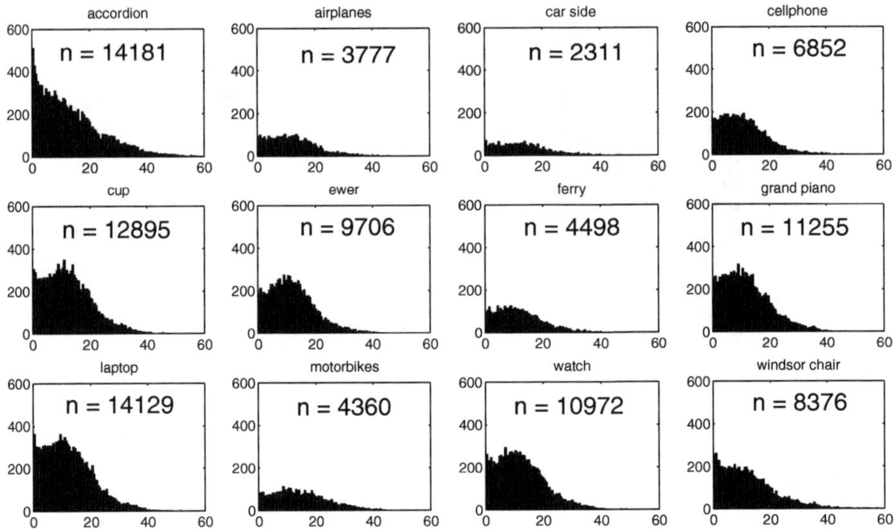

Fig. 3.2 Histograms of prototype leveraging coefficients α_j per category. The overall number of prototypes in each category is reported as well

histograms, i.e. those containing most of the prototypes with the largest coefficient values. We used all non-training images (2,039 overall) to test MLNN.

3.3.1.2 Classification

In order to obtain an overall classification score for a test image $\boldsymbol{X}$, each query descriptor $\boldsymbol{x} \in \boldsymbol{X}$ was first classified independently by our leveraged k-NN rule (3.2). Because prototype classes are highly imbalanced, as displayed in Fig. 3.2, we smoothed out aggregate scores with a standard technique [10]. Hence, we predict label $\hat{c}$ for a query image $\boldsymbol{Q}$ as follows:

$$\hat{c}(\boldsymbol{X}) \doteq \arg\max_c \frac{1}{m_c^{\mathscr{P}}} \sum_{\boldsymbol{x} \in \boldsymbol{X}} h_c^{\ell}(\boldsymbol{x}) \tag{3.13}$$

where $m_c^{\mathscr{P}}$ is the cardinal of retained prototypes of class c. In order to speed up the execution time we used a CUDA GPU implementation of Nearest Neighbor search [8].

Classification results are summarized in Fig. 3.3, where the mean Average Precision (mAP %) over all test images is shown for different prototype sets. We computed mAP as the average of diagonal entries in the confusion table, whereas the size of prototype set is reported as θ, that is the ratio of the number of retained prototypes and the overall size of training data. Fig. 3.3 also reports results of vanilla k-NN

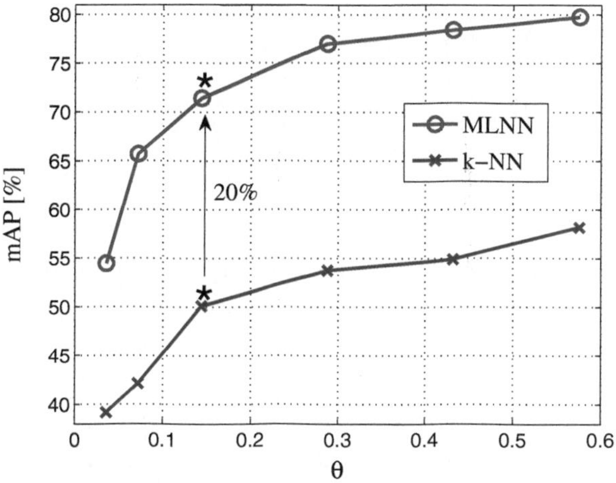

Fig. 3.3 MLNN classification performances in terms of MAP as a function of the proportion θ of retained prototypes

(with random sampling of the prototypes from the training data). We observe that the improvement over regular k-NN is dramatic, even when decreasing the prototype number. For example only 14 % of prototypes allow for 20 % improvement over classic k-NN. (See the marked points in the figure.) Besides this precision improvement, MLNN also enables to drastically reduce the computational complexity with respect to plain k-NN (gain up to a factor 4 when discarding half prototypes).

Finally, the confusion table reported in Fig. 3.4 highlights the difficulty of discriminating between couples of visually similar object categories, like "cup" and "ewer". Moreover, most of mistakes may be due to an insufficient representation of an object category in the prototype set. Namely, categories with few prototype descriptors, like "motorbikes", are more likely to be confused with over-represented categories (e.g. "accordion"). Normalizing the number of prototypes per class, e.g. by adapting the resolution of dense descriptors to the actual object size, is expected to improve classification rate in such categories [12].

3.3.2 Scene Categorization

In this section, we present experimental results of MLNN on the scene categorization task. Namely, we focused on evaluating how the average classification precision varies as a function of the number of prototypes that are used for testing. Indeed, one of the main features of our method is to allow to explicitly fix the number of data to be used at classification time, thus directly bounding the computational cost of the test phase. In particular, in all the reported experiments we carried out prototype selection by setting $T < m$, which corresponds to retaining at most T relevant prototypes (the

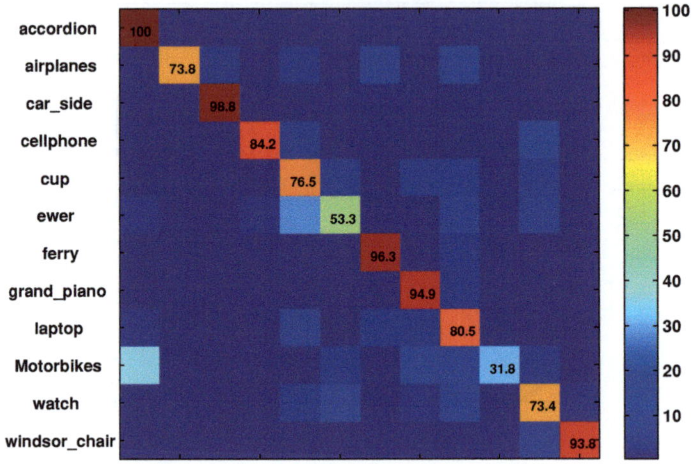

Fig. 3.4 Confusion table when retaining all prototypes with positive leveraging coefficients ($\theta = 0.58, k = 11$)

exact number of prototypes depends on which criterion is chosen for the WIC oracle, namely whether allowing a prototype to be selected at multiple steps or not). When running the baseline k-NN method, we carried out *random prototype selection*, which is the easiest strategy for data reduction, and averaged the classification results over a number of iterations.

We report image categorization results on the 15-cat database [12], which consists of the following 15 categories: coast (360 images), forest (328), mountain (374), open country (410), highway (260), inside of cities (308), tall buildings (356), street (292), suburb residence (241), bedroom (174), kitchen (151), living room (289), office (216), store (315), and industrial (311). In order to represent these images in terms of feature sets, we tested a common descriptor for natural image classification, i.e. Bag-of-Features (BoF) histograms computed from SIFT descriptors [20].

3.3.2.1 Settings

In the context of generic image categorization, the *Bag-of-Features* scheme is among the best performing feature representation methods. Besides its simplicity, the main advantage of this image representation approach is the *one-to-one* association between images and feature vectors, that allows for a straightforward use of discriminative learning tools like k-NN and SVM. Originally proposed for text categorization, the BoF descriptors have been successfully applied to image classification problems, with several implementations, ranging from using them as feature vectors for discriminative learning [5] to more sophisticated approaches like pyramid match kernels [12].

In this section we present and discuss results we obtained using MLNN for Bag-of-Features classification on the 15-cat database, which mixes outdoor scenes with more difficult indoor scenes. All the results presented in this section refer to 10-fold *cross-validation*. Thus we split the 15-cat database in 10 distinct random folds, in order to form 10 different training/test subsets, each one containing more than 4,000 training images and about 450 testing images. The following statistics refer to averaging over these ten folds. For each training/test combination we first built a *vocabulary* of visual words, that are SIFT descriptors densely extracted at four resolution levels on a fixed regular grid (typically 1,000÷10,000 SIFT per image). In order to build the visual vocabulary, we ran k-means with $k = 1500$, thus representing each image as a feature vector in dimension $d = 1500$.

3.3.2.2 Histogram Intersection Metrics

Although BoF descriptors are widely used in most state-of-the-art image classification techniques, some crucial issues are still unsolved and may significantly impact on classification performances. In particular, we consider the two following problems:

1. how to *normalize* such image descriptors in order to make comparison between different images as unbiased as possible;
2. which *distance metric* to use for measuring the dissimilarity between two descriptors.

Such problems mainly arise from the histogram-based nature of BoF descriptors and have been the object of much research effort in the computer vision community in the recent years. Indeed, on the one hand, their *normalization* is particularly crucial when images differ significantly from each other in terms of the local descriptors counts, thus resulting in largely variable descriptor norms. Thus it is common to pre-process BoF descriptors such that they have equal ℓ_1 norms. Less commonly, these descriptors have been ℓ_2-normalized, mostly when normalization is part of a pre-processing technique, like the squared root (sqrt) recently proposed by Perronnin et al [16]. The most common alternative to ℓ_1-normalization for BoF descriptors is represented by the TF-IDF schema, which was originally proposed in the context of text retrieval and then successfully applied to image indexing, in order to take into account the larger informative "power" of rare visual words [20].

On the other hand, defining the right *dissimilarity measure* between histograms (not necessarily ℓ_1-normalized) is challenging, and the resulting behaviour often strongly depends on the application. For example the Euclidean distance between ℓ_2-normalized descriptors [2] or TF-IDF-weighted descriptors [20] are still the most common choices for image classification. All the results we report in this section refer to normalizing Bag-of-Features descriptors in terms of the ℓ_1 norm and comparing them using the Manhattan distance. This choice was motivated by our evaluation of different normalization/metric combinations that we report in Table 3.1, which refer to 10-fold cross-validation using k-NN ($k = 10$). In particular, we tested some of the most suitable histogram distance metrics, as defined in a recent taxonomy

Table 3.1 Comparison of k-NN classification performances using different histogram normalization criteria and intersection metrics on the 15-cat database ($k = 10$)

metric	normalization	mAP
Euclidean	ℓ_1	54.55
Euclidean	none	57.56
Euclidean	ℓ_2	59.60
Manhattan	ℓ_1	**64.29**
Manhattan	none	62.09
Manhattan	ℓ_2	53.19
Canberra	ℓ_1	60.80
Canberra	none	59.57
Canberra	ℓ_2	59.81
Lorentzian	ℓ_1	**64.34**
Lorentzian	none	58.21
Lorentzian	ℓ_2	52.07
Manhattan	tf-idf	**64.47**
Euclidean	tf-idf	55.22
Euclidean	ℓ_2+sqrt	60.78
Euclidean	ℓ_1+sqrt	62.22

of histogram intersection measures [4], i.e. *Manhattan*, *Canberra*, *Lorentzian*, besides the baseline Euclidean distance. Furthermore, we evaluated different descriptor normalization/pre-processing methods, like the common ℓ_1, ℓ_2 and TF-IDF [20], as well as the most recent *squared root* pre-processing [16] and the baseline (absence of normalization). First of all, our results show that the ℓ_1 normalization always outperforms ℓ_2 and the baseline for a fixed metric (except for the Euclidean distance, for which ℓ_2-normalization is the best), and the gap is particularly significant for distances like Manhattan and Lorentzian (more than 11 % over ℓ_2-normalization). The best performances are obtained for Manhattan and Lorentzian with ℓ_1-normalization, and Manhattan with TF-IDF normalization, which still outperform the squared root pre-processing strategy. So as for the distance metric choice, our results clearly show that the Euclidean distance is generally *not* the optimal choice for comparing those histogram-like descriptors, thus suggesting intersection metrics as better alternatives. (See for instance the 10 % gap between Euclidean and Manhattan for ℓ_1-normalized BoF, or the 9 % gap between the same two metrics when using TF-IDF normalization.)

In Fig. 3.5 we quantify the gain provided by using the Manhattan (ℓ_1) distance over the Euclidean (ℓ_2) distance for our MLNN approach.Results of baseline k-NN classification are also shown for both metric distances. This plot shows that the choice of the k-NN metric significantly impacts on the precision of our method, with a 10 % gap between the Euclidean distance implementation and the Manhattan histogram intersection-based implementation.

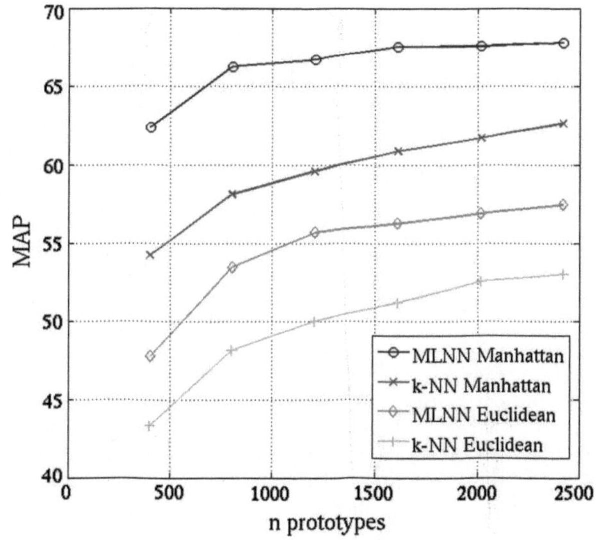

Fig. 3.5 Comparison between Manhattan (ℓ_1) and Euclidean (ℓ_2) distances for both MLNN and k-NN classification

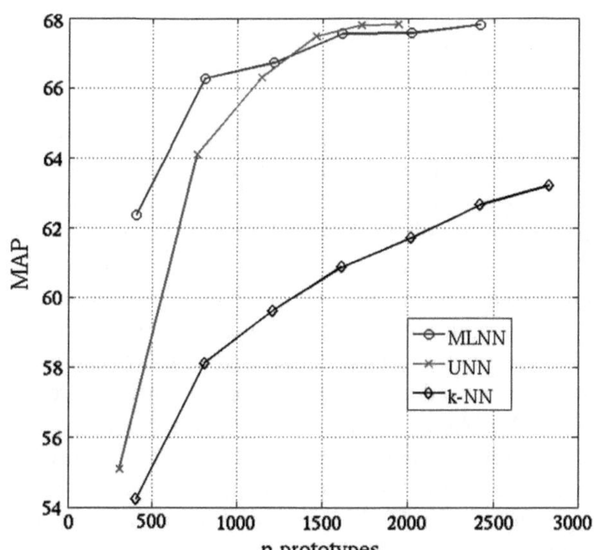

Fig. 3.6 MLNN with Histogram Intersection kernel compared to UNN (one-versus-all) and k-NN. All the three classification methods rely on the same distance metric, that is ℓ_1 distance

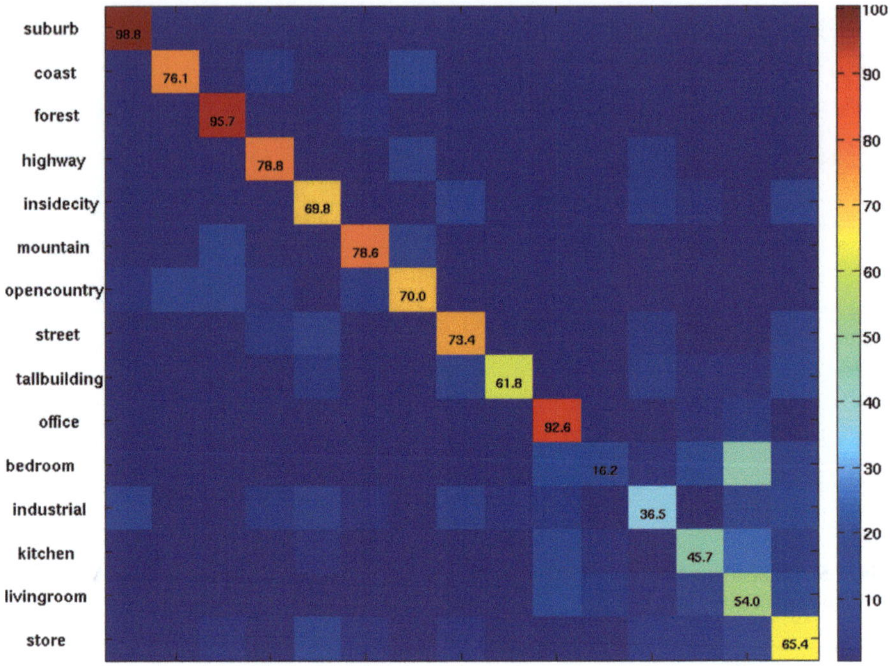

Fig. 3.7 Confusion table for MLNN on the 15-cat database

3.3.2.3 MLNN Performances

In order to evaluate the classification performances of our method, we report the trend of cross-validation mAP as a function of the overall number of prototypes used for testing (Fig. 3.6). We also compare these results, which were obtained using the MLNN implementation relying on the Manhattan distance, with our one-versus-all UNN method [18] and the baseline k-NN classification. Our method outperforms k-NN classification significantly (gap between 6 and 8 %) while reducing the prototype dataset, thus the computational cost, considerably. (See for instance the precision of MLNN for 400 prototypes, which equals that of k-NN using 2,400 prototypes, thus resulting in a dataset reduction by a factor 6.) Furthermore, the advantage of using our multiclass MLNN algorithm over its binary counterpart, UNN, is mainly concentrated at very low prototype set sizes, e.g. 10 % of the original set, where the multiclass learning allows for precision improvement up to 7 %. Notice that the number of prototypes for UNN is reported as the average over the multiple one-versus-all problems, i.e. it should be multiplied by the number of classes in order to compute the actual number of prototypes involved in classification. Hence, although UNN performances appear almost identical to those of MLNN, this latter still benefits a significant computational advantage over UNN, thus providing the best precision/cost trade-off.

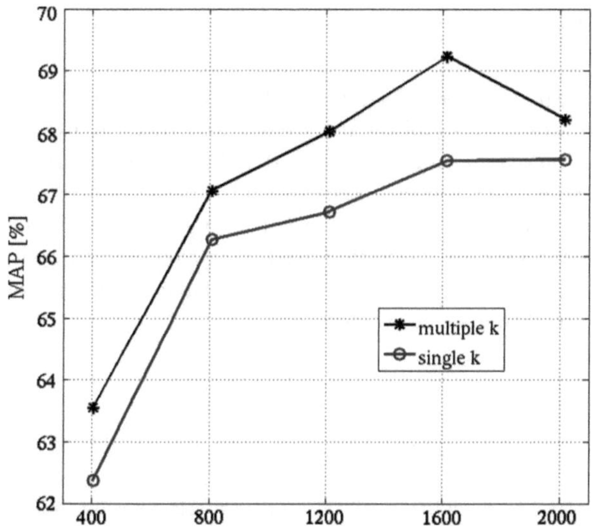

Fig. 3.8 Results of MLNN and k-NN obtained when combining multiple k values into the classification rule

Then, we look deeper into MLNN classification performances by analyzing the confusion matrix (Fig. 3.7). Overall, the method performs well on most outdoor scenes, as well as on the "office" category. Only some reasonable confusions are present, e.g. between "coast" and "open country" or "inside city" and "street". Thus, average performance (mAP under 70 %) drops off mainly because of the low recognition rate in a few more challenging categories, such as the indoor category "bedroom" (recognition rate of only 16.2 %), and the "industrial" category (36.5 %), that mixes outdoor and indoor images, thus making recognition very challenging. In particular, notice that "bedroom" images are more often misclassified into the "living room" category, due to their similar scene layouts and the presence of similar objects, e.g. paintings on the walls and sofas/beds. This drastically reduces the prototypical relevance of bedroom images during the learning phase, thus biasing misclassification into the "living room" category. This phenomenon is related to the well-known "semantic gap", which affects low-level visual descriptors that only collect statistics on the appearance information without any semantic interpretation. This problem is particularly critical for prototype-based methods when the inter-class variability is low, preventing them from learning reliable discriminating prototypes.

A very simple strategy for improving the overall precision of MLNN is to combine the scores from multiple MLNN tests, e.g. by summing the classification scores corresponding to different values of k for the same (truncated) Histogram Intersection kernel. An example of the performance increase enabled by this strategy is shown in Fig. 3.8, where MLNN improves by almost 2 % by summing the scores obtained for $k = 5, 10, 15$, with a best mAP of 69.23 %.

3.4 Conclusion

In this chapter, we have proposed a novel method (MLNN) for boosting k-NN voting in the context of object recognition and scene categorization, by minimizing a surrogate risk function over a training dataset. Results on benchmark image categories have shown considerable improvements over the classic uniform voting rule, both in precision and in computation time. Furthermore, since our method is completely independent on the kind of descriptor used, it is expected to conveniently apply to k-NN–based methods relying on other state-of-the-art descriptors, like VLAD descriptors [11] or Fisher kernel vectors [15].

References

1. Bartlett P, Jordan M, McAuliffe JD (2006) Convexity, classification, and risk bounds. J Am Stat Assoc 101:138–156
2. Bosch A, Zisserman A, Muñoz X (2008) Scene classification using a hybrid generative/discriminative approach. IEEE Trans Pattern Anal Mach Intell 30(4):712–727
3. Brighton H, Mellish C (2002) Advances in instance selection for instance-based learning algorithms. Data Min Knowl Disc 6:153–172
4. Cha SH (2008) Taxonomy of nominal type histogram distance measures. In: Proceedings of the American conference on applied mathematics, pp 325–330
5. Csurka G, Bray C, Dance C, Fan L (2004) Visual categorization with bags of keypoints. In: Workshop on statistical learning in computer vision (ECCV), pp 1–22
6. Davis J, Kulis B, Jain P, Sra S, Dhillon I (2007) Information-theoretic metric learning. In: International conference on machine learning (ICML), pp 209–216
7. Freund Y, Schapire RE (1997) A decision-theoretic generalization of on-line learning and an application to Boosting. J Comput Syst Sci 55(1):119–139
8. Garcia V, Debreuve E, Barlaud M (2008) Fast k nearest neighbor search using gpu. In: CVPR workshop on computer vision on GPU (CVGPU), Anchorage, Alaska, USA
9. Hastie T, Tibshirani R (1996) Discriminant adaptive nearest neighbor classification. IEEE Trans Pattern Anal Mach Intell 18(6):(6):607–616
10. Jegou H, Douze M, Schmid C (2008) Hamming embedding and weak geometric consistency for large scale image search. In: European conference on computer vision (ECCV), vol I, pp 304–317
11. Jégou H, Douze M, Schmid C, Pérez P (2010) Aggregating local descriptors into a compact image representation. In: IEEE Conference on computer vision & pattern recognition, pp 3304–3311. http://lear.inrialpes.fr/pubs/2010/JDSP10
12. Lazebnik S, Schmid C, Ponce J (2006) Beyond bags of features: spatial pyramid matching for recognizing natural scene categories. In: Computer vision and pattern recognition (CVPR), pp 2169–2178. http://dx.doi.org/10.1109/CVPR.2006.68
13. Lowe D (2004) Distinctive image features from scale-invariant keypoints. Int. J. Comput. Vision 60(2):91–110. http://dx.doi.org/10.1023/B:VISI.0000029664.99615.94
14. Nock R, Nielsen F (2009) Bregman divergences and surrogates for learning. IEEE Trans. Pattern Anal. Mach. Intell. 31:2048–2059
15. Perronnin F, Dance C (2007) Fisher kernels on visual vocabularies for image categorization. In: Computer vision and pattern recognition (CVPR), pp 1–8
16. Perronnin F, Sánchez J, Liu Y (2010) Large-scale image categorization with explicit data embedding. In: Computer vision and pattern recognition (CVPR), pp 2297–2304

17. Piro P, Nock R, Nielsen F, Barlaud M (2009) Boosting k-NN for categorization of natural scenes. ArXiv:1001.1221
18. Piro P, Nock R, Nielsen F, Barlaud M (2010) Leveraging *k*-NN for generic classification boosting. In: Proceedings of the 2010 IEEE international workshop on machine learning for signal Processing (MLSP)
19. Schapire RE, Singer Y (1999) Improved boosting algorithms using confidence-rated predictions. J Mach Learn 37:297–336
20. Sivic J, Zisserman A (2003) Video google: a text retrieval approach to object matching in videos. In: Proceedings of the international conference on computer vision, vol 2, pp 1470–1477. http://www.robots.ox.ac.uk/~vgg
21. Sivic J, Zisserman A (2006) Video google: efficient visual search of videos. pp 127–144. http://www.robots.ox.ac.uk/~vgg
22. Zou H, Zhu J, Hastie T (2008) New multicategory boosting algorithms based on multicategory fisher-consistent losses. Ann Appl Stat 2(4):1290–1306
23. Zhang H, Berg A, Maire M, Malik J (2006) SVM-KNN: discriminative nearest neighbor classification for visual category recognition. In: Computer vision and pattern recognition (CVPR'06), pp 2126–2136

Part II
Motion and Activity Analysis

Chapter 4
Semi-Automatic Object Tracking in Video Sequences by Extension of the MRSST Algorithm

Marko Esche, Mustafa Karaman and Thomas Sikora

Abstract The objective of this work is to investigate a new approach for segmentation of real-world objects in video sequences. While some amount of user interaction is still necessary for most algorithms in this field, in order for them to produce adequate results, these can be reduced making use of certain properties of graph-based image segmentation algorithms. Based on one of these algorithms a framework is proposed that tracks individual foreground objects through arbitrary video sequences and partly automates the necessary corrections required from the user. Experimental results suggest that the proposed algorithm performs well on both low- and high-resolution video sequences and can even, to a certain extent, cope with motion blur and gradual object deformations.

Keywords Image segmentation · Object tracking · Binary partition tree

4.1 Introduction

The segmentation of real world objects in still images and the tracking of such objects in video sequence have many applications especially in video processing and video editing [4]. While a lot of progress has been made concerning the extraction of binary object masks from single frames, automatic generation of such masks for an entire video remains mainly unsolved. Since dynamic foreground objects both need to be correctly identified and can also undergo sudden changes in shape and color, most video segmentation tools rely on user interaction in order to produce accurate results. In this chapter a new framework for the tracking of foreground objects in video sequences based on the *Modified Recursive Shortest Spanning Tree Algorithm*

M. Esche (✉) · M. Karaman · T. Sikora
Communication Systems Group Technische Universität Berlin Sekr. EN1,
Einsteinufer 17, 10587 Berlin, Germany
e-mail: esche@nue.tu-berlin.de

N. Adami et al. (eds.), *Analysis, Retrieval and Delivery of Multimedia Content*,
Lecture Notes in Electrical Engineering 158, DOI: 10.1007/978-1-4614-3831-1_4,

(MRSST) is presented. It incorporates the interactive Object Contour Extraction method proposed by Adamek and O'Connor in [1]. A new approach for identifying corresponding image regions in consecutive frames is also integrated using a *binary partition tree* (BPT) for each frame. In addition, an algorithm is described that automatically locates a tracked object in new frames even if the object shape, size or colour distribution changes. The remainder of the chapter is structured as follows. Section 4.2 briefly revisits the segmentation of color images using the MRSST and its applicability to the task of object extraction. Both a general outline of the proposed algorithm and detailed descriptions of two of its main components are given in Sect. 4.3. Experimental results and objective evaluation measures are provided in Sect. 4.4. Section 4.5 shows how future extensions can improve the algorithm, while Sect. 4.6 concludes the chapter with a short discussion.

4.2 Object Extraction and the MRSST

Among the various image segmentation approaches graph-based algorithms, such as the one presented by Cooray et al. in [6], have lately received significant attention. This is partly due to their ability to represent image segments of arbitrary sizes as nodes in a BPT. This property enables the user or an automatic algorithm to easily extract spatially connected regions or objects from the frame by labeling certain parts of the BPT. A detailed analysis of the BPT's suitability to object extraction and especially the detection of individual objects through using specialized merging criteria can be found in [10]. There Vilaplana et al. also introduce the notion of the BPT as a simplified search space that can used to identify certain image regions. One of these graph-based approaches is the MRSST presented in [3], which also introduces new so-called syntactic features. These features represent geometric porperties of image regions and their spatial configurations. The following descriptions shall be used to illustrate the basic idea behind syntactic features:

- Homogeneity: This feature controls the spatial color homogeneity of a region. When considering two neighboring regions the color difference along their common boundary is used as a measure of similarity.
- Complexity: Starting with the boundary length l_i of a region and its spatial area a_i, the complexity of the region is then given by $x_i = l_i / \sqrt{a_i}$. The least complex theoretical region, therefore, is a perfect circle.
- Compactness: A region is considered to be compact if all its constituent parts are adjacent to one another.

A mathematically sound definition of each these syntactic features together with a more detailed description may be found in [3]. Initially, the MRSST treats every pixel of an image as a node in a graph that is connected via weighted edges to its four direct neighbors. Here the weight of an edge depends on the colour of individual pixels, the spatial complexity of the regions to be merged and the potential complexity of the merged region. The segmentation of the image is achieved by iteratively

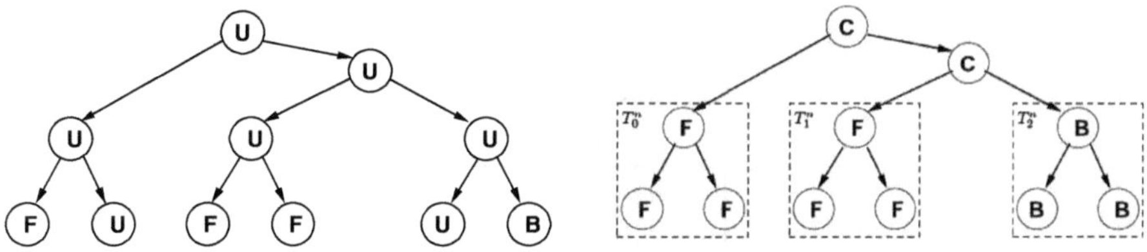

Fig. 4.1 *Top* labels added to the BPT through scribbles drawn on the image by the user. *Bottom* after the processing of the scribbles added by the user, local homogeneous subtrees of the BPT are formed (marked with *dashed rectangles*), denoted here by T_0^n, T_1^n and T_2^n

merging those nodes that are connected by the least-cost link, which also enforces the recalculation of associated graph edges and their weights. During this process the BPT is created by storing the order in which nodes are merged and establishing a father-child relationship between the newly created node and the two merged ones. The output of the MRSST in this case is a BPT, whose root node contains the entire image while the leafs are the individual pixels of the image. In [1] Adamek and O'Connor proposed a method that allows for quick extraction of arbitrary foreground shapes from a BPT by letting the user draw so-called scribbles on the image. Both foreground (**F**) and background (**B**) scribbles are passed to the leafs of the BPT in the form of competing labels. These are then iteratively propagated up the BPT. When the algorithm tries to assign both labels to the same node, that node is marked as a conflict node (**C**) and all its parent nodes are marked with the conflict label as well. This results in the formation of homogeneously labeled local subtrees that either belong exclusively to foreground or background. An example of a BPT before and after label propagation is given in Fig. 4.1, where each subtree represents an individual spatial region of its own. By adding more scribbles to the image the resulting foreground mask can be refined further.

4.3 Proposed Algorithm

The new object tracking algorithm, which is proposed in this article, consists of a three-stage approach as outlined by Fig. 4.2 for a video sequence of N frames. Initially a BPT is created for the first frame of the sequence which is then labeled by the user in order to correctly identify the foreground object to be tracked, which results in the extraction of a shape S^n. The labeling is done based on the method described in [1]. This initial step is illustrated by Fig. 4.3. Where the image on the left shows the scribbles drawn by the user. Background scribbles are shown in blue while all foreground labels are marked in red. In the actual implementation both the drawing of the scribbles and the propagation of labels in the graph are done simultaneously. This ensures that the user is at all times immediately presented with the resulting object share, which reduces the number of scribbles to be drawn and the amount of error

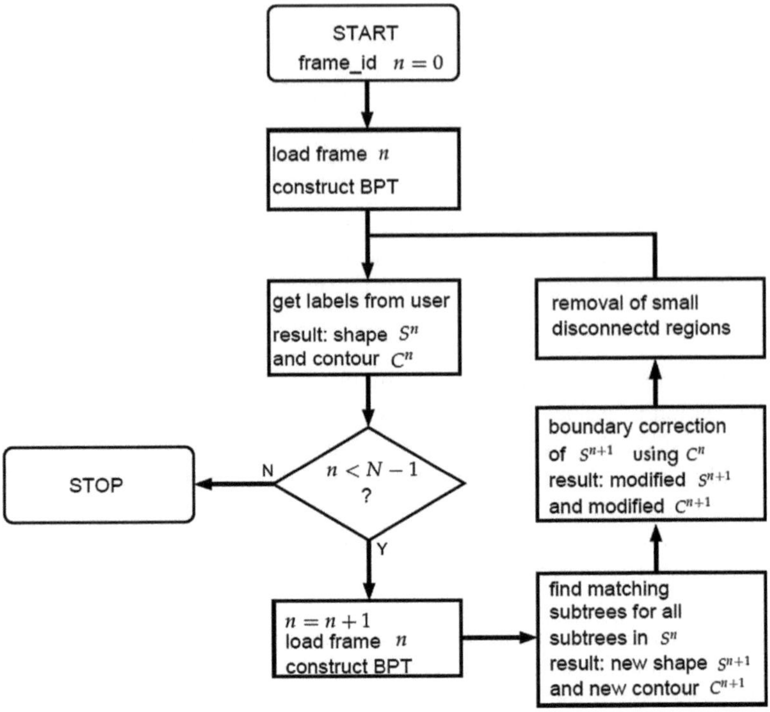

Fig. 4.2 The main part of the algorithm is a loop transferring shape silhouette and contour in form of background and foreground labels from frame to frame. After every iteration the intermediate result is presented to the user, who can then modify the labels by placing additional scribbles or by starting with an entirely new object mask

introduced by wrongly placed labels. The image on the right displays the extracted object shape after the processing of these scribbles. For every consecutive frame a BPT is also created. An initial estimate of the shape S^{n+1} of the tracked object in the next frame is determined by matching local subtrees of the BPT among neighboring frames. More details on this technique are provided in Sect. 4.3.1. Having obtained this estimate, the object contour C^n is transferred into the next frame and used to correct the object contour C^{n+1} of the predicted object shape S^{n+1} by automatically generating an independent set of scribbles. See Sect. 4.3.2 for details. Afterwards the predicted object shape is presented to the user who can now add further labels to refine the object shape or to make necessary corrections. An approach with a similar workflow, which however does not make use of a graph-based image representation and therefore relies solely on the object contour, has been proposed in [5]. Another approach based on a Mean Shift algorithm was for instance described in [7].

Fig. 4.3 Scribbles added by the user for the first frame of the *PlanetEarth* sequence are shown on the *left*. The extracted object is shown on the *right*

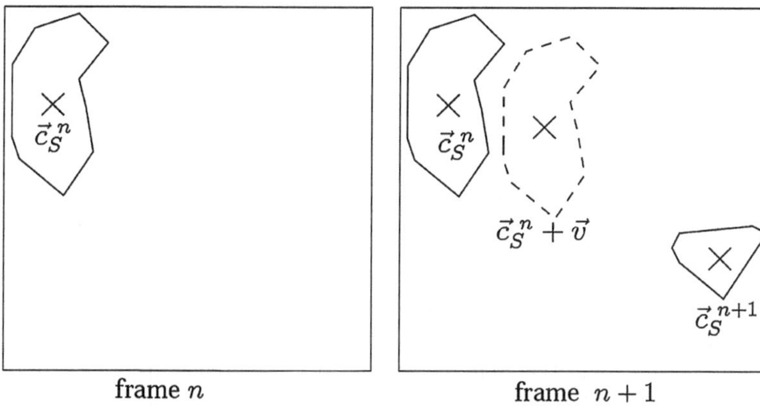

Fig. 4.4 *Left* object shape in the previous frame, *right* partially constructed object shape for the current frame with center $\mathbf{c}_S^{n+1}$ and the moved reference shape with center $\mathbf{c}_S^n + \mathbf{v}$

4.3.1 Identification of Corresponding Subtrees

Before describing the actual algorithm for transferring the current object shape S^n in frame n into the next frame $n+1$, a number of definitions have to be made. The areas in pixels occupied by shapes Sx^n and S^{n+1} are denoted by a_S^n and a_S^{n+1} respectively. The area associated with a subtree in the current frame is consequently denoted by $a_{T_i}^n$. A similar notation is used to describe the center pixel of the entire object shape S in the current frame $\mathbf{c}_S^n$ or the center pixel of a specific subtree T_i^{n+1} in the next frame $\mathbf{c}_{T_i}^{n+1}$. In order to evaluate the suitability of a subtree T_j^{n+1} for inclusion in the new object shape S^{n+1}, the previously determined object shape is transferred into the next frame using the motion vector $\mathbf{v}$ given in

$$\mathbf{v} = (\mathbf{c}_S^{n+1} \cdot a_S^{n+1} + \mathbf{c}_S^n \cdot (a_S^n - a_S^{n+1}))/a_S^n - \mathbf{c}_S^{n+1}. \tag{4.1}$$

The choice of this motion vector ensures that the predicted object shape S^{n+1} is initially placed at the same location $\mathbf{c}_S^n$ as the previous object shape, see Fig. 4.4 for a simple example. Once more local subtrees have been assigned to the new object shape, the location of the predicted object shape $\mathbf{c}_S^{n+1}$ is adapted to fit to the new object location. This results in an object center that gradually moves itself to the new position of the tracked object. The motion vector in Eq. 4.1 is only used as long as $a_S^{n+1} < a_S^n$. Here the areas a_S^{n+1} and a_S^n can be interpreted as a measure of confidence with which the tracked shape is at a certain location in the image. Based on the transferred version of S^n it is now possible to compute a relative overlapp $c_{ov}({T_j}^{n+1})$ between every subtree in the current frame and the entire object shape in the previous frame.

Additionally, a color cost $c_{co}(T_j^{n+1})$ is assigned to every local subtree T_j^{n+1} in the new frame. It is computed as the smallest euclidean distance in *CIEL*u*v** color space between the average color of T_j^{n+1} and any subtree T_i^n in the previous frame within a search radius $r(T_j^{n+1})$. The usage of this colorspace according to [8] guarantees a certain robustness against changing lighting conditions in the video sequence due to the fact that these mostly affect the L-component. Initially the search radius $r(T_j^{n+1})$ is set to $\frac{1}{5}\sqrt{a_{T_j}^{n+1}}$. The paradigm behind this choice is the following: During the first iterations of the algorithm, when the location of the tracked object in the next frame is still unknown, only larger subtrees are searched for as they are allowed more motion according to the search radius given above. Large connected regions that are represented by a single subtree in the previous frame are here expected to be more easily reidentifyable in the next frame, as there is a strong likelihood that at least some parts of these regions are again merged into a larger subtree. A greedy algorithm that tests every subtree in the new frame for inclusion in the new object shape is now developed: During every iteration the subtree with the smallest inclusion cost as given in Eq. 4.2 is chosen and included in the new object shape

$$c_{inc}(T_j^{n+1}) = \begin{cases} \infty, \text{ if } c_{co}(T_j^{n+1}) > c_{co}^{max} \vee c_{ov}(T_j^{n+1}) < c_{ov}^{min} \\ \frac{c_{co}(T_j^{n+1})}{c_{co}^{max}} + 1 - c_{ov}(T_j^{n+1}), \quad \text{otherwise.} \end{cases} \tag{4.2}$$

Additionally, the subtree T_k^n in the previous frame with the smallest color distance to the included subtree T_l^{n+1} is removed from the previous object shape to ensure that the color distributions of S^n and S^{n+1} remain identical. The thresholds c_{co}^{max}, c_{ov}^{max} and r^{max} used in Eq. 4.2 are needed to control the behaviour of the inclusion algorithm. When a subtree receives an infinite inclusion cost, it is split into its respective shildren which are then examined during the next iteration. Here the BPT's structure is employed as a binary search tree that ensures quick localization of individual nodes within the tree and their respective spatial representation as seperate regions in the image. Should it not be possible to include any candidate subtree during the current iteration because no subtree received a finite inclusion cost, then the three thresholds are adapted according to a three-stage schedule. The individual steps of this schedule are described in the following list:

1. Subtrees T_j^{n+1} with at least 50 % overlapp with the moved object shape from the last frame and small color difference and little individual segment motion $r(T_j^{n+1})$ are added to the new object shape S^n. For most tracked objects this produces a rough reconstruction around the object center, which moves along the motion trajectory of the entire object.
2. All subtrees T_j^{n+1} with at least 50 % overlapp with the moved object shape, a bigger color difference and more segment motion are merged with the object shape in the new frame. During this step object parts with a slowly changing color (i.e. due to lighting conditions) are included as well as slowly moving object parts such as arms and legs for human beings or animals.
3. All segments T_j^{n+1} that have a color distribution similar to T_i^n and lie somewhere within the entire region of interest are included without considering overlapp or individual segment motion. Assuming that the average size of an object does not change dramatically from frame to frame, this step results in a labeled group of subtrees in the next frame which are a good approximation of the new object shape. The region of interest around the predicted object center is a rectangular image region whose dimensions depends on the size of the object shape in the previous frame.

The resulting workflow of this step is illustrated by Fig. 4.5. As both color and spatial orientation play a strong role during the first stage of the algorithm, only such regions (marked in white) are added that undergo no change between the current and the next frame. After iteration 5, small parts of the added regions do not belong to the object to be tracked. Nevertheless, due to the overlapp criterion, such errors only occur at the object boundary, where they can easily be corrected by the next step of the algorithm, as will be shown in Sect. 4.3.2. The adding of new subtrees is automatically stopped once the area of the reconstructed object shape a_S^n is bigger than 99 % of the object shape a_S^{n+1} from the last frame. For sequences where the tracked object moves very fast either away from or towards the camera, this threshold can be modified by the user. The empirical value of 99 %, however, produced good results for all sequences the algorithm was tested on since small missing regions are automatically added again by the next step of the algorithm.

Two properties of most tracked objects that have not yet been taken into account now also need to be examined. Firstly, in most cases tracking will only be done here on non-dividable objects. That is to say, no tracked object shall be allowed to be split or to be merged with other objects. Secondly, changes in the objects appearance are to be expected at the object contour only while the central area of the two-dimensional representation of the object should not undergo noticable changes from frame to frame. Both of these properties as well as the errors introduced by the merging step described above are now examined.

Fig. 4.5 During every iteration of the algorithm that identifies corresponding subtrees between neighboring frames new regions (marked in *white*) are added to the predicted object shape in the next frame. These added regions are here shown in *white*. From *left* to *right* the merged object shape for the new frame after 4, 7 and 12 iterations are shown

4.3.2 Boundary Correction

Since object deformations usually manifest themselves as a modification of the object boundary, it is necessary to fit the object contour C^n from the previous frame to the new object shape S^{n+1} in the current frame. The strategy proposed here is to first move the previous object shape into the current frame. For this step the motion vector given in Eq. 4.1 is again utilized. Due to potential mismatch introduced by the merging algorithm and due to natural changes of the object shape, the original object contour C^n will not perfectly match the initially predicted object shape S^{n+1}. In order to overcome this discrepancy every pixel along the original object contour is now examined. For a square patch of size $l \times l$, where $l = \frac{1}{4}\sqrt{a_S^n}$, around the pixel location in the previous frame the contrast measure (abbreviated by *con*) given in Eq. 4.3 is computed. Here L_F, u_F, v_F and L_B, u_B, v_B are the three components of the average colors of foreground and background inside the patch, respectively. The value of $\frac{1}{4}\sqrt{a_S^n}$ is chosen based on empirical data and represents a good compromise between computational complexity and the amount of variability allowed for the object contour, since a large value l increases the size of the square patch to be compared quadratically, while at the same time allowing for bigger modifications of the boundary. Should S^n be a circular region with radius r for instance, then l is roughly $0.44r$

$$con = \frac{(L_F - L_B)^2 + (u_F - u_B)^2 + (v_F - v_B)^2}{3 \cdot 255 \cdot 255}. \tag{4.3}$$

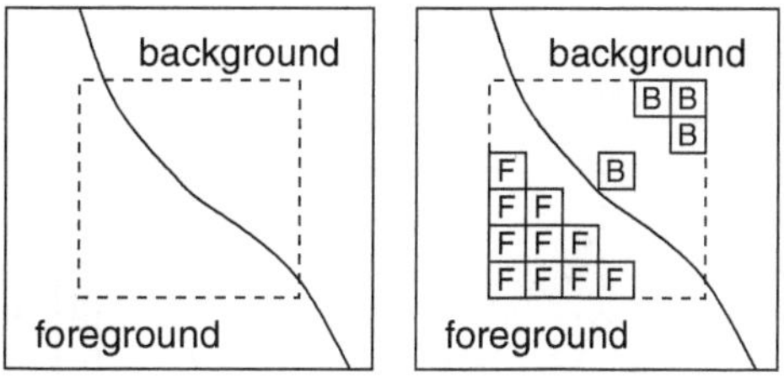

Fig. 4.6 Original object contour (*left*) and adapted automatic scribbles (*right*). **B** and **F** indicate background and foreground labels, respectively

The contrast measure in turn indicates the confidence with which a contour pixel can be located in the new frame. Should the contrast between background and foreground be higher than 0.1 % then a local scribble model as shown in Fig. 4.6 is built. The scribble model is basically a distribution of labels over the square patch mentioned above that is assumed to extract the correct object shape if these automatic scribbles were added to an unlabeled BPT. The actual distribution is chosen with respect to possible contour deformations and only the one background pixel closest to the center of the patch is expected to remain a background pixel. A best match for the $l \times l$ patch is now found by performing a fullsize blocksearch around the initial contour location. Once the best match has been identified, the previously obtained automatic scribbles are added to the BPT of the new frame around the corrected location of the considered contour pixel. An example for such a boundary correction step is provided in Fig. 4.7. Of particular interest in the displayed frame is the flag that occupies the lower right part of the image and is correctly identified as part of the background despite having the same color as the foreground object. In order to enforce the compactness of the reconstructed shape all disconnected segments with a size smaller than 10 % of the entire labeled region are removed from the object shape and are treated as background.

4.4 Experimental Results

The propsed algorithm has been implemented in *C++* and tested on the MPEG CIF test sequences *House*, *Highway* and *Group*. It was also tested on one sequence each from the following movies in DVD resolution: *Harry Potter and the Sorcerer's Stone*, *Planet Earth* (BBC Documentary) and *Star Wars-Episode IV*. Some of these also include global camera motion or moving background objects. Keyframes and extracted foreground objects are shown in Fig. 4.8. Detailed examples for four successive frames may be found in Fig. 4.9. In order to objectively evaluate the algorithm all automatically generated object masks were compared with manually segmented groundtruth masks. For each frame precision (p), recall (r) and f-measure (f) were computed. In addition, the number of user interactions (u) and the number of manually labeled pixels (l) per frame were recorded. In this context, a user interaction is defined as a single connected scribble in either foreground or background color. The

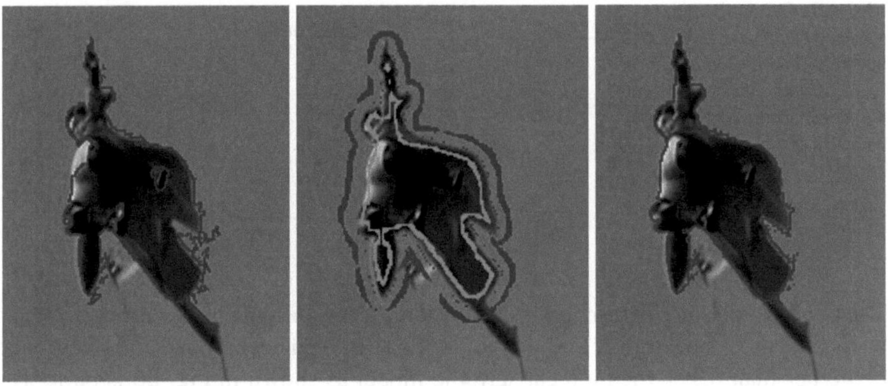

Fig. 4.7 *Left* initially predicted object shape (outlined in *red*) which was constructed using the subtree matching approach; *center* automatically generated scribbles (*green* for foreground, *red* for background); *right* corrected object shape, after the application of the scribbles

Table 4.1 Average precision, recall and f-measure per sequence for the automatically generated object masks. rel_4 and rel_6 indicate the percentage of frames for which less than 4 respectively less than 6 scribbles where necessary

Sequence	p (%)	r (%)	f (%)	rel_4 (%)	rel_6 (%)
Group	93.0	89.3	91.0	70.7	92.7
Highway	92.7	89.2	90.1	88.2	94.1
House	93.3	86.4	88.7	70.3	94.6
Harry Potter	87.8	94.2	90.8	54.2	80.6
Planet Earth	91.8	93.1	92.4	67.5	95.8
Star Wars	97.6	98.3	97.9	100	100

per-frame values for p, r, f, u and l for the *House* sequence are given in Fig. 4.10 in the left column. Due to the foreground object (a person walking from right to left) entering the scene during the first frames of the sequence satisfactory results are only achieved from frame five onwards, when the person is visible in its entirety. The right column shows the respective measures for the *PlanetEarth* sequence. Here, high precision and recall are achieved from the first frame onward, a quality decrease of the predicted masks can only be observed for frames 15 to 17, where foreground and background of the displayed scene have almost identical color. The average measures for all sequences are given in Table 4.2. In addition, the percentage of frames for which less than 4 or less than 6 scribbles were needed (rel_4 and rel_6 respectively) are provided. Of particular interest is the *Harry Potter* sequence throughout which motion blur is frequently present. Nevertheless, the dual approach still performs as well as for the other sequences. A comparison with similar interactive tracking approaches has not been conducted yet since, to the knwoledge of the authors, a general framework for objectively measuring the amount of user interaction still needs to be established (Fig. 4.10).

Fig. 4.8 Sample frames and extracted foreground objects for all tested sequences. For the high-resolution sequences displayed in the right columns only a magnified region of interest is shown

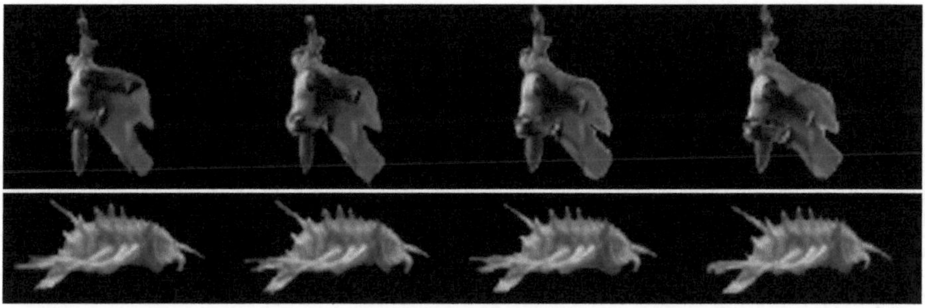

Fig. 4.9 In the *top-row* the extracted foreground objects for the first four frames of the *Harry Potter* sequence are shown. The respective foreground objects of the *PlanetEarth* sequence are shown in the *bottom row*

That none of the values shown in Table 4.2 are identical to 100 % is mainly due to the fact that for most video sequences the definition of the object boundary will be strongly dependent on the user providing the segmentation mask. This accounts for a slight discrepancy between the manually segmented groundtruth masks and the masks produced by the algorithm.

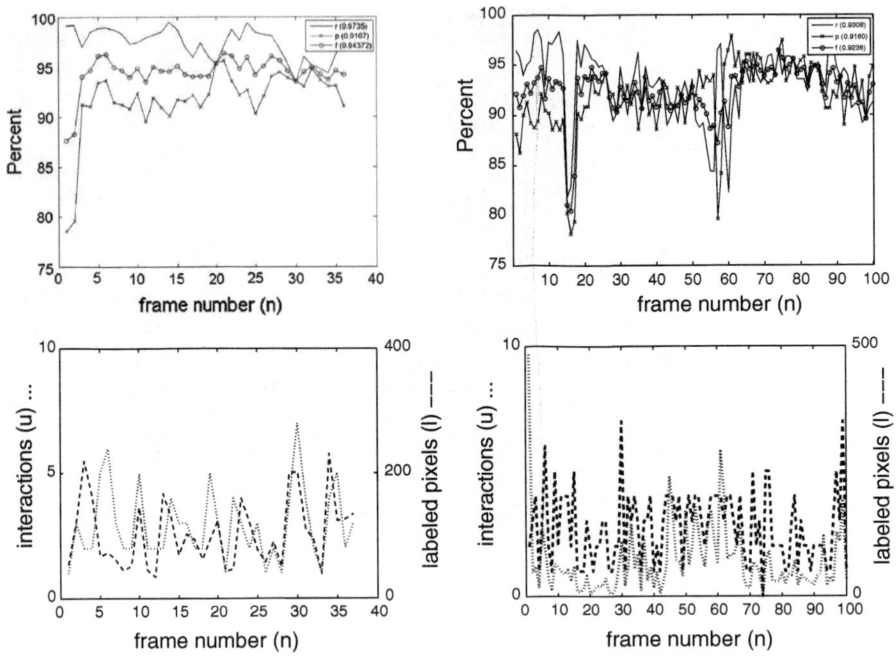

Fig. 4.10 *Left column* the average recall (r), precision (p) and f-measure f (*top row*) for the *House* sequence as well as the number of interactions (u) and the number of labeled pixels per frame l (*bottom row*) that were needed to produce these values are shown. *Right column* The respective measures for the *PlanetEarth* are shown

Table 4.2 Average precision, recall and f-measure per sequence for the corrected object masks after the application of user corrections. $\overline{N}$ denotes the average number of user interactions per frame which were required to achieve the respective measures in columns 1 to 3

Sequence	p (%)	r (%)	f (%)	$\bar{N}$
Group	94.1	93.8	93.9	1.92
Highway	95.4	93.5	94.4	0.93
Harry Potter	89.6	97.4	93.3	2.25
Planet Earth	93.1	94.2	93.6	1.73
Star Wars	97.8	98.0	97.9	0.33

4.5 Possible Extensions and Future Work

One advantage of the algorithm proposed here is the fact that additions to the *MRSST* algorithm such as the one proposed in [2] can easily be integrated without affecting the rest of the work flow. As described in [2] Dempster-Shafer theory can be used to further improve the segmentations produced by the *MRSST* by making use of a probabilistic model to describe new merging criterions during the formation of the

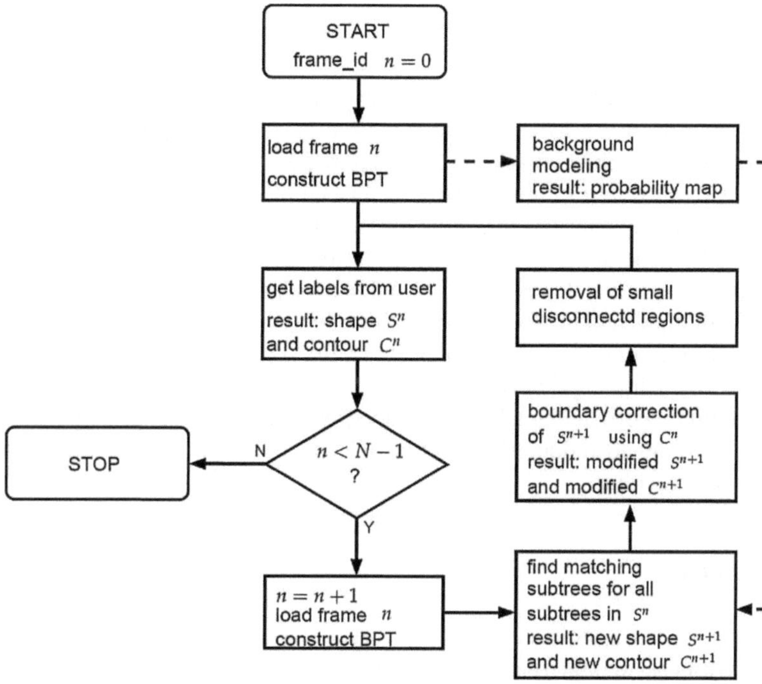

Fig. 4.11 Background modeling could be used to remove certain subtrees in the new frame from the inclusion process, which would speed up the subtree matching algorithm significantly

BPT. This addition, ensures a higher robustness against the unwanted merging of foreground and background regions. Equivalently a more robust formation of segments firstly guarantees that spatial regions in neighboring frames are similar and hence more easily identifyable. This could ensure that subtrees are identified with greater precision. Another extension, which could be used to improve the effectiveness of the approach discussed here, is the integration of a background subtraction step. As shown for example in [9] background modelling itself can be used to segment moving foreground objects, or even static objects if a moving camera has been used.

One of the problems with these approaches is the fact that they produce good segmentations of foreground objects for certain sequences, but are unable to identify matching segmented regions between neighboring frames. Especially when two foreground objects overlapp, no distinction is possible, concerning which pixels are part of which object. Here the advantages of tree-based segmentation and background modeling could be combined: In the approach outlined in Fig. 4.11 background modelling is used to generate a probability map for the current frame, giving for every pixel the probability with which it is either part of foreground or background. This information can then be used by the subtree matching step, to use only those subtrees that are considered likely foreground candidates. Other subtrees would equivalently only be considered, once all expected foreground subtrees have been tested.

4.6 Summary

The main objective of this work was to develop a new interactive object tracking approach based on the MRSST that represents a good compromise between the quality of extracted object masks and the required amount of user interaction. In this chapter a two stage algorithm has been described that tracks individual foreground objects in video sequences by matching local subtrees of the BPT among neighboring frames and by generating an automatic set of foreground and background labels for each consecutive frame. It has been shown that the algorithm performs comparatively well for both high- and low-resolution videos. In addition, no restrictions have been placed on movement, shape or variability of the tracked foreground object which makes the algorithm applicable for arbitrary videos and a wide range of real-world objects. Future work will include the incorporation of a background subtraction approach in order to reduce the amount of misclassification done during the first stage of the algorithm.

References

1. Adamek T, O'Connor NE (2006) Interactive object contour extraction for shape modeling. In: 1st international workshop on shapes and semantics, vol 1(1). pp 31–39
2. Adamek T, O'Connor NE (2007) Using dempster-shafer theory to fuse mulitple information sources in region-based segmentation. In: Proceedings of the 14th international converence on image processing ICIP, vol 2. pp 269–272
3. Adamek T, O'Connor NE, Murphy N (2005) Region-based segmentation of images using syntactic visual features. In: Proceedings of the 6th international workshop on image analysis for multimedia interactive services (WIAMIS)
4. Alatan A, Onural L, Wollborn M, Mech R, Tuncel E, Sikora T (1998) Image sequence analysis for emerging interactive multimedia services—the european cost 211 framework. IEEE Trans. Circuits Syst. Video Technol. 8:802–813
5. Bai X, Wang J, Simons D, Sapiro G (2009) Video snapcut: robust video object cutout using localized classifiers. In: SIGGRAPH'09: ACM SIGGRAPH 2009 papers, vol 28(3)
6. Cooray S, O'Connor N, Marlow S, Murphy N, Curran T (2001) Semi-automatic video object segmentation using recursive shortest spanning tree and binary partition tree. In: Proceedings of the 3rd international workshop on image analysis for multimedia interactive services (WIAMIS)
7. Corrigan D, Robinson S, Kokaram A (2008) Video matting using motion extended grabcut. In: 5th european conference on visual media production (CMVP), pp 1–9
8. Freixenet J, Munoz X, Raba D, Marti J, Cufi X (2002) Yet another survey on image segmenation: region and boundary information integration. Lect Notes Comput Sci 2352/2002:21–25
9. Krutz A, Glantz A, Borgmann T, Frater M, Sikora T (2009) Motion-based object segmentation using local background sprites. In: Proceedings of the IEEE international conference on acoustics, speech and signal processing (ICASSP), pp 1221–1224
10. Vilaplana V, Marqués F, Salembier P (2008) Binary partition trees for object detection. IEEE Trans. Image Process. 17:11

Chapter 5
A Multi-Resolution Particle Filter Tracking with a Dual Consistency Check for Model Update in a Multi-Camera Environment

Yifan Zhou, Jenny Benois-Pineau and Henri Nicolas

Abstract In this chapter, we present a novel tracking method with a multi-resolution approach and a dual model check to track a non-rigid object in an uncalibrated static multi-camera environment. It is based on particle filter methods using color features. The major contributions of the method are: multi-resolution tracking to handle strong and non-biased object motion by short term particle filters; stratified model consistency check by Kolmogorov-Smirnov test and object trajectory based view corresponding deformation in multi-camera environment.

Keywords Particle filter· multi-resolution· dual consistency check· Kolmogorov-Smirnov test· iteratively reweighted least-squares method

5.1 Introduction

Tracking of moving non-rigid objects is a key issue for efficient analysis of video streams in the framework of multimedia applications. Both mono-camera and multi-camera tracking of objects remain open issues despite a rich literature on the subject [1]. In this chapter, we propose a tracking method for monocular camera which is based on the family of particle filter methods [2]. Then, an extension of this method for multiple camera tracking will be developed in the uncalibrated and unconstrained environment.

Y. Zhou (✉) · J. Benois-Pineau · H. Nicolas
Laboratoire Bordelais de Recherche en Informatique (LaBRI), CNRS (UMR 5800), Université Bordeaux 1, 351 cours de la Libération, 33405 Talence cedex, France
e-mail: yifanzhou67@yahoo.fr

J. Benois-Pineau
e-mail: jenny.benois@labri.fr

H. Nicolas
e-mail: henri.nicolas@labri.fr

N. Adami et al. (eds.), *Analysis, Retrieval and Delivery of Multimedia Content*, Lecture Notes in Electrical Engineering 158, DOI: 10.1007/978-1-4614-3831-1_5,

The inspiration of our work came color distribution model [3]. It is a Sequential Monte Carlo method [4] based on Particle Filter (PF) whose color feature is used to evaluate importance function. We chose it for its capacity to solve non-linear and non-Gaussian state-space model, e.g., human tracking. Unfortunately, when applied for tracking a non-rigid object in low and/or variable frame-rate videos, their tracking quickly loses object because of its strong motion magnitude. Hence, we proposed a multi-resolution technique with PF method [5] to reduce this magnitude as well as computational time. The essential idea was to first quickly locate object by prediction at the lowest resolution level and then refine it gradually at higher levels. Instead of tracking only on full resolution [3] or a certain optimal level [6], our method involves all the levels. Therefore, it offered a better accuracy of tracking result especially in low and/or variable frame-rate videos in mono-camera environment.

A Consistency Check was employed to alarm the degeneracy phenomenon [7] in PF methods so as to conduct a object reinitialization step. Instead of using an effective sample size [7] to evaluate the effectiveness of a set of particles, the Kolmogorov-Smirnov test was applied to control object appearance change. It is proved to give a better mono-camera tracking result. Thus in this chapter, a Dual Consistency Check is presented to adapt to multi-camera environment.

Nevertheless, the reinitialization in mono-camera tracking could not provide an efficient solution to occlusion problems. Therefore, the reinitialization by an interaction of cameras in multi-camera tracking was proposed to relocate the object in one camera by those in other cameras. Our transformation matrix between cameras was estimated by least-squares (LS) method. The particularity of our approach is to use object center position in each camera view as observations in place of static scene elements. However, the position error can be relatively strong. Therefore, in this chapter, we propose a robust (reweighted) least-squares estimation for the transformation matrix with accumulation of observations along the time.

This chapter is organized as follows. The details of our tracking method will be presented in Sect. 5.2. Some examples will be illustrated and discussed in Sect. 5.3. The conclusion and perspectives will be shown in Sect. 5.4.

5.2 Tracking Algorithm

We first developed the method for one non-rigid object tracking in mono-camera environment [5]. Figure 5.1 depicts the general scheme on 3 resolution levels and Fig. 5.2 illustrates an example. The object estimate is firstly predicted on the lowest resolution on the time level based on a short term temporal particle filter (red line in Fig. 5.1) and refined gradually on the spatial level based on a short term spatial particle filter (blue line in Fig. 5.1). By using this meander strategy (red line in Fig. 5.2), the important motion change of objects in videos of low and variable frame-rate can be largely alleviated.

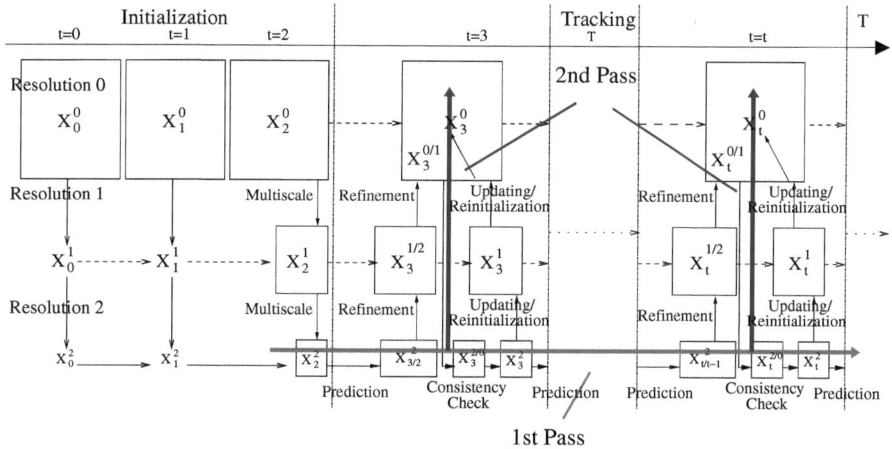

Fig. 5.1 General scheme of multi-resolution particle filter tracking with consistency check in mono-camera environments

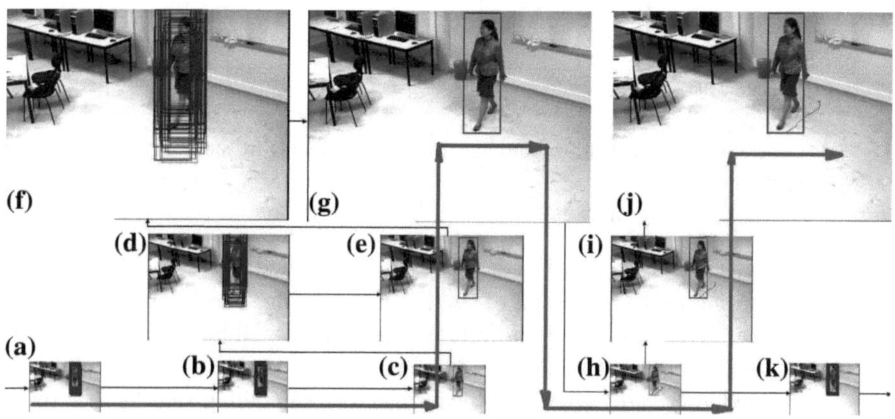

Fig. 5.2 Illustration of multi-resolution particle filter tracking with consistency check in mono-camera environments

In this chapter, we further extend the method to multi-camera tracking. Suppose $\mathbf{X}_t^{l,c}$ is an estimate of object state at time t of resolution l in camera c computed by the mean of a predefined number N^l of particles $\mathbf{x}_{i,t}^l$ with $i = 1, 2, \ldots, N^l, t = 0, 1, \ldots, T, l = 0, 1, \ldots, L$. Figure 5.3 and Table 5.1 shows the general scheme and process of our method Multi-resolution Particle Filter Tracking by Multiple Cameras in two-camera environments. The method consists fundamentally of Single Camera Tracking stage and Interaction of Cameras stage.

During the single camera tracking stage, for every camera $c = 1, 2$, the estimate of object state $\mathbf{X}_{t/t-1}^{L,c}$ is rapidly located by the prediction at lowest resolution $l = L$. It is gradually refined at every higher level. The refined estimate $\mathbf{X}_t^{0/1,c}$ at full

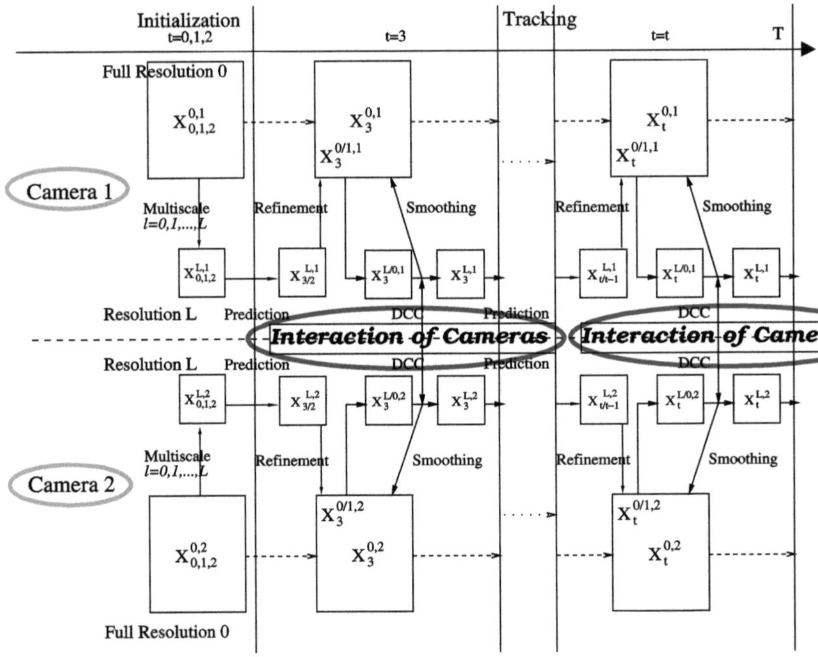

Fig. 5.3 General scheme of multi-resolution particle filter tracking by multiple cameras in two-camera environments

resolution $l = 0$ is then repassed to the lowest level for the Dual Consistency Check (DCC). According to the test result of $\mathbf{X}_t^{L/0,c}$, an updating, a self reinitialization or an adaptive iteratively weighted multiple reinitialization will be carried out during the Interaction of Cameras (IoC) stage. The final object estimate of $\mathbf{X}_t^{l,c}$ on all levels can be obtained afterwards after the object trajectory smoothing step. A new set of particles is generated and resampled based on this final estimate in each camera in order to prepare for the tracking to the next frame. The whole process is repeated until the last frame $t = T$.

5.2.1 Single Camera Tracking

In our method, a particle $\mathbf{x}$ is composed of a motion $\mathbf{M}(\mathbf{x})$ and an appearance state $\mathbf{H}(\mathbf{x})$. The motion state consists of a static $\mathbf{E}$, a velocity $\mathbf{E}'$ and an acceleration state $\mathbf{E}''$ which are used for an adaptive 1st/2nd order propagation. The appearance state contains a joint color histogram **HA** in a chosen 3D color space, a 2D cumulative histogram **HC**, a color weight w based on **HA** and a color distance d based on **HC**. In practice, a particle is a rectangle centered at coordinates (x, y) and of size (sx, sy), i.e., $\mathbf{E}=(x, y, sx, sy)$. The object is included in this rectangle. Thus, a

Table 5.1 General process of multi-resolution particle filter tracking by multiple cameras in two-camera environments

For time $t(t = 0, \ldots, T)$:

I. **Single camera tracking** ($l = 0, \ldots, L, c = 1, 2$):
1. **Initialization**:
(a) Multiscaling: scale frame $\mathbf{I}_t^{0,c}$ into $\mathbf{I}_t^{l,c}$;
(b) Color space: change RGB to HSV if necessary;
(c) Quantization: quantize frame into $\mathbf{I}_t^{q,l,c}$;
(d) Detection ($t = 0, 1, 2$):initial object state as $\mathbf{X}_0^{l,c}, \mathbf{X}_1^{l,c}, \mathbf{X}_2^{l,c}$;

For every camera $c(c = 1, 2)$:
2.**Prediction I** ($l = L$):
(a) Propagation: propagate $\mathbf{B}_{t-1}^{'L,c}$ by $\mathbf{M}(\mathbf{X}_{t-1}^{L,c})$ into $\mathbf{B}_{t/t-1}^{L,c}$;
(b) Estimation: estimate object state, noted as $\mathbf{X}_{t/t-1}^{L,c}$;
3. **Refinement** ($l = L - 1, \ldots, 0$):
(a) Generation: generate $\mathbf{B}_t^{l,c}$ on $\mathbf{X}_t^{'l/l+1,c}$;
(b) Resampling: resample $\mathbf{B}_t^{l,c}$ to $\mathbf{B}_t^{'l,c}$;
(c) Estimation: estimate object state, noted as $\mathbf{X}_t^{l/l+1,c}$;
(d) Repeat (a)–(c) until the full resolution 0;
4.**Dual Consistency Check** ($l = L, \ldots, 0$):
(a) Pass $\mathbf{M}(\mathbf{X}_t^{0/1,c})$ to $\mathbf{MX}_t^{L/0,c}$;
(b) DCC Test ($l = L$);
End for, go to II;

For every camera $c(c = 1, 2)$:
5.**Smoothing** ($l = L, \ldots, 0$):
smooth the object corrected estimate at the lowest resolution, pass it to other levels, noted as $\mathbf{X}_t^{l,c}$;
6.**Prediction II** ($l = L$):
(a)Generation: generate a set of particles $\mathbf{B}_t^{l,c}$ on $\mathbf{X}_t^{l,c}$;
(b)Resampling: resample $\mathbf{B}_t^{l,c}$ to $\mathbf{B}_t^{'l,c}$;
End for, go to I1;

II. **Interaction of Cameras** ($l = L, \ldots, 0, c = 1, 2, c' = 2, 1$):
1. Pass $\mathbf{MX}_t^{L/0,c}$ to $\mathbf{MX}_t^{'0/L,c}$;
2. If $d(\mathbf{X}_t^{L/0,c}) \leq \Gamma_0^c$: Updating;
3. If $\Gamma_0^c < d(\mathbf{X}_t^{L/0,c}) < \Gamma_1^c$: Self-Reinitialization;
4. If $d(\mathbf{X}_t^{L/0,c}) \geq \Gamma_1^c$ and $d(\mathbf{X}_t^{L/0,c'}) \leq \Gamma_0^{c'}$:
Adaptive Iteratively Weighted Multiple Reinitialization;
5. If $d(\mathbf{X}_t^{L/0,c}) \geq \Gamma_1^c$ and $d(\mathbf{X}_t^{L/0,c'}) \geq \Gamma_0^{c'}$: Hidden Tracking;
6. In all cases, $\mathbf{X}_t^{'l,c}$ is obtained, go to I5;

End for

particle descriptor is presented as [2]:

$$\mathbf{x} = (\mathbf{M}, \mathbf{H}) = ((\mathbf{E}, E', E''), (\mathbf{HA}, \mathbf{HC}, w, d). \tag{5.1}$$

During the single camera tracking stage, the object is tracked individually in parallel in both cameras. For the sake of simplicity, we ignore the index c in this section.

5.2.1.1 Initialization

The initialization step (cf. Fig. 5.3 and Table 5.1, Initialization) scales the video frames to several resolution levels by a Gaussian pyramid decomposition [6], quantizes them in RGB/HSV color space as well as initializes the object initial state. It also detects the object initial state on the first three frames.

Thus, the velocity and acceleration state are calculated as:

$$\mathbf{E}'(\mathbf{X}_t^0) = \frac{\mathbf{E}(\mathbf{X}_t^0) - \mathbf{E}(\mathbf{X}_{t-\Delta t}^0)}{\Delta t}; \mathbf{E}''(\mathbf{X}_t^0) = \frac{\mathbf{E}'(\mathbf{X}_t^0) - \mathbf{E}'(\mathbf{X}_{t-\Delta t}^0)}{\Delta t}. \tag{5.2}$$

with $\Delta t = 1$ in our case.

A simple down scaling based on the multi-resolution pyramid is used to determine the object motion state at lower resolution levels:

$$\mathbf{M}(\mathbf{X}_t^{l+1}) = \frac{1}{\varepsilon}\mathbf{M}(\mathbf{X}_t^l). \tag{5.3}$$

with ε the scaling coefficient.

The object appearance state $\mathbf{H}(\mathbf{X}_t^l)$ at different resolution levels is calculated individually based on its own static state at that resolution. In our method, two histograms are required. The joint histogram **HA** is calculated as [3]:

$$\mathbf{HA}(\mathbf{X}_t^l)^u = g_n \sum_{j=1}^{J} g\left(\frac{\|\mathbf{E}(\mathbf{X}_t^l) - \mathbf{E}(\mathbf{X}_{t(j)}^l)\|}{a}\right)\delta[h(\mathbf{HA}(\mathbf{X}_{t(j)}^l) - u]. \tag{5.4}$$

where $U = 256/\eta * 256/\eta * 256/\eta$, the total bin number in the histogram.

One of the novelties in our tracking method is that a Dual Consistency Check is applied to evaluate the consistency of the estimate. It is realized by comparing the marginal cumulative histogram of the current estimate with that in the previous frame. The marginal cumulative histogram **HC** first integrates **HA** to one of the 3 components (**HM** marginal histogram) and then accumulates itself along each component (**HC** cumulative histogram):

$$\mathbf{HC}(\mathbf{X}_t^l)^z = \begin{pmatrix} \mathbf{HC}(\mathbf{X}_t^l)^z_{R/H} \\ \mathbf{HC}(\mathbf{X}_t^l)^z_{G/S} \\ \mathbf{HC}(\mathbf{X}_t^l)^z_{B/V} \end{pmatrix}$$

$$= \begin{pmatrix} \mathbf{HC}(\mathbf{X}_t^l)^{z-1}_{R/H} + (\sum_{a=0}^{A})_{G/S}(\sum_{b=0}^{B})_{B/V}\mathbf{HA}(\mathbf{X}_t^l)^z_{R/H} \\ \mathbf{HC}(\mathbf{X}_t^l)^{z-1}_{G/S} + (\sum_{a=0}^{A})_{R/H}(\sum_{b=0}^{B})_{B/V}\mathbf{HA}(\mathbf{X}_t^l)^z_{G/S} \\ \mathbf{HC}(\mathbf{X}_t^l)^{z-1}_{B/V} + (\sum_{a=0}^{A})_{R/H}(\sum_{b=0}^{B})_{G/S}\mathbf{HA}(\mathbf{X}_t^l)^z_{B/V} \end{pmatrix} \quad (5.5)$$

with $z = 1, \ldots, 256/\eta - 1$ and $A = B = 256/\eta - 1$.

5.2.1.2 Prediction

The prediction stage is realized only at lowest level L (cf. Fig. 5.3 and Table 5.1, Prediction). It consists of generation, resampling, propagation and estimation steps.

A set of particles $\mathbf{x}_{i,t-1}^L$ is generated in the neighborhood of object final estimate $\mathbf{X}_{t-1}^L$ on the previous frame $t-1$ during the generation step. They are resampled in a Sequential Importance Sampling way [7] during the resampling step. In the next propagation step, they are then propagated to the current frame t by an adaptive 1st/2nd order way where the object velocity and acceleration are updated with time (cf. Eq. 5.2):

$$\mathbf{M}(\mathbf{x}_{i,t}^L) = \mathbf{D} \cdot [\mathbf{M}(\mathbf{X}_{t-1}^L)]^T + [\mathbf{M}(\mathbf{x}_{i,t-1}^L)]^T + \mathbf{V}_p^L. \quad (5.6)$$

where $\mathbf{x}_{i,t}^L$ is a particle in the set $\mathbf{B}_{t-1}^{\prime L}$ with $i = 1, \ldots, N^L$, $\mathbf{D}$ is the propagation matrix and $\mathbf{V}_p^L$ is a Gaussian white noise vector $\mathbf{V}_p^L \sim \begin{pmatrix} N(0, (\sigma_p^{2L}) \\ \ldots \quad) \\ N(0, (\sigma_p^{2L}) \end{pmatrix}_{12\times 1}$.

Here, we consider a particle incomplete motion state in the vector $\mathbf{M}(\mathbf{x}_{i,t-1}^L) = (x, y, sx, sy, 0, 0, 0, 0, 0, 0, 0, 0)$ (cf. Eq. 5.1). The new particle set is noted as $\mathbf{B}_{t/t-1}^L$.

There are two propagation ways: an adaptive first order (AFO) way **D1** and an adaptive second order(ASO) way **D2**:

$$\mathbf{D1} = \begin{pmatrix} \mathbf{0} & \Delta t\mathbf{I} & \mathbf{0} \\ \mathbf{0} & \mathbf{I} & \mathbf{0} \\ \mathbf{0} & \mathbf{0} & \mathbf{I} \end{pmatrix}_{12\times 12} ; \mathbf{D2} = \begin{pmatrix} \mathbf{0} & \Delta t\mathbf{I} & \frac{\Delta t^2}{2}\mathbf{I} \\ \mathbf{0} & \mathbf{I} & \mathbf{0} \\ \mathbf{0} & \mathbf{0} & \mathbf{I} \end{pmatrix}_{12\times 12} \quad (5.7)$$

with $\mathbf{I}_{4\times 4}$, an identity matrix and $\mathbf{0}_{4\times 4}$, a zero matrix. Here, $\Delta t = 1$.

Once the particles have been propagated into the current frame, their joint histogram $\mathbf{HAx}_{i,t}^L$ is calculated individually based on their new motion state $\mathbf{Mx}_{i,t}^L$ during the estimation step.

The color weight of each particle is computed as:

$$w(\mathbf{x}_{i,t}^{L}) = \frac{1}{\sqrt{2\pi}\sigma_e{}^L} e^{-\left(\frac{1-\rho[\mathbf{HA}(\mathbf{x}_{i,t}^{L}),\mathbf{HA}(\mathbf{X}_{t-1}^{L})]}{2\sigma_e^{2L}}\right)}. \tag{5.8}$$

with σ_e^{2L} the variance of a white noise and ρ the Bhattacharyya coefficient [3]. A weight normalization of particles is carried out to ensure that $\sum_{i=1}^{N_t^L} w(\mathbf{x}_{i,t}^{L}) = 1$.

Thus, the object state can be estimated as [3]:

$$\mathbf{M}(\mathbf{X}_{t/t-1}^{L/L}) = \sum_{i=1}^{N^L} w(\mathbf{x}_{i,t}^{L})\mathbf{M}(\mathbf{x}_{i,t}^{L}). \tag{5.9}$$

Then $\mathbf{M}$ $(\mathbf{X}_{t/t-1}^{L})$ is passed immediately to higher resolution levels for the refinement.

5.2.1.3 Refinement

During the refinement stage, the estimate is passed to higher levels and refined gradually (cf. Fig. 5.3 and Table 5.1, Refinement). That is, the estimate $\mathbf{X}_{t/t}^{l/l+1}$ at level $l+1$ is passed to the next successive higher level l by image pyramid [6]. A new set of particles is generated in the neigborhood of the passed estimate and resampled. The object state is re-estimated.

5.2.1.4 Dual Consistency Check

The refined estimate is repassed to lowest level $\mathbf{X}_{t/t}^{L/0}$ for DCC (cf. Fig. 5.3 and Table 5.1, DCC). Because the Particle Filter methods are predictive methods, an efficient metric has to be proposed to evaluate the reliability of the particle set. In most of the PF methods, the effective sample size is applied to check the effectiveness of a particle set [7]. In our method, there is a repetitive generation of particle set in the neighborhood of the last estimate. The effectiveness of the particle set is always very high. Therefore, a new evaluation has to be brought about. We propose the Dual Consistency Check based on Kolmogorov-Smirnov (KS) test. It checks directly the reliability of the object estimate in the place of the particle set.

In statistics, the Kolmogorov-Smirnov test is a goodness of fit test to determine whether two hypothesized distributions are generated by the same underlying probability distribution based on the finite samples [8]. Both samples are described by their cumulative distribution discrete functions, i.e., cumulative histogram, which is given by the "Smirnov test" [9]. It calculates the similarity of 2 histograms by building and comparing the cumulative distribution function of each histogram.

We use the KS test to decide the reliability of the checked estimate. It is conducted only at the lowest resolution level L. First, the KS statistics is calculated. In our case, it is the color distance $d(\mathbf{X}_t^{L/0})$ of the marginal cumulative histogram between the object checked estimate after the refinement stage and the final estimate in the reference frame:

$$d(\mathbf{X}_t^{L/0}) = \max_z |\mathbf{HC}(\mathbf{X}_\tau^L)^z - \mathbf{HC}(\mathbf{X}_t^{L/0})^z|. \tag{5.10}$$

with $z = 0, 1, \ldots, 256/\eta - 1$ and τ the index of the reference frame. The result is actually the biggest difference of the marginal cumulative histogram on the 3 components between $\mathbf{X}_\tau^L$ and $\mathbf{X}_t^{L/0}$.

Kolmogorov and Smirnov proved that if the observed distance $d(\mathbf{X}_t^{L/0})$ is greater than a threshold Γ_0, the 2 experimental distributions are not from the same hypothesized distribution:

$$\Gamma_0 = \sqrt{-\frac{1}{2}(\frac{1}{g} + \frac{1}{h}) \cdot \ln \lambda_0}. \tag{5.11}$$

where g and h are the cardinality of statistical samples. In our case, $g = h = (256/\eta) \times 3$, hence:

$$\Gamma_0 = \sqrt{-((256/\eta) \times 3)^{-1} \cdot \ln \lambda_0}. \tag{5.12}$$

λ_0 is a parameter depending on the probability β_0 that a tracking failure happens. It is calculated by:

$$\lambda_0 \sim \sqrt{-\frac{\ln \beta_0}{2}}. \tag{5.13}$$

A second parameter β_1 is also applied, defined as the chance that the occlusion on object happens, the object disappears from the scene and the scene changes. It decides the second threshold λ_1 as the same way as Eq. 5.13. Therefore, the reliability of the object estimate is actually depended on these two parameters $\beta_{0,1}$.

According to the area that the color distance of object estimate falls in, the object consistency is evaluated as:

- $d(\mathbf{X}_t^{L/0}) < \Gamma_0$: a good estimate;
- $\Gamma_0 < d(\mathbf{X}_t^{L/0}) < \Gamma_1$: a right estimate;
- $d(\mathbf{X}_t^{L/0}) > \Gamma_1$: a bad estimate;

When a bad estimate is announced, the tracking is interrupted. However, simply stop the tracking can not principally solve these problems. Thus, an Interaction of Cameras stage in a multi-camera environment is proposed in order to relocate the "bad" estimate in the current camera by the "good" estimates in the other camera in case of tracking degeneracy.

Cam 1 \ Cam 2	$d_t^2 < T_0^2$	$T_0^2 < d_t^2$, $d_t^2 < T_1^2$	$T_1^2 < d_t^2$
$d_t^1 < T_0^1$	(1) Cam 2: UP; Cam 1: UP	(2) Cam 2: SR; Cam 1: UP	(3) Cam 2: AIWMR; Cam 1: UP
$T_0^1 < d_t^1$, $d_t^1 < T_1^1$	(2) Cam 2: UP; Cam 1: SR	(4) Cam 2: SR; Cam 1: SR	(5) Cam 2: HT; Cam 1: SR
$T_1^1 < d_t^1$	(3) Cam 2: UP; Cam 1: AIWMR	(5) Cam 2: SR; Cam 1: HT	(6) Cam 2: HT; Cam 1: HT

With:

a. $T_0^1 < T_1^1$, $T_0^2 < T_1^2$

b. UP: Updating (mono–camera)

c. SR: Self–Reinitialization(mono–camera)

d. AIWMR: Adaptive Iteratively Weighted Multiple Reinitialization (multi–camera)

e. HT: Hidden Tracking (mono–camera)

Fig. 5.4 Table of interaction of cameras in two-camera environments

5.2.2 Interaction of Cameras

During the Interaction of Cameras stage (cf. Fig. 5.3 and Table 5.1, Interaction of Cameras), the view correspondence between cameras are calculated based on the object trajectory in each camera by the Iteratively Reweighted Least-Squares (IRLS) method [10]. It is conducted at the full resolution $l = 0$ since the number of resolution levels used for tracking in 2 cameras might not be the same. The possible combinations of each camera action based on the "good", "right" or "bad" estimate are shown in Fig. 5.4 and listed below:

1. $UP \leftrightarrow UP$: if a **good** estimate is obtained in both cameras, an updating (UP) step is conducted in each camera separately (cf. (1) in Fig. 5.4);
2. $UP \leftrightarrow SR$: if a **good** estimate is found in one camera while a **right** one is found in the other, an updating and a self-reinitialization (SR) step are carried out respectively (cf. (2) in Fig. 5.4);
3. $UP \leftrightarrow AIWMR$: if a **good** estimate is found in one camera while a **bad** one is found in the other, an updating and an adaptive iteratively weighted multiple reinitialization (AIWMR) step are carried out respectively (cf. (3) in Fig. 5.4);
4. $SR \leftrightarrow SR$: if a **right** estimate is obtained in both cameras, a self-reinitialization step is conducted in each camera separately (cf. (4) in Fig. 5.4);
5. $SR \leftrightarrow HT$: if a **right** estimate is found in one camera while a **bad** one is found in the other, a self-reinitialization and a hidden tracking (HT) step are carried out respectively (cf. (5) in Fig. 5.4);
6. $HT \leftrightarrow HT$: if a **bad** estimate is obtained in both cameras, a hidden tracking step is conducted in each camera separately (cf. (6) in Fig. 5.4).

5.2.2.1 Updating

The updating (UP) step is conducted when the object is considered consistent, where object joint color histogram at every level is updated recursively [5]:

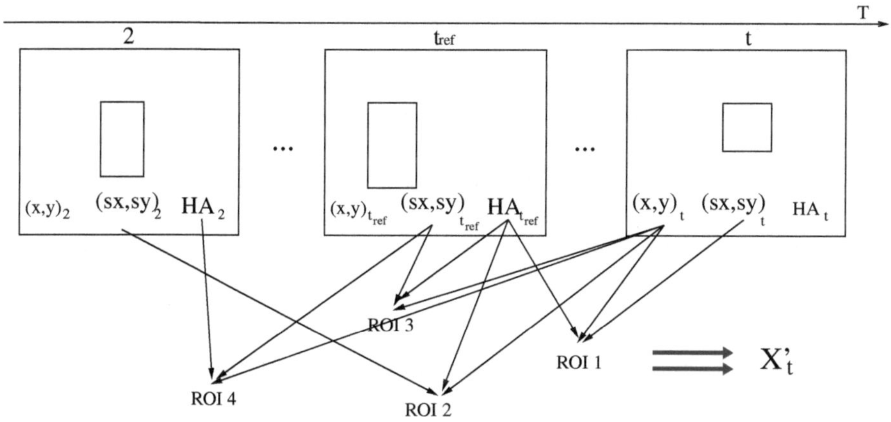

Fig. 5.5 Illustration of the four ROI reinitialization ways

$$\mathbf{HA}(\mathbf{X}_t^{l,c}) = \alpha_0 \mathbf{HA}(\mathbf{X}_t^{l,c}) + (1 - \alpha_0)\mathbf{HA}(\mathbf{X}_{t-1}^{l,c}). \quad (5.14)$$

with $\alpha_0 \in [0, 1]$ a predefined update coefficient.

5.2.2.2 Self-Reinitialization

The self-reinitialization (SR) step occurs when a "right" estimate is announced. The object estimate is reinitialized by its past state in the same camera in region-of-interest (ROI) way. Four different ROI ways are proposed here according to the different failure situations:

1. ***Reinitialization by ROI 1***: use its own position, its own size and the appearance in the nearest past estimate: $\mathbf{E}(\mathbf{X}_t^{'L}) = \mathbf{E}(\mathbf{X}_t^{L/0})$; $\mathbf{H}(\mathbf{X}_t^{'L}) = \mathbf{H}(\mathbf{X}_{t_{ref}}^{L})$
2. ***Reinitialization by ROI 2***: use its own position, the size in the initial estimate and the appearance in the nearest past estimate: $\mathbf{E}(\mathbf{X}_t^{'L}) = \{(x, y)(\mathbf{E}(\mathbf{X}_t^{L/0})), (sx, sy)(\mathbf{E}(\mathbf{X}_2^{L}))\}$; $\mathbf{H}(\mathbf{X}_t^{'L}) = \mathbf{H}(\mathbf{X}_{t_{ref}}^{L})$
3. ***Reinitialization by ROI 3***: use its own position, the size in the nearest past estimate and the appearance in the nearest past estimate: $\mathbf{E}(\mathbf{X}_t^{'L}) = \{(x, y)(\mathbf{E}(\mathbf{X}_t^{L/0})), (sx, sy)(\mathbf{E}(\mathbf{X}_{t_{ref}}^{L}))\}$; $\mathbf{H}(\mathbf{X}_t^{'L}) = \mathbf{H}(\mathbf{X}_{t_{ref}}^{L})$
4. ***Reinitialization by ROI 4***: use its own position, the size in the nearest past estimate and the appearance in the initial estimate: $\mathbf{E}(\mathbf{X}_t^{'L}) = \{(x, y)(\mathbf{E}(\mathbf{X}_t^{L/0})), (sx, sy)(\mathbf{E}(\mathbf{X}_{t_{ref}}^{L}))\}$; $\mathbf{H}(\mathbf{X}_t^{'L}) = \mathbf{H}(\mathbf{X}_2^{L})$

Here, t_{ref} is the frame index. This frame is one frame after the frame where last reinitialization has occurred. $\mathbf{X}_2^{L}$ is the initial object state on the frame $t = 2$ where the tracking started. Figure 5.5 illustrates the four different ROI ways.

5.2.2.3 Hidden Tracking

The hidden tracking (HT) occurs when a "bad" estimate in the current camera **and no** "good" estimate in the other camera is announced. In this case, there is no need to save the tracking result since it is surely inaccurate. However, the tracking is continued in a hidden way. That is, the tracking continues, only the estimate is abandoned (not saved/shown) on the current frame. The tracker stays in the neighborhood of the last estimate where the object is occluded, it disappears from the scene or the scene changes.

5.2.2.4 Adaptive Iteratively Weighted Multiple Reinitialization

Contrarily to the hidden tracking, Adaptive Iteratively Weighted Multiple Reinitialization (AIWMR) is conducted when a "bad" estimate in the current camera **and a** "good" estimate in the other camera is announced.

Since no prior knowledge on the scene is assumed in our method, the only way to establish this relationship is then by using the object state already estimated, i.e., the object trajectory in both cameras. Hence, the Iteratively Reweighted Least-Squares (IRLS) method in MRPFMC is used for its ability to handle the regression situations in which the data points are of varying quality [10]. More specifically, it is employed to calculate the Adaptive Iteratively Weighted Transformation Matrix (AIWTM) between two cameras at the full resolution.

Suppose the object trajectory matrix in the current and the other camera are:

$$\mathbf{Tr}_t^c = \begin{pmatrix} x(\mathbf{X}_0^{0,c}) & y(\mathbf{X}_0^{0,c}) & 1 \\ x(\mathbf{X}_1^{0,c}) & y(\mathbf{X}_1^{0,c}) & 1 \\ \cdots & \cdots & \cdots \\ x(\mathbf{X}_{v_t}^{0,c}) & y(\mathbf{X}_{v_t}^{0,c}) & 1 \end{pmatrix}_{t\times 3} ; \mathbf{Tr}_t^{c'} = \begin{pmatrix} x(\mathbf{X}_0^{0,c'}) & y(\mathbf{X}_0^{0,c'}) & 1 \\ x(\mathbf{X}_1^{0,c'}) & y(\mathbf{X}_1^{0,c'}) & 1 \\ \cdots & \cdots & \cdots \\ x(\mathbf{X}_{v_t}^{0,c'}) & y(\mathbf{X}_{v_t}^{0,c'}) & 1 \end{pmatrix}_{t\times 3} . \tag{5.15}$$

with $v_t = 0, 1, \ldots t-1$, the number of frames having already been processed at time t.

Therefore, the view correspondence between cameras can be computed by the view correspondence matrix AIWTM [11]:

$$\mathbf{Tr}_t^c \cdot \Lambda_t = \mathbf{Tr}_t^{c'}. \tag{5.16}$$

where the AIWTM is defined as $\Lambda_t = \begin{pmatrix} a & d & 0 \\ b & e & 0 \\ c & f & 1 \end{pmatrix}_{3\times 3}$ with a, b, e, d, c, f, the coefficients of affine transformation. Figure 5.6 illustrates the relationship between object trajectory matrix and the view correspondance matrix.

A new AIWTM is calculated whenever the AIWMR step is required based on the object trajectory already estimated at that instant. The reason that the object trajectory

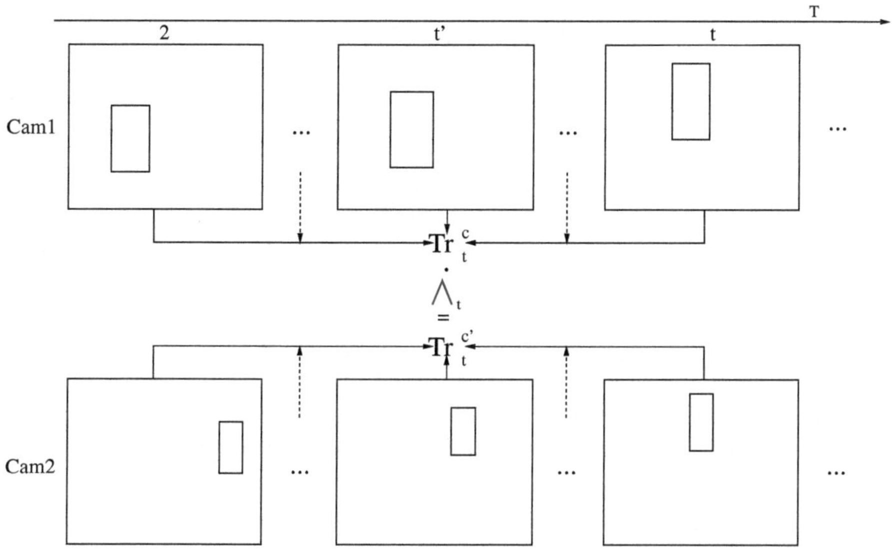

Fig. 5.6 Relationship between object trajectory matrix and the view correspondance matrix

can be used to compute the 2D transformation between cameras is that there is one but only one final estimate at every frame (and the cameras are stationary). The estimates in the two cameras are surely correspondent. However, those estimates contain the tracking error whose standard deviation at each frame is not constant. By applying the IRLS method, the influence of this error can be reduced for its ability of assigning different importance to the estimates of different tracking accuracy. Therefore, the AIWTM in our method can be computed as [12]:

$$\Lambda_t = ((\mathbf{Tr}_t^c)^T \cdot \mathbf{WT}_t \cdot \mathbf{Tr}_t^c)^{-1} \cdot (\mathbf{Tr}_t^c)^T \cdot \mathbf{WT}_t \cdot \mathbf{Tr}_t^{c'}. \quad (5.17)$$

with the weight matrix defined as:

$$\mathbf{WT}_t = \begin{pmatrix} wt_0 & 0 & \dots & 0 & 0 \\ 0 & wt_1 & \dots & 0 & 0 \\ \dots & \dots & \dots & \dots & \dots \\ 0 & 0 & \dots & 0 & wt_{v_t} \end{pmatrix}_{t \times t} . \quad (5.18)$$

Thus, finding the best view correspondence matrix AIWTM is actually to find the best weight matrix. In [13], the authors report on the use of various robust estimators to the problem of view correspondence for stereo imaging. The Lorentz and German-McClure estimators yield good results. From the general theory of robust parameter estimation presented in the fundamental work by [14], the weights can be found from the derivation of estimators. Hence, for Lorentz and German-McClure estimators, they will be as follows:

$$wt_{v_t,L} = \frac{2}{2\delta^2 + r_{v_t}^2}. \tag{5.19}$$

$$wt_{v_t,G} = \frac{2\delta}{(\delta^2 + r_{v_t}^2)^2}. \tag{5.20}$$

with δ^2, a predefined weight distribution variance.

The r_{v_t} is the Euclidean distance between the object estimate in the current camera and the projection of the estimate from the other camera to the current camera. Suppose the projected estimate is:

$$(x, y, 1)(\mathbf{X}_{v_t}^{0,c/c'}) = (x, y, 1)(\mathbf{X}_{v_t}^{0,c'}) \cdot \Lambda_t^{-1}. \tag{5.21}$$

Then, the Euclidean distance between the estimate in the current camera at the time v_t and its projection from the other camera is calculated as [15]:

$$r_{v_t} = \sqrt{(x(\mathbf{X}_{v_t}^{0,c/c'}) - x(\mathbf{X}_{v_t}^{0,c}))^2 + (y(\mathbf{X}_{v_t}^{0,c/c'}) - y(\mathbf{X}_{v_t}^{0,c}))^2}. \tag{5.22}$$

The calculation of the AIWTM is repeated m_{v_t} times until the weight matrix is comparatively stable:

$$|\sum_{v_t=0}^{t-1} wt_{v_t}^{m_{v_t}} - \sum_{v_t=0}^{t-1} wt_{v_t}^{m_{v_t}-1}| < \epsilon_0. \tag{5.23}$$

with ϵ_0, a random generated threshold close to 0.

Once the AIWTM is obtained, it can be used to project the "good" estimate of the other camera to the current camera and relocate the "bad" estimate. The size of the projected estimate is reinitialized by the object initial size in the current camera and its appearance is computed correspondingly. The tracking then goes to the single camera tracking stage for the object trajectory smoothing.

5.3 Examples and Discussion

In this section, several tracking examples will be shown and discussed. The system implementation is written in C++ and runs on a standard single-core 3.00 GHz CPU Linux computer. To avoid the initialization error, the object initial state is initialized manually. The different parameters are predefined experimentally.

5.3.1 Multi-Resolution Improvement

The performance of a tracking method depends on many factors. Among them, observation frame-rate, background clutters and illumination changes do have an important impact, even for a single object tracking [16]. As a consequence, it is quite important to fully examine each application problem formulation.

Our objective is to realize a real-time non-rigid object tracking in a complex indoor environment. The term "non-rigid" indicates that the object changes its shape and histogram along time. When it moves in a complex indoor environment, its speed varies, the illumination changes and the occlusion happens from time to time. Besides, a real-time tracking requires a system with a considerable computational ability. All of these have to be taken into account in system design in order to get a satisfactory tracking accuracy.

The inspiration of using Multi-resolution technique comes from our applications with videos of low and variable frame-rate. Some videos in our disposal have only 8 fps. The object motion between successive frames in some of our videos has more than 60 pixels. By passing the prediction step to lower resolution levels, the object motion can be largely reduced, therefore, make it possible to use particle filter methods.

Moreover, we not only track the object on the lowest resolution, we also refine the estimate on higher resolutions. Figure 5.7 shows a comparison of tracking only on full resolution, only on lowest resolution and on all the resolution levels. The object motion magnitude between two successive frames is about 30 pixels. Tracking only on full resolution (first line) uses a simple version of the method developed based on [3]. Tracking only on lowest resolution (second line) uses a simple version of the method developed based on [6]. And the tracking on all the resolution levels actually uses our tracking method presented in this chapter. It can be seen that the tracking degeneracy happens quickly if only track on the full resolution (method in [3]). However, if the object is tracked only on the lowest resolution $L = 2$, the tracking degeneracy does not happen, but the object size is not well estimated (method in [6]). By adding the refinement at higher resolution levels, our method not only succeeds in tracking, but also has a good accuracy of the estimate size (third line).

Another advantage of using multi-resolution technique is the reduction of computational time. A large number of particles is needed in order to get a good approximation of the object probability distribution. However, this number is always a bottleneck for calculation, especially when tracking the object of large size [6]. By passing thc prediction stage onto the lowest resolution, the neighborhood of the particle generation become smaller. Thus, the necessary number of particles can be greatly reduced. For example, if 600 particles are needed to succeed in tracking only on the full resolution 0, only 200 particles are needed on the level 2, plus the 150 particles for the refinement on resolution 1 and 100 on resolution 0. Thus, the total necessary number for tracking in our method will be 450. In this way, the computational time is reduced.

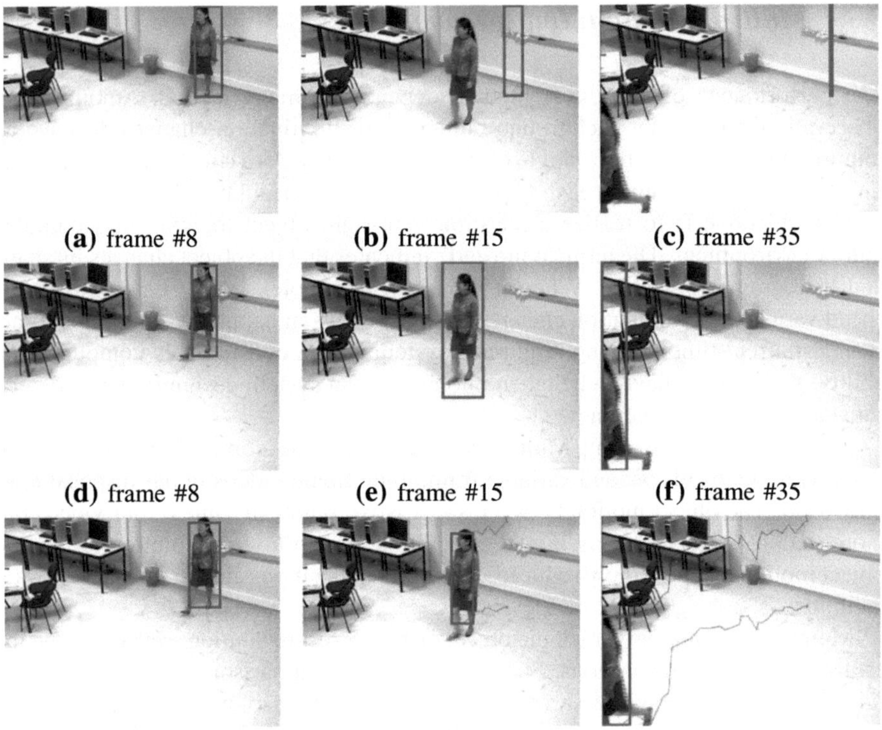

(a) frame #8 (b) frame #15 (c) frame #35

(d) frame #8 (e) frame #15 (f) frame #35

(g) frame #8 (h) frame #15 (i) frame #35

Fig. 5.7 Comparison of tracking on 3 resolution levels, result shown on the full resolution. *First line*: tracking only on the full resolution; *second line*: tracking only on the lowest resolution; *third line*: tracking on all the 3 resolutions. (Sequence "Yifan2Cam1", corpus LaBRI, image size on full resolution 640 × 480)

Finally, only the estimate (mean state) of a particle set is passed between time or resolution levels. In other words, a repetitive particle generation is carried out in our tracking method. The original thought of doing this is that as the prediction stage is only conducted at the lowest resolution, it is unnecessary to propagate this set to higher resolutions just for the refinement purpose as the object state has already been well pre-located. Therefore, when refining the object estimate at high resolutions, a new particle set is always generated on the same level. The biggest benefit of this repetitive particle generation is that the well-known degeneracy problem and the sample impoverishment problem in Particle Filters methods are somehow avoided.

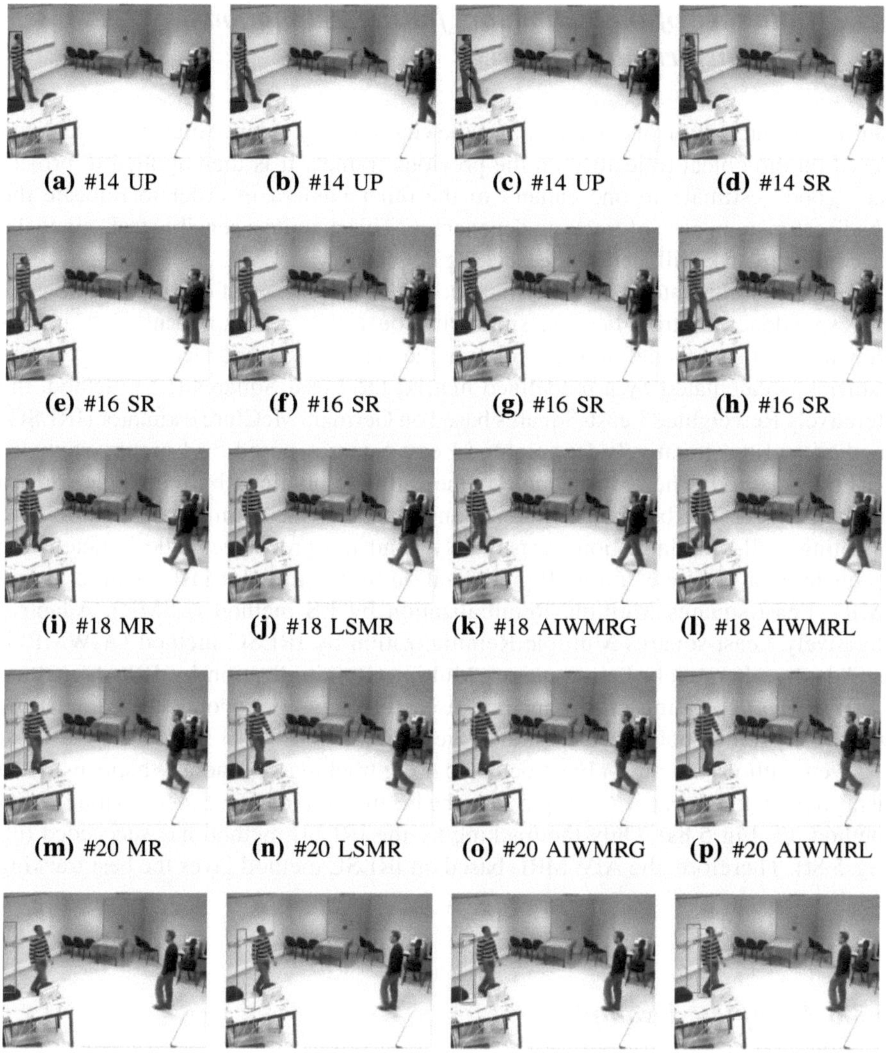

(a) #14 UP **(b)** #14 UP **(c)** #14 UP **(d)** #14 SR

(e) #16 SR **(f)** #16 SR **(g)** #16 SR **(h)** #16 SR

(i) #18 MR **(j)** #18 LSMR **(k)** #18 AIWMRG **(l)** #18 AIWMRL

(m) #20 MR **(n)** #20 LSMR **(o)** #20 AIWMRG **(p)** #20 AIWMRL

(q) #22 MR **(r)** #22 LSMR **(s)** #22 AIWMRG **(t)** #22 AIWMRL

Fig. 5.8 Comparison of different ways of calculating the view correspondence matrix. *First column*: by a predefined matrix; *second column*: by LS method; *third column*: by IRLSG method; *fourth column*: by IRLSL. (Sequence "RonanChris", corpus LaBRI, image size on full resolution 640 × 480)

5.3.2 *Influence of the Estimators for the Estimation of Transformation Matrix*

During the interaction of cameras stage, the view correspondence matrix is calculated based on the object trajectory on the previous frames. It is then applied to project the "good" estimate in one camera to the other camera in order to relocate the "bad" estimate in case of tracking degeneracy. This is quite a sensible task, since the inaccurate matrix will lead to the tracking degeneracy.

Figure 5.8 illustrates a comparison among different ways of calculating the view correspondence matrix. For the simplicity, only the tracking result in Cam2 is shown here. From the left to right column, the view correspondence matrix between cameras is calculated by a predefined matrix, the Least-Squares (LS) method, the Iteratively Reweighted Least-Squares based on German-McClure estimator (IRLSG) method and the Iteratively Reweighted Least-Squares based on Lorentz estimator (IRLSL) method. The red rectangle is the estimate obtained by the single camera tracking stage. The black and blue rectangle present the estimate obtained by the updating, self-reinitialization, respectively. And the green rectangle indicates the estimate obtained by Multiple Reinitialization by the initial transformation matrix (MR), Least-squares Multiple Reinitialization by LS method (LSMR), Adaptive Iteratively Least-squares Multiple Reinitialization by IRLSG method (AIWMRG) or Adaptive Iteratively Least-squares Multiple Reinitialization by IRLSL method (AIWMRL) depending on the method used for view correspondence calculation, which is specified below each frame in the figure.

Even without the inter-object occlusion, the tracking degeneracy happens when using a predefined matrix (cf. Fig. 5.8q), the LS method (cf. Fig. 5.8r) and the IRLSG method (cf. Fig. 5.8s). Only the tracking by the IRLSL method has succeeded (cf. Fig. 5.8t). Therefore, the AIWMRL based on IRLSL method gives the best tracking result.

5.3.3 *Complete Examples*

Figure 5.9 shows an example of tracking a car in an outdoor environment. The left column is the tracking result in Cam1 while the right column is that in Cam2. The red rectangle is the estimate obtained by the single camera tracking stage. The black, blue and green rectangle present the estimate obtained by UP, SR and AIWMRL, respectively. The object final tracking trajectory is illustrated in Fig. 5.9o, p.

Our system succeeds in tracking the car in both two cameras. In particularly, the system once locates the head of the car in Cam2 (cf. black rectangle in Fig. 5.9j). The estimate which is at the back of the car is considered incorrect by the system (cf. red rectangle in Fig. 5.9i); and relocated by the estimate in Cam2 which is exactly at the head of the car (cf. green rectangle in Fig. 5.9i). However, it can be noticed that

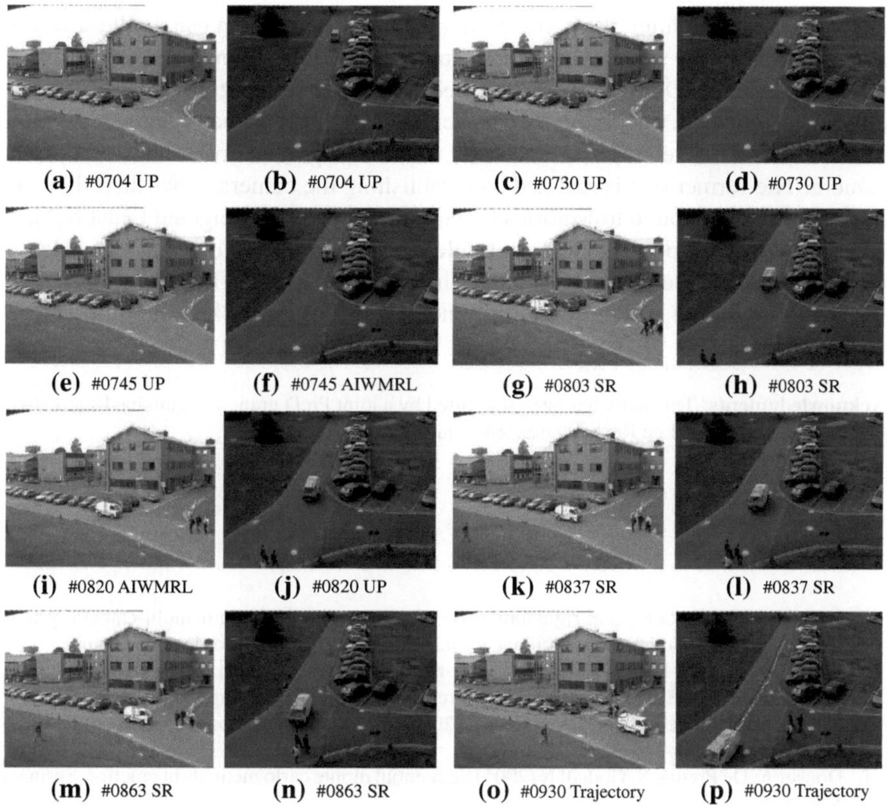

Fig. 5.9 Example of tracking a car in a parking. (Sequence "Data1Testing", corpus PETS2001, image size on full resolution 768 × 576, 231 frames tested)

the object shape is not well estimated when the car is zooming. This is due to the constraint of color-based particle filter method.

Finally, the processing rate of this sequence is 2.69 f/s versus 25 f/s observation rate.

5.4 Conclusion and Perspectives

In this chapter, a novel Multi-resolution Particle Filter Tracking with a Dual Consistency Check for Model Update in an environment of two uncalibrated static cameras is presented. The cameras track a non-rigid object in parallel in a meander strategy by using multi-resolution particle filter method. In this way, the object strong non-linear articulated motion change can be greatly reduced, therefore, make it possible for the prediction step of particle filter methods. However, due to this short term temporal and spatial particle filter, the particle filter degeneracy problem can not be solved in a conventional way, such as Sequential Importance Resampling. That is the reason why we propose a Dual Consistency Check based on Kolmogorov-Smirnov test to directly evaluate the tracking result in each camera. Once a "bad" estimate is found in one camera, it is relocate/self locate by the correct/past estimate in the other/same camera. The former one is realized by establishing the camera views directly using the object already found trajectories based on Iteratively Reweighted Least-Squares method. However, our system is not able to achieve a real-time tracking. Hence, one of our future work is to accelerate the processing rate. The other future work includes developing system to multiply non-rigid objects tracking in multi-camera environment.

Acknowledgments This work has been supported by a joint Ph.D grant of Aquitaine Region from CNRS (Centre National de Recherche Scientifique).

References

1. Zhou Y (2010) Tracking non-rigid multi-object based on particle filter in multi-camera systems for the application of video surveillance. Ph.D. Thesis, University Bordeaux 1, France
2. Zhou Y, Nicolas H, Benois-Pineau J A multi-resolution particle filter tracking in a multi-camera environment. In: IEEE International Conference on Image Processing, 2009
3. Nummiaro K, Koller-Meier E, Van Gool L (2003) An adaptive color-based particle filter. Image Vis Comput 21:99–110
4. Doucet A, De Freitas N, Godsill N (2001) Sequential monte carlo methods in practice. Springer, New York
5. Zhou Y, Benois-Pineau J, Nicolas H Multi-resolution tracking of a non-rigid target with particle filters for low and variable frame-rate videos. In: 10th International Workshop on Image Analysis for Multimedia Interactive Services. IEEE, 2009
6. Dutta Roy S, Tran SD, Davis LS, Sreenivasa Vikram B (2008) Multi-resolution tracking in space and time. In: IEEE ICVGIP, pp 352–358
7. Arulampalam MS, Maskell S, Gordon N, Clapp T (2002) A tutorial on particle filters for online nonlinear/non-gaussian bayesian tracking. IEEE Trans. Sig. Process 50(2):174–188
8. Li M, Vitányi P (1997) An introduction to kolmogorov complexity and its applications. Springer, New York
9. Van Der Waerden BL (1967) Statistique mathématique. Dunaud press
10. Carroll RJ, Ruppert D (1988) Transformation and weighting in regression. Chapman and Hall, London
11. Black J, Ellis T (2006) Multi camera image tracking. Image Vis. Comput. 24:1256–1267
12. D'Apuzzo N (2000) Motion capture by least squares matching tracking algorithm. In AVATARS, Lausanne, Switzerland
13. Hasler D, Sbaiz L, Süsstrunk S, Vetterli M (2003) Outlier modeling in image matching. IEEE Trans Pattern Anal Mach Intell 25(3):301–315
14. Odobez J-M, Bouthemy P (1995) Robust multiresolution estimation of parametric motion models. J Vis Comm Image Represent 6(4):348–365
15. Del Bimbo A, Dini F, Grifoni A, Pernici F (2008) Uncalibrated framework for on-line camera cooperation to acquire human head imagery in wide areas. AVSS, pp 252–258
16. Lanz O (2007) An information theoretic rule for sample size adaptation in particle filtering. ICIAP 317–322

Chapter 6
Activity Detection Using Regular Expressions

Mattia Daldoss, Nicola Piotto, Nicola Conci and Francesco G. B. De Natale

Abstract In this chapter we propose a novel method to analyze trajectories in surveillance scenarios by means of Context-Free Grammars (CFGs). Given a training corpus of trajectories associated to a set of actions, a preliminary processing phase is carried out to characterize the paths as sequences of symbols. This representation turns the numerical representation of the coordinates into a syntactical description of the activity structure, which is successively adopted to identify different behaviors through the CFG models. The obtained model is the basis for the classification and matching of new trajectories *versus* the learned templates and it is carried out through a parsing engine that enables the online recognition of human activities. An additional module is provided to recover parsing errors (i.e., insertion, deletion, or substitution of symbols) and update the activity models previously learned. The proposed system has been validated in indoor, in an assisted living context, demonstrating good capabilities in recognizing activity patterns in different configurations, and in particular in presence of noise in the acquired trajectories, or in case of concatenated and nested actions.

Keywords Activity analysis · Context-free grammar · Regular expressions · Activity classification · Anomaly detection

6.1 Introduction

In the recent years there has been a relevant attention towards the implementation of automatic systems for activity detection and analysis in several application areas, including environmental monitoring and video surveillance [13]. The analysis

M. Daldoss · N. Piotto · N. Conci (✉) · F. G. B. De Natale
Multimedia Signal Processing and Understanding Lab, DISI—University of Trento,
Via Sommarive 14, 38123 Trento, Italy
e-mail: conci@disi.unitn.it

N. Adami et al. (eds.), *Analysis, Retrieval and Delivery of Multimedia Content*,
Lecture Notes in Electrical Engineering 158, DOI: 10.1007/978-1-4614-3831-1_6,

consists of the extrapolation of meaningful information about the event occurring in the observed scene by interpreting and classifying low level features such as objects trajectories [18, 19]. There are though situations in which the trajectory analysis tools may be misleading, due to their connection to physical and geometrically referenced displacements. In fact, geometric displacements can be of great help in a lot of situations where the personalization does not play a crucial role, such as for example traffic monitoring (vehicles are driven by humans, but they have to comply with a set of rules) [17], or in case the structure and the semantic of the environment are not known a priori (monitoring in outdoor [20]). On the contrary, in other and more defined circumstances, the structure of the environment is well known, typically facilitating the interpretation of the context in which a specific activity is carried out. This is the case for example of human behavior analysis in homes or offices. In these situations the topology of the rooms is typically kept constant over a considerably long range of time. However, although the observed space is static, subjects move freely and tend to perform also the most common actions in slightly different manners each time. This personalization factor is clearly not voluntary and relies on a number of factors that cannot easily be measured. In practice, it consists of collecting sets of trajectories, in which the point-to-point displacement of the moving subject is different each time, thus making it impossible to apply conventional curve matching tools. The activity detection can though be achieved by analyzing the event at a higher level of abstraction, establishing connections between the moving subject and the environment, and trying to categorize the activities on the basis of interactions. Interactions can be of different forms, but in general, and considering trajectories as the principal source of information, they can be modeled as the permanence for a specific amount of time in the neighborhood of one or more meaningful spots in the room.

The proposed framework stems from a preliminary work [4] the authors have carried out in this area, and concerns a video analysis tool that exploits regular expressions to automatically associate an observed activity pattern to a template of activities learned a priori. In this work actions are modeled through an abstraction of the top-view trajectory, which is summarized into a symbolic stream of *Hot Spots*, each of them corresponding to specific and relevant areas in the observed environment. The derived expressions corresponding to the activity models, are automatically learned as separate CFGs using a set of training sequences. In the test phase, the resulting symbolic strings are parsed using the Earley-Stolcke algorithm to determine the similarity against all available CFGs. The main contribution with respect to the previous work consists of the progressive update of the activity database and therefore the capability of the system in adapting to the changes in the environment, as well as users' habits.

The chapter is structured as follows. After a quick overview of the algorithms available in literature to capture and recognize activities (Sect. 6.2), Sect. 6.3 provides a concise description of the CFG formalism. Section 6.4 introduces the proposed framework focusing on the representation and discovery of activities, together with the presentation of the corresponding matching strategy. The experimental validation is presented in Sect. 6.6 for an indoor scenario.

6.2 Related Work

The research in the area of analysis and modeling of complex activities is attracting a lot of researchers also thanks to the reliability of motion tracking algorithms, which allow locating the moving objects in the video with constantly increasing precision. Once the spatio-temporal object displacement is extracted for each time instant, data processing can be performed at different abstraction levels. Treating the trajectories as generic series of coordinates, one could apply low-level matching techniques [5, 2] to classify incoming samples and match them with pre-stored templates. Although these techniques can effectively work in some simple situations, their application to trajectories associated to complex activities, may give unreliable results due to different factors, including noise or other uncertainties that typically affect the low-level data.

For these reasons, higher level reasoning is often applied on top of the raw data processing modules. The available literature is quite rich with this respect and some of the most relevant approaches are summarized in the following paragraphs.

A common and widely used way to model the structure of human behaviors relies on purely probabilistic approaches exploiting, for example, Hidden Markov Models (HMM), related modifications [6], as well as Dynamic Bayesian Networks (DBN) [12]. The general idea of these approaches is to extract sets of features from the low-level data and feed them into the probabilistic graphical model used to define the event structure.

As an example, the work in [6] implements a strategy to learn and recognize human activities through a special type of Hidden Markov Models (Switching Hidden Semi-HMM). A two-layer representation is proposed: in the bottom layer a sequence of concatenated Hidden Semi-Markov Model (generalization of HMM with random state duration) defines the atomic activities; the upper layer handles the temporal structure of the activities composing the event by means of a sequence of switching variables. In the same spirit, the authors of [16] proposed a Hierarchical HMM, in order to exploit both the hierarchical structure and the shared semantics contained in the movement trajectories. Moreover, they introduce a Rao-Blackwellised particle filter in the recognition engine in order to cope with real-time recognition constraints. Exploiting such a representation, the method first learns the actions of a subject from an unsegmented training data set, and successively performs an online activity classification, segmentation and anomaly detection. Among the purely probabilistic approaches, an alternative is proposed in [12] where a scalable approach for complex activity recognition is described. The system includes three major modules: a low-level action detector, for the extraction of sub-events from the low-level data, a Dynamic Bayesian Network (DBN) that encodes the prior knowledge of sub actions ordering constraints, and a Viterbi-based inference algorithm, used to maintain the most likely activity given the DBN status and the output of the low-level detectors.

The main advantage of these methods is the capability of handling the uncertainties generated during the low-level processing. On the other hand, as the event complexity increases, the recognition performance dramatically drops, due to a combination of factors including insufficient training data, semantic ambiguity in the model of the process, or temporal ambiguity in competing hypothesis. Although some methods for unsupervised parameter estimation of the graphical model have been proposed [3], the major problem remains the definition of the network topology, which is usually too complex to be learned from a sparse dataset, and is commonly pre-defined by human operators.

Some other approaches perform the activity recognition in a symbolic domain. In particular, in these works an intermediate step is introduced between the low-level feature extraction and the high level reasoning. The low-level primitive processing is carried out in different ways (e.g., HMM or similar), while for the high level behavior modeling a common approach is to adopt the Context-Free Grammar formalism. In [9], for example, the authors propose to split the problem into two parts, using a statistical approach to detect primitives (low-level activities), and a syntactic approach to detect the high-level structures. In the first phase, HMMs are employed to propose candidates for low-level temporal features; these features serve then as input for the Stochastic Context-Free Grammar (SCFG), providing longer range temporal constraints, disambiguating uncertain low-level detections, and allowing the inclusion of a priori knowledge about the structure of temporal events. In [14] a system is proposed to generate detailed annotations of complex behaviors of humans performing the Towers of Hanoi through a parameterized and manually-defined stochastic grammar, able to identify both single operations as well as more complex tasks. In [15] the authors also use SCFG to extract high-level behaviors from video sequences, in which multiple subjects can perform different separable activities. An alternative approach is proposed in [10]. Here, the so-called attribute grammars [11] are employed as descriptors for features that are not easily represented by finite symbols. They provide, in other words, a formal way to define attributes for the production rules of a formal grammar. The final goal of the proposed work is to recognize activities and signal potential anomalies. In particular, the proposed framework can handle concurrent behaviors involving multiple entities, as well as uncertainties in semantic conditions on the attributes which are used to express a confidence measure over the recognized events.

A common drawback of the systems relying on formal grammars is in the definition and update of the production rules. In fact, an exhaustive formalization and structuring of the observable activities a person can perform in everyday life, is in practice not available, since all possible actions cannot be defined a priori. For this reasons, in [8] a computational framework is proposed, able to recognize behaviors in a minimally supervised manner, relying on the assumption that everyday activities can be encoded through their local event subsequences, and assuming that this encoding is sufficient for activity discovery and classification. In this work, the authors introduce the concept of *Motif*, defined as the most frequent subsequences that appeared in the data collection phase, that may be associated to relevant atoms to be recognized as behaviors. The activity recognition is then based on the discovery and matching of the *Motif* elements. However, since behaviors are modeled using rigid variable-length event subsequences, the method is sensitive to the noise introduced for example by changing the order of the sub-events. Another major limitation of SCFG based system that prevents their use in real applications is that the parsing strategy can handle only sequential relations between sub-events, with no capability in catching the parallel temporal relations often existing in complex events. Recently, to overcome this issue, the authors of [21] have proposed to extract the terminal symbols of a SCFG from motion trajectories. In particular, the motion trajectories are transformed into a set of basic motion patterns (*primitives*) that are taken as terminals for the formal grammar. Then, a rule induction algorithm based on the Minimum Description Length (MDL) is proposed to automatically derive the spatio-temporal structure of the event from the primitive stream. The complex temporal logic between atomic events is modeled through a combination of SCFG and Allen's temporal logic, while a Multi-Thread Parsing algorithm with Viterbi-like error recovering is developed in order to recognize interesting events in the stream.

6.3 Overview on Context-Free Grammars

As a brief introduction to CFGs [7], we define a *language* as a set of strings over a finite set of symbols. The CFG is used as a formal mean to specify which strings belong to the language. A CFG is defined as in (6.1):

$$G = (N, \Sigma, P, S) \tag{6.1}$$

where N is a finite set of *non-terminal* symbols, Σ is a finite set of *terminal* symbols ($N \cap \Sigma = 0$), P is a finite set of rules of the form $A \rightarrow \alpha$ ($A \in N$ and $\alpha \in (N \cup \Sigma)^*$), and S is the starting symbol ($S \in N$).

In order to decide whether a given string X is compatible with a grammar G, a *parser* scans it from left to right: for each symbol X_i a set of states is constructed, representing the conditions of the recognition process. The algorithm is composed of three stages, recursively executed:

- *Prediction*: estimates the possible continuation of the input, based on the current position in the parsing process
- *Scanning*: reads the next input symbol and matches it against all pending states. The states not confirmed by the read symbol are discarded
- *Completion*: updates the states confirmed by the scanning phase

If during *Scanning* the procedure encounters symbols that do not match any of the predicted pending states (i.e., the input does not verify any of the available rules), the procedure is aborted; the algorithm terminates successfully otherwise.

Before going into the details of the proposed solution, a toy example that describes how the grammars work is provided in order to highlight the power and the flexibility of the matching through CFGs. The flowchart of Fig. 6.1 summarizes the steps required to implement the action *Watching TV*: the activity includes three main parts, i.e., the initial stage (b), the core of the action (c) and an end phase (d). Every section, represented by a *non-terminal* symbol, is composed by one or more symbols, each of

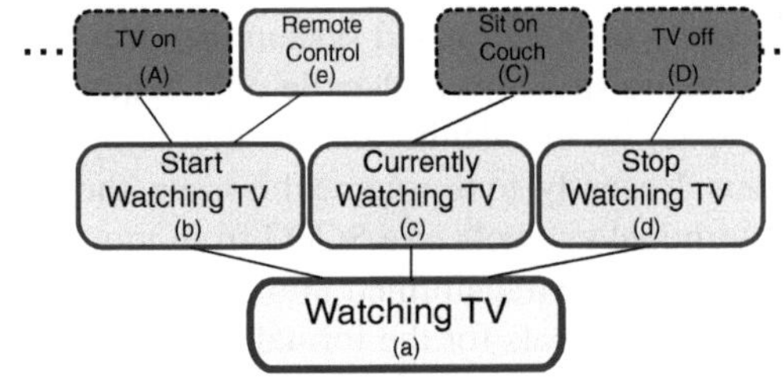

Fig. 6.1 Flowchart for "Watching TV"

Table 6.1 Complete CFG rules for example in Fig. 6.1

$act \rightarrow$	a
$a \rightarrow$	$b\ c\ d$
$b \rightarrow$	$A\ e$
$e \rightarrow$	$\varepsilon \mid B$
$c \rightarrow$	C
$d \rightarrow$	D

them being either a *terminal* (i.e., dark grey boxes) or a *non-terminal* symbol(light grey boxes), leading to the tree structure reported in Table 6.1. Some symbols can also represent *wildcard* elements, as for example the block corresponding to the *Remote Control (e)*, making the presence of one or more specific symbols optional (ε). As a consequence, the sequences reported in 6.2 and 6.3 are both recognized as the action *Watching TV*:

$$act_1 \rightarrow ACD \tag{6.2}$$

$$act_2 \rightarrow ABCD \tag{6.3}$$

6.4 Proposed Framework

Observing and tracking an object allows the characterization of its motion by means of a trajectory. The trajectory consists of the frame-by-frame displacement of the moving object with respect to a reference system (the ground plane is typically adopted). In order to fulfill the requirements of the CFG framework, it is necessary to convert the numerical representation of the trajectory into a symbolic stream. This process can be carried out at different abstraction layers, such as for example the quantization of the ground floor into a coarser grid, in order to avoid sudden fluctuations of the object centroid projection. In our implementation we have chosen to synthesize the trajectory as the succession of significant spots as pointed out also in [8], therefore describing an activity as the corresponding symbol string resulting from the concatenation of these elements. At first, a number of areas of interest in the observed room is selected, and then each raw trajectory is converted into an event

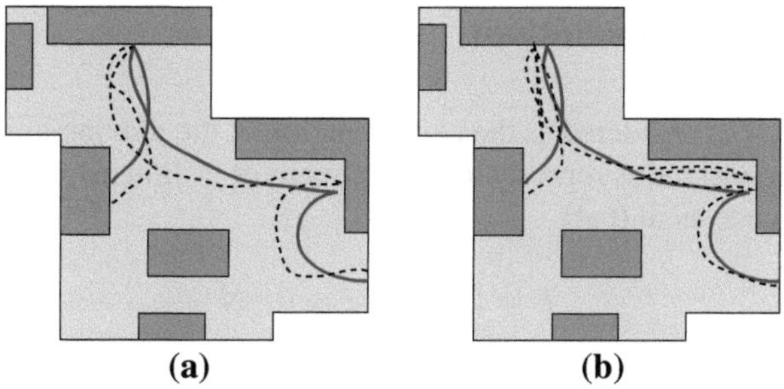

Fig. 6.2 Two different instances (*dashed line*) of the same action (*solid line*)

stream of concatenated spanned regions. This representation, although much coarser than the storage of the frame-by-frame coordinates, allows gathering an exhaustive overview of the path the moving object has performed. The motion rules identifying the training paths are automatically extracted and a specific CFG is computed for each activity, allowing a parallel and online classification of specific behaviors.

After the initial setup, the method we propose operates as follows:

(i) *pre-processing* of the incoming paths and representation in a symbolic domain;
(ii) *automatic discovery* of the grammar sets encoding the motion rules for the training activities (learning);
(iii) *verification*, through parsing, of whether an incoming trajectory fits any of the available rules (classification);
(iv) *update* of the grammar rules corresponding to the learned activities.

The choice of a framework, purely based on syntactical elements, compared to more traditional approaches for trajectory analysis, arises from the fact that people perform actions by introducing each time a certain level of personalization, which makes any instance of the same activity different one from each other. Therefore, the application of techniques that rely only on the spatial displacements, may be in these situations less appropriate. For the same reason it can happen that the connection among relevant elements for a specific actions is at some point interrupted due external events such as for example, the phone ringing, or someone knocking at the door. In this situation, the resulting trajectory would be strongly altered and the match with the pre-stored template would be missed, even though the global action is the same. As an example, in Fig. 6.2 it is possible to observe two different instances of the same action, each of them showing non-negligible spatiotemporal differences compared to the original template.

6.4.1 Activity Representation

Each trajectory T, considered as the concatenation of the moving object locations at successive time instant, is represented as a stream of 2-dimensional samples with a temporal reference as in (6.4)

$$T = \{P_i, t_i\};\ \ i = 0...N \tag{6.4}$$

where each sample $P_i = (x_i, z_i)$ is the projection on the ground plane of the moving object centroid at time i. As anticipated, it is worth noting that apart from the noise, object traces associated to the same activity may evolve in very different ways. This is the main motivation behind our choice of defining a number of *Hot Spots* instead of considering the whole motion trajectory, allowing for the extrapolation of a sort of *signature* of the activity. Finally, the activities are represented as the stream returned by the concatenation of *Hot Spots* the actor has interacted with, where the term interaction is reduced here to the proximity of the actor to the specific *Hot Spot* for a predefined temporal interval. This allows simplifying the representation in (6.4) to a stream of indexed regions associated to a timestamp, as in (6.5).

$$T' = \{R_j, t_j\};\ \ j = 0...M \tag{6.5}$$

In (6.5), R_j is an indexed *Hot Spot* and t_j is the corresponding temporal reference. Sampling the path in *Hot Spots* rather than at fixed time intervals allows preventing the potential issues arising from the acquisition phase, such as noise and outliers, yet preserving the general spatial evolution of the activity.

6.4.2 CFG Rules Discovery

For any of the considered activities, a set of symbolic sequences is employed in the grammar induction phase to discover its specific CFG rules. The strategy employed in this work relies on [1], a tool originally employed for NLP applications: here, each sentence (i.e., each symbolic sequence) is iteratively decomposed in *expressions* and *contexts*. For instance, in the sentence *Jack (drinks) juice*, *drinks* is the *expression* while *Jack (-) juice* is the *context*. Intuitively, given the entire set of training sentences, the algorithm searches for frequent combinations of *expressions* and *contexts* and interprets them as a grammatical type. Types are then extended, whenever possible, and the derivation rules based on the types are formulated as a CFG. Operatively, two distinct stages are carried out: initially, all the *expression/context* pairs are arranged in a matrix form, then a 2D-clustering algorithm is run to group the expressions appearing in the same context, allowing the effective grammar definition.

As a simple example taken from NLP let us consider 4 sentences:

- (s1) *Jack drinks juice*;

	(-) drinks juice	Jack (-) juice	Jack drinks (-)	(-) juice	Jack (-)	(-)	(-) likes juice	Jack likes (-)	(-) drinks water	Jack (-) water	(-) water	(-) likes water
Jack	●						●		●			●
drinks		●								●		
juice			●					●				
Jack drinks				●							●	
drinks juice					●							
Jack drinks juice						●						
likes		●								●		
Jack likes				●							●	
likes juice					●							
Jack likes juice						●						
water			●					●				
likes water					●							
Jack likes water						●						
drinks water					●							
Jack drinks water						●						

Fig. 6.3 List of all the *expression context* pairs for s1, s2, s3 and s4

	Jack (-) juice	Jack drinks (-)	(-) juice	Jack (-)	(-)	Jack likes (-)	Jack (-) water	(-) water
drinks	●						●	
juice		●				●		
Jack drinks			●					●
drinks juice				●				
Jack drinks juice					●			
likes	●						●	
Jack likes			●					●
likes juice				●				
Jack likes juice					●			
water		●				●		
likes water				●				
Jack likes water					●			
drinks water				●				
Jack drinks water					●			

(a)

["Jack (.) juice", {"drinks","likes"}]
["Jack drinks (.)", {"water","juice"}]
["(.) juice", {"Jack drinks","Jack likes"}]
["Jack (.)", {"drinks juice", "likes juice", "drinks water", "likes water"}}]
["(.)", {"Jack drinks juice", "Jack likes juice", "Jack drinks water", "Jack likes water"}]
["Jack likes(.)", {"water","juice"}]
["Jack (.) water", {"drinks","likes"}]
["(.) water", {"Jack drinks","Jack likes"}]

(b)

Fig. 6.4 **a** List of all the *expression context* pairs for s1, s2, s3 and s4 after initial clustering; **b** explicit rules

- (s2) *Jack likes juice*;
- (s3) *Jack drinks water*;
- (s4) *Jack likes water*.

All the initial *expression/context* pairs are reported in tabular format in Fig. 6.3. In an initial stage, a clustering of the context appearing in the same expressions is carried out, leading to the pairs in Fig. 6.4a and to the rules in Fig. 6.4b. A second clustering is successfully performed leading to the final pairs shown Fig. 6.5. The final grammar rules for the considered sequences are shown in Fig. 6.5b.

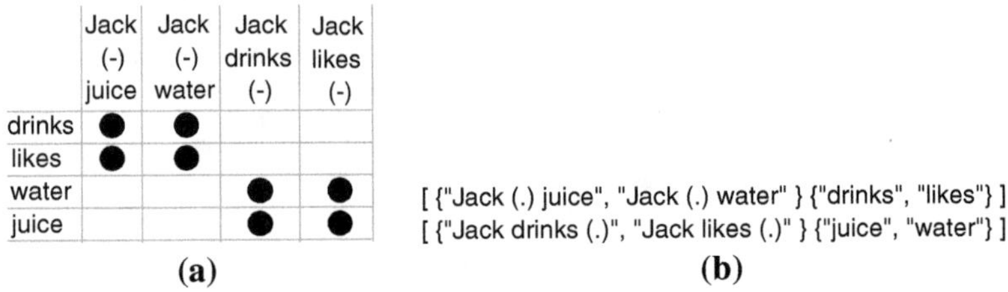

Fig. 6.5 **a** List of all the *expression context* pairs for s1, s2, s3 and s4 after the final clustering; **b** explicit final rules

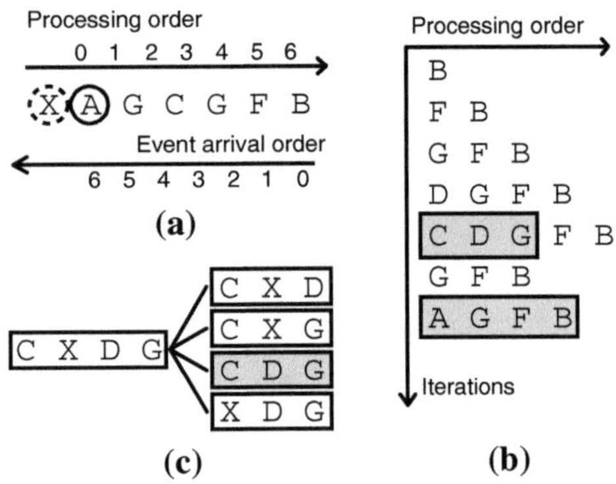

Fig. 6.6 **a** Sequence layout at the parser; **b** detection of nested actions; **c** detection of 1-corrupted sequence

6.4.3 CFG-Based Parsing for Activity Recognition

In order to recognize predefined activities in a stream of events, we developed an online algorithm to parse and analyze the incoming symbols. The stream of *Hot Spots* (referred to as *events*) as described in Sect. 6.4.1, is taken as input and the presence of symbol chunks (i.e., activities) satisfying the learned CFGs (as in Sect. 6.4.2) are verified. Symbols are processed as soon as they are acquired (see Fig. 6.6a) and combined with the previous events to verify the pertinence of the chunk against the learned CFGs. For computational reasons, the string is parsed backwards, removing the detected chunks as soon as they are identified. In this way, an activity is detected when it ends, without the need of generating multiple hypothesis. Moreover, the temporal reference of each event allows recovering both the duration and the hierarchy among possibly nested activities. In Fig. 6.6b the procedure is illustrated, considering *C D G* and *A G F B* as activities: according to the parsing order, the most recent symbol at each iteration is on the left hand side of the string. In order to maintain a finite list of possible events, a time-to-live constraint is also applied to the symbols, such that the oldest events are dropped from the stream and not considered for further processing.

Besides the detection of nested actions, an additional problem consists of the activity detection in presence of noise, mainly due to sparse events not related to any of the known activities. To overcome this issue, we perform a second level of analysis. We call $G(n)$ a set of n grammars, and $G_{max}(i)$ the maximum length of any of the possible sequence belonging to the i-th grammar. From the original sequence S of length l, we compute

$$S' = S(l-1)...S(l-q) \tag{6.6}$$

with

$$q = max_i\{G_{max}(i)\} + f \tag{6.7}$$

and f the number of corrupted symbols we choose to tolerate. Starting from S', we generate m new sequences S'', removing at each step one symbol $S(x)$ from the original string, as shown in Fig. 6.6c. The quantity m is obtained as follows:

$$m = \sum_{i=0}^{f-1} \prod_{j=0}^{i} (q-j) \tag{6.8}$$

We apply $G(n)$ on S'', in order to verify whether the sequences belong to any of the available grammars. If so, the symbols involved in the detected action are removed from the sequence, and $S(x)$ is restored.

6.4.4 CFG Rule Update

In order to maintain a coherent model for the learned activities, able to take into account the temporal evolution, an update procedure is required to include potential modifications. The choice we have adopted is to concentrate on a temporal window and rebuild the grammar rules considering the most frequent observations matching a given activity within that window. To this aim, a separate database is built to keep track of all the incoming sequences fulfilling the rules of each grammar. For every sequence matching the grammar, a set of features is stored, such as the corresponding activity (i.e., the grammar) identifier and the list of symbols belonging to it, together with the temporal reference that informs about the interaction time with the specific *hotspots*. Moreover, for every grammar, the system keeps track of the most recent observation, as well as, the list of all possible activity variations according to the error tolerance strategy.

The system is updated on a regular basis and in accordance with the application requirements. When an update takes place, the grammar rules for the models are recalculated following the procedure illustrated in Sect. 6.4.2 and considering the most recent set of observations for every particular grammar. Two combined criteria are considered before updating:

- the number of occurrences of the single instance;

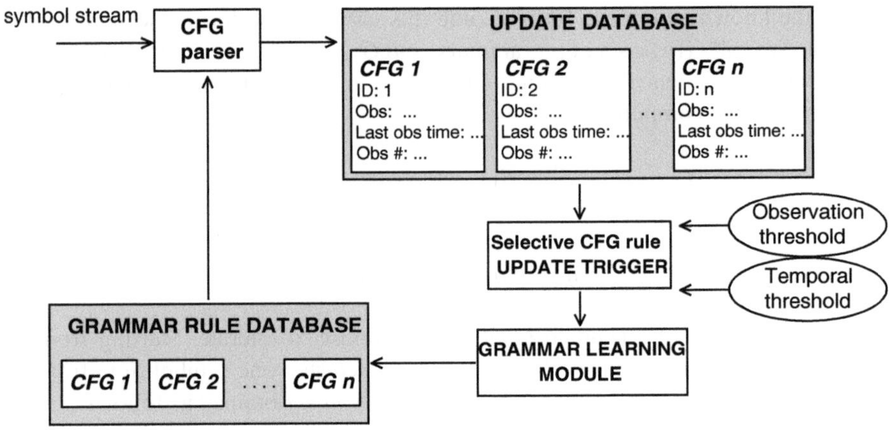

Fig. 6.7 Flowchart of the CFG rule update strategy

- the timestamp of the last recognized instance.

In other words, an update is operated only when the number of observations for a specific grammar exceeds a given occurrence threshold, and the time instant of the last observation is within the given temporal window. The occurrence threshold and the temporal window extension are tunable parameters regulating the update process. As a consequence of the update, the system can successfully adapt the detection of the defined actions to the variations that may occur due to changes of the environment (e.g., new environmental objects to interact with) or in the habits of the user (Fig. 6.7).

6.5 Results

In order to validate the proposed approach, we acquired a set of video sequences and extracted the top-view trajectories of each moving subject by means of calibrated camera. Being the tracking algorithm out of the scope of this work, the experimental validation only concerns the trajectory analysis module. All tests have been carried out in an assisted living lab (Fig. 6.8a) and concern people monitoring while performing three classes of activities, namely *Cooking*, *Serving food*, and *Taking a break*. In Fig. 6.9 a sample instance of each activity is reported for completeness.

Given the nature of the considered activities, the pre-processing phase described in Sect. 6.4 is implemented considering the *Hot Spots* in Fig. 6.8b. For the grammar induction phase, we selected and segmented five instances of each activity obtaining three different CFGs. For conciseness reasons, only the rules for the *Taking a break* activity are reported in Fig. 6.10a. The choice of these actions among all possible scenarios is due to the consideration that they can occur in sequential or nested configurations, thus slightly complicating the recognition process.

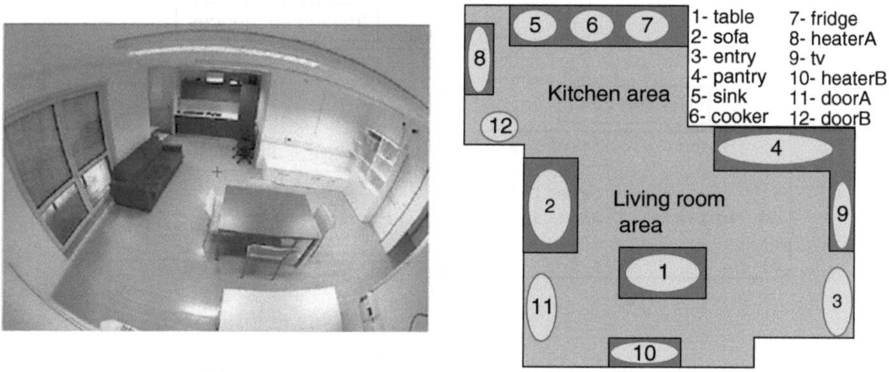

(a) (b)

Fig. 6.8 **a** Snapshot and **b** map of the environment. *Hot Spots* are provided with the corresponding legend

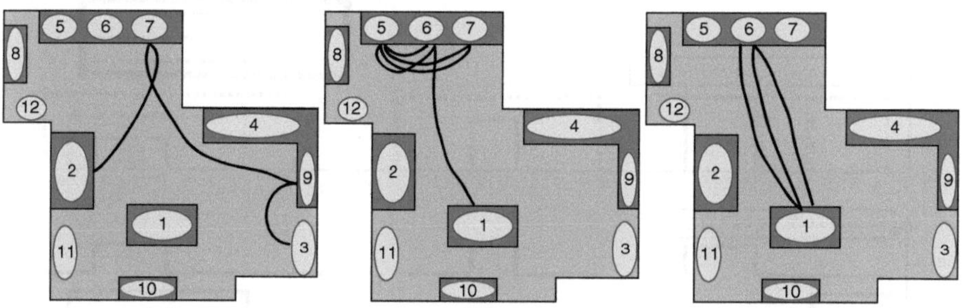

(a) (b) (c)

Fig. 6.9 Sample activity for **a** taking a break, **b** cooking, and **c** serving food

The test phase is divided in two different stages: we will first show how the learned grammars can generalize, by recognizing activity instances not defined in the training set; we will then show the capability of spotting activity sequences from a symbolic stream, also in a nested configurations.

For the first tests, let us take as an example the CFG learned from one simple activity (e.g., *Taking a break*), even though the reasoning can be extended to more complex grammars. As it can be seen from the rules reported in Fig. 6.10a, the structural skeleton of the activity is modeled through the first three rules: each rule provides for three possible variations, leading to a total number of sequences satisfying the CFG rules, equal to 9. In general, the number of sequences belonging to a given grammar depends on the grammar structure, and thus on the complexity of the sequences used in the training. In this specific case, the rules have been extracted from five training sequences (Fig. 6.10b) and the grammar returns a total number of nine candidates. The generalized sequences are shown in Fig. 6.10c, and result in

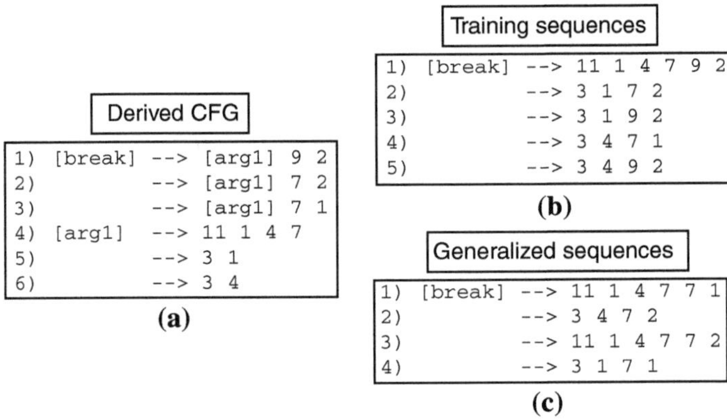

Fig. 6.10 **a** CFG for *Taking a break* (regions as in Fig. 6.8); **b** training sequences; **c** generalized realizations

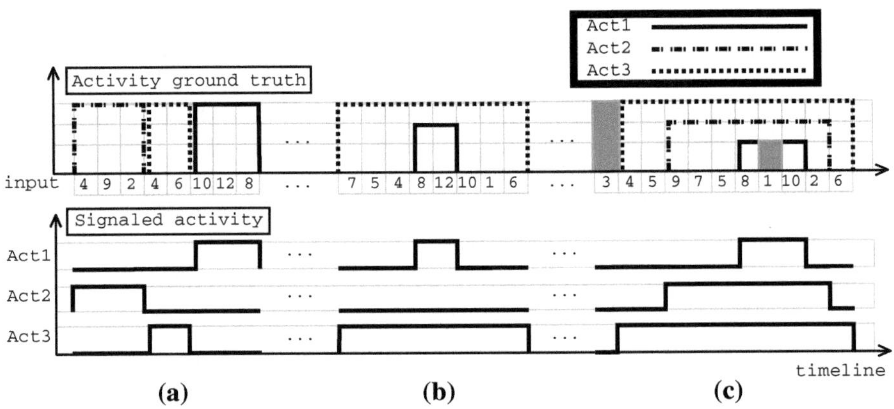

Fig. 6.11 Activity spotting example: **a** 3 consecutive sequences; **b** hierarchy between 2 activities; **c** 3 nested activities with noisy events

semantically legitimate examples of the activity *Taking a break*, even though they were not available in the original training set.

The second test phase aims at demonstrating the capability of the parsing strategy in spotting known activity patterns from a continuous event stream. In particular, given the backward parsing paradigm, we show how the proposed engine can recognize activities also in a nested form. To this aim, we randomly selected one activity instance for each of the three learned grammars and composed them in different nested configurations, as shown in Fig. 6.11. We first consider three consecutive activities (a), then two simply nested sequences (b), and, finally, a complex hierarchy including 3 activities with noisy symbols (i.e., grey box in the figure). From top to bottom we indicate (i) the ground truth for the activity stream, (ii) the event input stream, and (iii) the signaled activity. As it can be noticed, the system is fully capable of disclosing chunks of activities even if the incoming data stream is corrupted by noisy symbols.

The algorithm we have implemented is able to carry out the analysis of the incoming data, in compliance with the real-time constraints, in order to raise alarms in case an anomaly is detected.

6.6 Conclusions

In this chapter we have described a reconfigurable tool for activity detection in indoor scenarios based on CFGs. Starting from the trajectory acquired by the cameras installed in the environment, the algorithm takes as input a set of training trajectories. The data is processed in order to extract the meaningful information of the path, namely the interaction with a number of relevant areas in the observed environment. Through the identification of these *Hot Spots*, it is then possible to construct the CFG-based representation of the activities. During the test phase, the parsing algorithm is run to evaluate the CFG that best matches the current trajectory with respect to the acquired prototypes. The algorithm has been validated in an experimental test site targeted at monitoring daily activities of people living in the environment and it is able to recognize the activities, both as standalone, as well as in consecutive or nested hierarchies, also handling noisy events.

Acknowledgments The research has been developed under the project ACube funded by the Provincia Autonoma di Trento (Italy).

References

1. Adriaans PW, Vervoort M (2002) The emile 4.1 grammar induction toolbox. In: International colloquium on grammatical inference, Springer-Verlag GmbH, pp 293–295
2. Berndt D, Clifford J (1994) Using dynamic time warping to find patterns in time series. In: Workshop on knowledge discovery and databases, pp 229–248
3. Bilmes J (1998) A gentle tutorial of the em algorithm and its application to parameter estimation for gaussian mixture and hidden markov models. Technical Report
4. Daldoss M, Piotto N, Conci N, De Natale FGB (2010) Learning and matching human activities using regular expressions. In: IEEE international conference on image processing, pp 4681–4684
5. Das G, Gunopulos D, Mannila H (1997) Finding similar time series. In: Proceedings of the European symposium on principles of data mining and knowledge discovery, Springer-Verlag GmbH, pp 88–100
6. Duong TV, Bui H, Phung DQ, Venkatesh S (2005) Activity recognition and abnormality detection with the switching hidden semi-markov model. In: IEEE international conference on computer vision and pattern recognition, vol 1, pp 838–845
7. Earley J (1970) An efficient context-free parsing algorithm. Commun ACM 13(2):94–102
8. Hamid R, Maddi S, Johnson A, Bobick A, Essa I, Isbell C (2009) A novel sequence representation for unsupervised analysis of human activities. J Artif Intell 173(14):1221–1244

9. Ivanov YA, Bobick A (2000) Recognition of visual activities and interactions by stochastic parsing. IEEE Trans Pattern Anal Mach Intell 22(8):852–872
10. Joo SW, Chellappa R (2006) Attribute grammar-based event recognition and anomaly detection. In: IEEE international conference on computer vision and pattern recognition workshop, pp 107–107
11. Knuth DE (1968) Semantics of context-free languages. Theory Comput Syst 2(2):127–145
12. Laxton B, Lim J, Kriegman D (2007) Leveraging temporal, contextual and ordering constraints for recognizing complex activities in video. In: IEEE international conference on computer vision and pattern recognition, pp 1–8
13. Morris B, Trivedi M (2008) A survey of vision-based trajectory learning and analysis for surveillance. IEEE Trans Circuits Syst Video Technol 18(8):1114–1127
14. Minnen D, Essa I, Starner T (2003) Expectation grammars: leveraging high-level expectations for activity recognition. In: IEEE international conference on computer vision and pattern recognition, vol 2, pp 626–632
15. Moore D, Essa I (2001) Recognizing multitasked activities using stochastic context-free grammar. In: Proceedings of AAAI conference
16. Nguyen NT, Phung DQ, Venkatesh S, Bui H (2005) Learning and detecting activities from movement trajectories using the hierarchical hidden markov models. In: IEEE international conference on computer vision and pattern recognition, vol. 2, pp 955–960
17. Piciarelli C, Micheloni C, Foresti G (2008) Trajectory-based anomalous event detection. IEEE Trans Circuits Syst Video Technol 18(11):1544–1554
18. Piotto N, Conci N, De Natale F (2009) Syntactic matching of trajectories for ambient intelligence applications. IEEE Trans Multimedia 11(7):1266–1275
19. Prati A, Calderara S, Cucchiara R (2008) Using circular statistics for trajectory shape analysis. In: IEEE international conference on computer vision and pattern recognition, pp 1–8
20. Stauffer C, Grimson W (2000) Learning patterns of activity using real-time tracking. IEEE Trans Pattern Anal Mach Intell 22(8):747–757
21. Zhang Z, Tan T, Huang K (2011) An extended grammar system for learning and recognizing complex visual events. IEEE Trans Pattern Anal Mach Intell 33(2):240–255

Chapter 7
Shape Adaptive Mean Shift Object Tracking Using Gaussian Mixture Models

Katharina Quast and André Kaup

Abstract GMM-SAMT, a new object tracking algorithm based on a combination of the mean shift principal and Gaussian mixture models (GMMs) is presented. GMM-SAMT stands for Gaussian mixture model based shape adaptive mean shift tracking. Instead of a symmetrical kernel like in traditional mean shift tracking, GMM-SAMT uses an asymmetric shape adapted kernel which is retrieved from an object mask. During the mean shift iterations the kernel scale is altered according to the object scale, providing an initial adaptation of the object shape. The final shape of the kernel is then obtained by segmenting the area inside and around the adapted kernel into object and non-object segments using Gaussian mixture models.

Keywords Object tracking · Mean shift tracking · Gaussian mixture models

7.1 Introduction

There has been an increasing interest in object tracking as it is one of the most important and challenging tasks in computer vision. Among the many different methods developed for object tracking the mean shift algorithm [1, 2] is one of the most famous tracking techniques, because of its ease of implementation, computational speed, and robust tracking performance. Mean shift is a nonparametric statistical method which iteratively shifts each data point to the average of data points in its neighborhood [3]. It has been applied to several computer vision tasks such as segmentation [2] and object tracking [1, 4].

In spite of its advantages traditional mean shift tracking has two main drawbacks. The first problem is the fixed scale of the kernel or the constant kernel bandwidth. In

K. Quast (✉) · A. Kaup
Chair of Multimedia Communications and Signal Processing,
University of Erlangen-Nuremberg, Cauerstr. 7, 91058 Erlangen, Germany
e-mail: quast@lnt.de

N. Adami et al. (eds.), *Analysis, Retrieval and Delivery of Multimedia Content*,
Lecture Notes in Electrical Engineering 158, DOI: 10.1007/978-1-4614-3831-1_7,

order to achieve a reliable tracking result of an object with changing size an adaptive kernel scale is necessary. The second drawback is the use of a radial symmetric kernel. Since most objects are of anisotropic shapes a symmetric kernel with its isotropic shape is not a good representation of the object shape. In fact the symmetric kernel shape leads to an inclusion of background information into the target model, which can cause tracking failures. In order to achieve a reliable tracking result of an object with changing size and shape an adaptive kernel is necessary.

An intuitive approach of solving the first problem is to run the algorithm with three different kernel bandwidths, former bandwidth and former bandwidth ±10 %, and to choose the kernel bandwidth which maximizes the appearance similarity (±10 % method) [5]. A more sophisticated method using difference of Gaussian mean shift kernel in scale space has been proposed in [6]. The method provides good tracking results, but is computationally very expensive. And both methods are not able to adapt to the orientation or the shape of the object. Mean shift based methods which are adapting the kernel scale and the orientation of the kernel are presented in [4, 7]. In [4] scale and orientation of a kernel are obtained by estimating the second order moments of the object silhouette, but that is of high computational costs. Combining the mean shift method with adaptive filtering as in [7] is another possibility to achieve adaptive kernel scale and orientation. But even if the estimation of kernel scale and orientation are good, due to the use of a symmetric kernel, both methods do not achieve an adaptation of the kernel to the actual object shape.

Methods working with adaptive and asymmetric kernels are described in [8–10]. The method of [8] focuses on face tracking and uses ellipses as basic face models, thus it can not easily be generalized for tracking other objects since adequate models are required. In [9] asymmetric kernels are generated using implicit level set functions. Since the search space is extended by a scale and an orientation dimension, the method simultaneously estimates the new object location, scale and orientation. However, the method can only estimate the orientation of the object for in-plane rotations. In case of out-of-plane rotations this algorithm is also not capable to adapt to the objects orientation and therewith to the object shape. A first approach for mean shift tracking with an adaptive asymmetric kernel being able to deal with out-of-plane rotations was presented in [10]. However, the technique for adapting the kernel to the shape of an object is rather heuristic, since image segments produced by a segmentation process are mainly assigned as object segments if more than 50 % of a segment is included in an inital mask.

In this chapter we propose a new model-based method for kernel shape adaptation which provides a more reliable discrimination between object and background and therefore improves the tracking performance of the shape adaptive mean shift tracking [10]. The proposed method takes advantage of two Gaussian mixture models (GMMs) modeling the color histogram of the object and the histogram of the background as usually done in applications for video matting or compositing [11]. The scale adapted kernel, given after running the mean shift iterations in an extended search space, is fully adapted to the object shape by a maximum a posteriori estimation considering the GMM of the object and of the background. Thus, a good fit of the object shape is retrieved even if the object is performing out-of-plane rotations.

The rest of the chapter is organized as followed. Section 7.2 gives an overview of the mean shift tracking algorithm. In Sect. 7.3 the proposed method is described explaining the construction of the object shaped kernel, the execution of the mean shift iterations in the spatial-scale-space and the final estimation of the kernel shape using using GMMs. Experimental result and the evaluation of the tracking algorithm is then described and given in Sect. 7.4. Finally conclusions are drawn in Sect. 7.5.

7.2 Mean Shift Tracking Overview

Mean shift tracking discriminates between a target model in frame n and a candidate model in frame $n + 1$. For tracking purposes the target model is mostly defined as the color density distribution of the object, but any other object feature than the color of the object could also be used. Thus, the target model is estimated from the discrete density of the weighted objects feature histogram $q(\hat{\boldsymbol{x}}) = \{q_u(\hat{\boldsymbol{x}})\}_{u=1\ldots m}$ with $\sum_{u=1}^{m} q_u(\hat{\boldsymbol{x}}) = 1$. The probability of a certain feature belonging to the object with the centroid $\hat{\boldsymbol{x}}$ is expressed as the probability of the feature $u = 1 \ldots m$ occurring in the target model. Which is

$$q_u(\hat{\boldsymbol{x}}) = C \sum_{i=1}^{N} k\left(\left\|\frac{\boldsymbol{x}_i - \hat{\boldsymbol{x}}}{h}\right\|^2\right) \delta[b(\boldsymbol{x}_i) - u], \tag{7.1}$$

where δ is the impulse function, h is the kernel bandwidth, N is the number of pixels of the target model and normalization constant C is the reciprocal of the sum of values of the kernel profile $k(z)$. Basically Eq. (7.1) estimates the weighted feature bin of the feature u, where the feature is weighted according to the positions of the pixels containing that feature. The kernel K with kernel profile $k(z)$ makes the density estimation more reliable, because it provides pixels farther away from the center of the ellipse with a smaller weight. Hence, the least reliable outer pixels don't influence the density estimation too much. Figure 7.1 shows the Epanechnikov kernel, which is typically used in mean shift tracking, and an object marked by an ellipse. The axes h_x and h_y of the ellipse are referred to as the bandwidth of the kernel. Usually the ellipse is scaled such that h_x and h_y have a length of one and only one bandwidth parameter h is necessary. Thus, the radial symmetric kernel can be applied for ellipses of different size and shape.

The candidate model $\boldsymbol{p}(\hat{\boldsymbol{x}}_{new}) = \{p_u(\hat{\boldsymbol{x}}_{new})\}_{u=1\ldots m}$ (whereas $\sum_{u=1}^{m} p_u = 1$) in the following frame and the probability of a certain feature appearing in the candidate model

$$p_u(\hat{\boldsymbol{x}}_{new}) = C \sum_{i=1}^{N} k\left(\left\|\frac{\boldsymbol{x}_i - \hat{\boldsymbol{x}}_{new}}{h}\right\|^2\right) \delta[b(\boldsymbol{x}_i) - u] \tag{7.2}$$

are defined similarly to the target model.

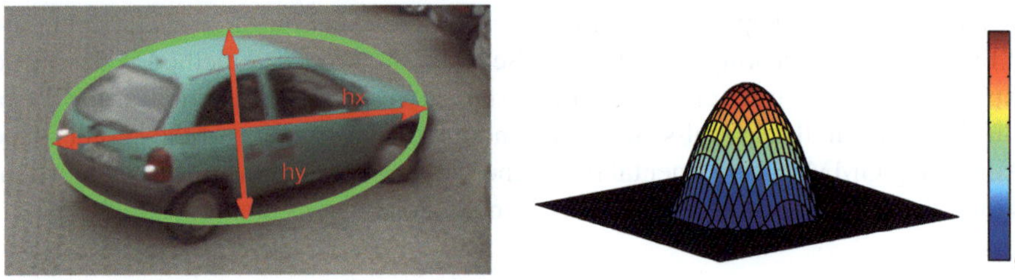

Fig. 7.1 Object in image marked by an ellipse (*left*) and radial symmetric Epanechnikov kernel (*right*), which is used to weight the pixel features according to the position of the pixels inside the ellipse

The core of the mean shift method is the computation of the offset from an old object position $\hat{\boldsymbol{x}}$ to a new position $\hat{\boldsymbol{x}}_{new} = \hat{\boldsymbol{x}} + \Delta\boldsymbol{x}$ by estimating the mean shift vector

$$\Delta\boldsymbol{x} = \frac{\sum_i K(\boldsymbol{x}_i - \hat{\boldsymbol{x}})w(\boldsymbol{x}_i)(\boldsymbol{x}_i - \hat{\boldsymbol{x}})}{\sum_i K(\boldsymbol{x}_i - \hat{\boldsymbol{x}})w(\boldsymbol{x}_i)}, \tag{7.3}$$

where $K(\cdot)$ is a symmetric kernel with bandwidth h defining the object area and $w(\boldsymbol{x}_i)$ is the weight of $\boldsymbol{x}_i$ which is defined as

$$w(\boldsymbol{x}_i) = \sum_{u=1}^{m} \delta[b(\boldsymbol{x}_i) - u]\sqrt{\frac{q_u(\hat{\boldsymbol{x}})}{p_u(\hat{\boldsymbol{x}}_{new})}}. \tag{7.4}$$

An important property of the weighting function $w(\boldsymbol{x}_i)$ is, that it sets all feature bins of the current candidate model to zero if they are not contained in the target model. Thus, the mean shift vector as given in Eq. (7.3) will shift the ellipse to a feature centroid where the new candidate model has a feature histogram which is much more similar to the one of the target model. In detail, the problem of localizing the candidate model in the next frame $n + 1$ is formulated as the derivation of the estimate that maximizes the Bayes error between the reference distribution of the target model and the distribution of the candidate model. Or in other words, to find the candidate model, which has the most similar feature distribution compared to the distribution of the target model. For the similarity measure the discrete formulation of the Bhattacharyya coefficient is chosen as in [1], since we have discrete feature distributions on the one hand and the Bhattacharyya coefficient is nearly optimal and imposes a metric structure on the other hand. The Bhattacharyya coefficient and the distance between the two color distributions of target and candidate model are defined as follows

$$\rho[\boldsymbol{p}(\hat{\boldsymbol{x}}_{new}), \boldsymbol{q}(\hat{\boldsymbol{x}})] = \sum_{u=1}^{m} \sqrt{p_u(\hat{\boldsymbol{x}}_{new})q_u(\hat{\boldsymbol{x}})}, \tag{7.5}$$

$$d(\hat{\boldsymbol{x}}_{new}) = \sqrt{1 - \rho[\boldsymbol{p}(\hat{\boldsymbol{x}}_{new}), \boldsymbol{q}(\hat{\boldsymbol{x}})]}. \tag{7.6}$$

The distance defined in Eq. (7.6) is also known as Hellinger distance in probability theory and measures the similarity between the distribution of the target model and of the candidate model. The aim is to minimize the Hellinger distance between the two color distributions as a function of $\hat{\boldsymbol{x}}_{new}$ in the neighborhood of a given position $\hat{\boldsymbol{x}}_0$ by using the mean shift algorithm. Starting with the Taylor expansion around $p_u(\hat{\boldsymbol{x}}_0)$ the Bhattacharyya coefficient is approximated as

$$\rho[\hat{\boldsymbol{p}}(\hat{\boldsymbol{x}}_{new}), \hat{\boldsymbol{q}}(\hat{\boldsymbol{x}})] \approx \frac{1}{2} \sum_{u=1}^{m} \sqrt{p_u(\hat{\boldsymbol{x}}_0) q_u(\hat{\boldsymbol{x}})} + \frac{C}{2} \sum_{i=1}^{N} w(\boldsymbol{x}_i) k\left(\left\| \frac{\boldsymbol{x}_i - \hat{\boldsymbol{x}}_{new}}{h} \right\|^2\right). \tag{7.7}$$

In Eq. (7.7) only the second term is dependent on $\hat{\boldsymbol{x}}_{new}$. Hence, for minimizing the distance it is sufficient to maximize the second term of (7.7). This term corresponds to the density estimate computed with kernel profile k at location $\hat{\boldsymbol{x}}_{new}$ in frame $n+1$, whereas the data is weighted with $w(\boldsymbol{x}_i)$. The maximization can be achieved using the mean shift algorithm. By running this algorithm the kernel is recursively moved from $\hat{\boldsymbol{x}_0}$ to $\hat{\boldsymbol{x}_1}$ according to the mean shift vector.

7.3 Shape Adaptive Mean Shift Tracking

7.3.1 Asymmetric Kernel Generation

Traditional mean shift tracking is working with a symmetric kernel. But a symmetric kernel can not describe an object shape properly. Hence, the use of isotropic kernels will always cause an influence of background information on the target model, which can lead to tracking errors and even to tracking failure. To solve this problem we are using an asymmetric and anisotropic kernel, which is adapted to the contour of the object.

As the mean shift tracker cannot initialize the object by itself, it either requires some user input or the result from a detection process which provides an object mask like [12] or [13]. Based on such an object mask our asymmetric kernel is constructed by estimating for each pixel inside the mask $\boldsymbol{x}_i = (x, y)$ its normalized distance to the object boundary:

$$K_s(\boldsymbol{x}_i) = \frac{d(\boldsymbol{x}_i)}{d_{max}}, \tag{7.8}$$

where the distance from the boundary is estimated using morphological operations. In Fig. 7.2 an object, its mask and the mask based asymmetric kernel are shown.

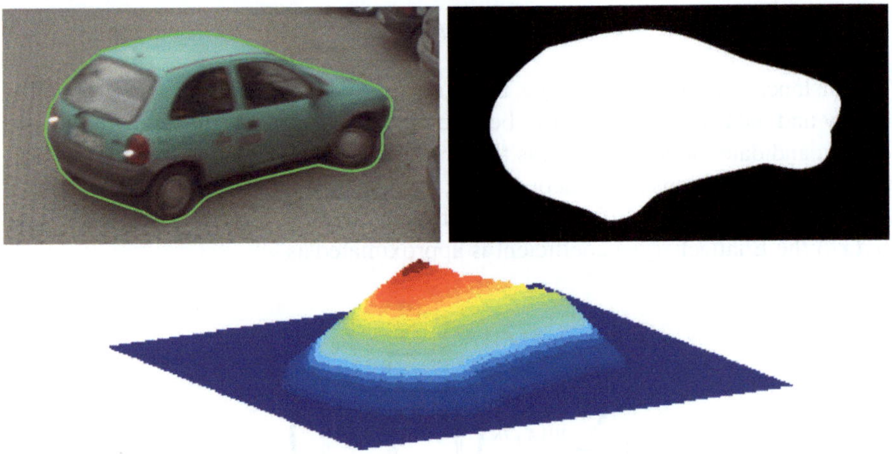

Fig. 7.2 Object in the original image with *green* marked contour (*left*), object mask (*right*) and asymmetric object kernel retrieved from object mask (*bottom*)

7.3.2 Mean Shift Tracking in Spatial-Scale-Space

Instead of running the algorithm only in the local space the mean shift iterations are performed in an extended search space $\Omega = (x, y, \sigma)$ consisting of the image coordinates (x, y) and a scale dimension σ as described in [9]. Since the mean shift iterations estimate a scale update $\Delta\sigma$ and a bandwidth update factor $d = 1+\sqrt(2)\Delta\sigma$, the new kernel bandwidth $r_{new}(\alpha) = dr(\alpha)$ at angle α can be computed from the product of the former bandwidth $r(\alpha)$ and the bandwidth update factor d. Thus, the object's changes in position and scale can be evaluated through the mean shift iterations simultaneously. To run the mean shift iterations in the spatial-scale-space a 3D kernel consisting of the product of the spatial object based kernel from Sect. 7.3.1 and a kernel for the scale dimension

$$K(x, y, \sigma_i) = K(x, y)K(\sigma) \tag{7.9}$$

is defined. The kernel for the scale dimension is a 1D Epanechnikov kernel with the kernel profile $k(z) = 1 - |z|$ if $|z| < 1$ and 0 otherwise, where $z = (\sigma_i - \hat{\sigma})/h_\sigma$. The mean shift vector given in (7.3) can now be computed in the joint space as

$$\Delta\Omega = \frac{\sum_i K(\Omega_i - \hat{\Omega})w(\boldsymbol{x}_i)(\Omega_i - \hat{\Omega})}{\sum_i K(\Omega_i - \hat{\Omega})w(\boldsymbol{x}_i)}, \tag{7.10}$$

with $\Delta\Omega = (\Delta x, \Delta y, \Delta\sigma)$.

Given the object mask for the initial frame the object centroid $\hat{\boldsymbol{x}}$ and the target model are computed. To make the target model more robust the histogram of a specified neighborhood of the object is also estimated and bins of the neighborhood histogram are set to zero in the target histogram to eliminate the influence of colors which are contained in the object as well as in the background. In case of an object mask with a slightly different shape than the object shape too many object colors might be suppressed in the target model, if the direct neighbored pixels are considered. Therefore, the directly neighbored pixels are not included in the considered neighborhood.

Taking the distribution $\{q_u(\hat{\boldsymbol{x}})\}_{u=1\ldots m}$ of the target model at location $\hat{\boldsymbol{x}}$ in frame n the algorithm iterates as follows:

1. Initialize the location of the candidate model in frame $n+1$ with $\hat{\boldsymbol{x}}_0 = \hat{\boldsymbol{x}}$ and set $d_0 = 1$.
2. Subsequently compute $\boldsymbol{p}(\hat{\boldsymbol{x}}_0)$ and $\rho[\boldsymbol{p}(\hat{\boldsymbol{x}}_0), \boldsymbol{q}(\hat{\boldsymbol{x}})]$.
3. Compute the weights $w(\boldsymbol{x}_i)$ according to Eq. (7.4).
4. According to the mean shift vector (7.10) estimate
   - the new position of the candidate model $\hat{\boldsymbol{x}}_1 = \hat{\boldsymbol{x}}_0 + \Delta\boldsymbol{x}$
   - the bandwidth update factor $d_1 = d_0(1 + \sqrt(2)\Delta\sigma)$
   - $\boldsymbol{p}(\hat{\boldsymbol{x}}_1)$
   - $\rho[\boldsymbol{p}(\hat{\boldsymbol{x}}_1), \boldsymbol{q}(\hat{\boldsymbol{x}})]$.
5. If $\|\hat{\boldsymbol{x}}_1 - \hat{\boldsymbol{x}}_0\| < \varepsilon$ stop, else $\hat{\boldsymbol{x}}_0 \leftarrow \hat{\boldsymbol{x}}_1$, $d_0 \leftarrow d_1$ and go to step 2.

The algorithm uses the mean shift vector in step 4 to maximize the Bhattacharyya coefficient. The termination threshhold ε in step 5 implies that the vectors $\hat{\boldsymbol{x}}_0$ and $\hat{\boldsymbol{x}}_1$ point at the same pixel in image coordinates. Therefore, the algorithm terminates for one thing if the same or a larger value for the Bhattacharyya coefficient is found and for the other thing if the candidate model does not change its position in two subsequent iterations.

7.4 Shape Adaptation Using GMMs

After the mean shift iterations have converged the final shape of the object is evaluated from the first estimate of the scaled object shape S_i. Therefore, the image is segmented using the mean shift method according to [2]. Figure 7.3 shows the segmented object area and its surrounding neighborhood indicated by the black outline.

Segments which are fully contained in the first shape estimate are assigned as object segments. For each segment being only partly included in the found object area we have to decide if it still belongs to the object shape or to the background. Therefore, we learn two Gaussian mixture models, one modeling the color histogram of the background and one the histogram of the object. The surrounding background area H of the object is estimated by

$$H = S_i \cdot k - S_i, \quad \text{with} \quad k \geq 1, \tag{7.11}$$

where k is usually chosen between 1.1 and 1.3. The GMMs are learned at the beginning of the sequence based on the binary mask already being used for the

Fig. 7.3 The object area and its surrounding neighborhood are segmented into segments of similar colors. The neighborhood H is the narrow band between the outline of the first shape estimate S_i (*green outline*) and the increased area of it (*black outline*)

initial kernel generation. Since we are working in RGB color space the multivariate normal density distribution of a color value $\boldsymbol{c} = (c_r, c_g, c_b)^T$ is given by

$$p(\boldsymbol{c}|\boldsymbol{\mu}_k, \boldsymbol{\Sigma}_k) = \frac{1}{(2\pi)^{\frac{3}{2}}|\boldsymbol{\Sigma}_k|^{\frac{1}{2}}} e^{-\frac{1}{2}(\boldsymbol{c}-\boldsymbol{\mu}_k)^T \boldsymbol{\Sigma}_k^{-1}(\boldsymbol{c}-\boldsymbol{\mu}_k)}, \tag{7.12}$$

where $\boldsymbol{\mu}_k$ is the mean and $\boldsymbol{\Sigma}$ is a 3×3 covariance matrix, while $|\boldsymbol{\Sigma}_k|$ is its determinant and $\boldsymbol{\Sigma}^{-1}$ its inverse. The Gaussian mixture model for an image area is given by

$$P(\boldsymbol{c}) = \sum_{k=1}^{K} P_k \cdot p(\boldsymbol{c}|\boldsymbol{\mu}_k, \boldsymbol{\Sigma}_k), \tag{7.13}$$

where P_k is the a priori probability of distribution k, which can also be interpreted as the weight for the respective Gaussian distribution.

To fit the Gaussians of the mixture model to the corresponding color histogram as illustrated in Fig. 7.4 the parameters $\Theta_k = \{P_k, \mu_k, \boldsymbol{\Sigma}_k\}$ have to be estimated. A very common method to solve the parameter estimation problem is the expectation maximization (EM) algorithm [14]. The EM algorithm is an iterative technique which can be used for finding the maximum likelihood parameter estimates when fitting a distribution to a given data set. A good description of the EM algorithm applied for fitting the parameters of a Gaussian mixture model can be found in [15]. During the EM iterations, first the probability (at iteration step t) of all N data samples $\boldsymbol{c}_n$ to belong to the kth Gaussian distribution is calculated by Bayes' theorem

$$p(k|\boldsymbol{c}_n, \Theta) = \frac{P_{k,t}\, p(\boldsymbol{c}_n|k, \boldsymbol{\mu}_{k,t}, \boldsymbol{\Sigma}_{k,t})}{\sum_{k=1}^{K} P_{k,t}\, p(\boldsymbol{c}_n|k, \boldsymbol{\mu}_{k,t}, \boldsymbol{\Sigma}_{k,t})}, \tag{7.14}$$

which is known as the expectation step. In the subsequent maximization step the likelihood of the complete data is maximized by re-estimating the parameters Θ:

$$P_{k,t+1} = \frac{1}{N} \sum_{n=1}^{N} p(k|\boldsymbol{c}_n, \Theta), \tag{7.15}$$

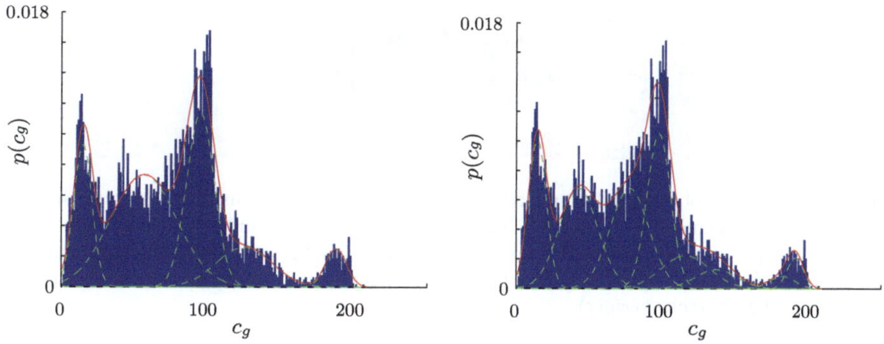

Fig. 7.4 Modeling the histogram of the *green* color channel of the car in sequence *parking_lot* with $K = 5$ (*left*) and $K = 8$ Gaussians (*right*)

$$\boldsymbol{\mu}_{k,t+1} = \frac{1}{NP_{k,t+1}} \sum_{n=1}^{N} p(k|\boldsymbol{c}_n, \Theta)\boldsymbol{c}_n, \tag{7.16}$$

$$\boldsymbol{\Sigma}_{k,t+1} = \frac{1}{NP_{k,t+1}} \sum_{n=1}^{N} p(k|\boldsymbol{c}_n, \Theta)(\boldsymbol{c}_n - \mu_{t+1})(\boldsymbol{c}_n - \mu_{t+1})^T. \tag{7.17}$$

The updated parameter set is then used in the next iteration step $t + 1$. The EM algorithm iterates between these two steps and converges to a local maximum of the likelihood. Thus, after convergence of the EM algorithm the Gaussian mixture model will be fitted to the discrete data and a nice representation of the histogram will be given by the fitted mixture model, see Fig. 7.4. Since the visualization of a GMM modeling a three-dimensional histogram is rather difficult to understand, Fig. 7.4 shows two GMMs modeling only the histogram of the green color channel of the car in sequence *parking_lot*.

The accuracy of a GMM depends on the number of Gaussians. Hence, the GMM with $K = 8$ Gaussian distributions models the histogram more accurate than the model with $K = 5$ Gaussians. Of course, depending on the histogram in some cases a GMM with a higher number of Gaussian distributions might be necessary, but for our purpose a GMM with $K = 5$ Gaussians showed to be a good trade-off between modeling accuracy and parameter estimation.

To decide for each pixel if it belongs to the GMM of the object $P_{obj}(\boldsymbol{c}) = P(\boldsymbol{c}|\alpha = 1)$ or to the background GMM $P_{bg}(\boldsymbol{c}) = P(\boldsymbol{c}|\alpha = 0)$ we use maximum a posteriori (MAP) estimation. Using log-likelihoods the typical form of the MAP estimate is given by

$$\hat{\alpha} = \underset{\alpha}{\operatorname{argmax}}(\ln p(\alpha) + \ln P(\boldsymbol{c}|\alpha)), \tag{7.18}$$

where $\alpha\hat{} \in [0, 1]$ indicates that a pixel, or more precise its color value $\boldsymbol{c}$, belongs to the object ($\hat{\alpha} = 1$) or the background class ($\hat{\alpha} = 0$), and $p(\alpha)$ is the corresponding

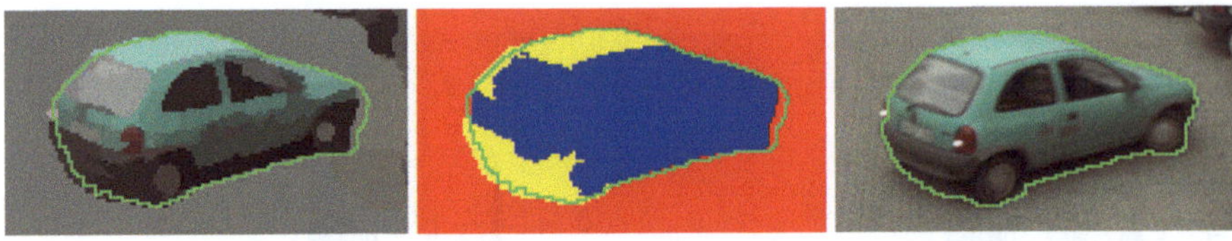

Fig. 7.5 The initial object mask retrieved from the mean shift iterations in spatial-scale-space is shown above the segmented object (*left*). The segments are classified either as one of the two possible object segment types (*blue* and *yellow* segments) or as background segments (*red* segments). According to the object segments the contour of the final object mask is estimated and displayed on the object being tracked (*right*)

a priori probability. To set $p(\alpha)$ to an appropriate value the object area and the background area of the initial mask are considered.

Based on the number of its object and background pixels, a segment is assigned as an object or background segment. If more than 50 % of the pixels of a segment belong to the the object class, the segment is assigned as an object segment, otherwise the segment is considered to belong to the background. Figure 7.5 shows from left to right the segmented object and its neighborhood, a color coding of the different types of segments and the final shape estimate after the segmentation process overlayed on the original image of the object. The first shape estimate given by running the mean shift iteration in the extended search space is also shown by the contour in Fig. 7.5 (left) and (middle). In Fig. 7.5 (middle) the three different types of segments are shown: segments which are completely included in the initial mask are shown in blue, segments which are partly included and are containing enough color information of the target model are marked in yellow and the red segments are background segments. The next object based kernel can now be obtained from the final shape and the next mean shift iterations can be initialized. To summarize the new shape adaptive mean shift tracking method GMM-SAMT a pseudo code of the tracking algorithm is given by Algorithm 1.

7.5 Experimental Results

For modeling the histogram of the object region and the background histogram we used $K = 5$ Gaussian distributions for each GMM, where K is determined empirically. We found a maximum of 30 iterations of the EM algorithm sufficient to fit the parameters of each GMM. The GMMs were estimated only once at the beginning of the sequence. Given the first object mask the object centroid and the mask based asymmetric kernel are estimated. The masked based kernel is then used for computing the histogram in the RGB space with $32 \times 32 \times 32$ bins. For the scale dimension the Epanechnikov kernel with a bandwidth of $h_\sigma = 0.4$ is used. For mean shift segmentation a multivariate kernel defined according to Eq. (35) in [2] as the product of two Epanechnikov kernels, one for the spatial domain (pixel coordinates) and one for the range domain (color), is used. The bandwidth of the Epanechnikov

Algorithm 1: Gaussian Mixture Model based Shape Adaptive Mean Shift Tracking (GMM-SAMT)

Input: frame n, frame $n+1$, mask n, $\boldsymbol{q}(\hat{\boldsymbol{x}})$, $\hat{\boldsymbol{x}}$, K Gaussian distributions
Output: mask $n+1$, $\hat{\boldsymbol{x}}_{new}$

```
GMM_obj ← expectation maximization(object histogram, K)
GMM_bg ← expectation maximization(background histogram, K)
x̂_0 ← x̂

begin
    K(x, y, σ) ← Eq. (7.9)
    p_u(x̂_0) ← Eq. (7.2)
    p(x̂_0) ← {p_u(x̂_0)}_{u=1...m}
    ρ[p(x̂_0), q(x̂)] ← Eq. (7.5)
    w(x_i) ← Eq. (7.4)
    (Δx, Δσ) ← ΔΩ ← Eq. (7.10)
    x̂_1 ← x̂_0 + Δx;
    d_1 = d_0(1 + √(2)Δσ);
    if ||x̂_1 − x̂_0|| > ε then
        to begin
    else
        x̂_0 ← x̂_1
        d_0 ← d_1
        x̂_new ← x̂_1
        mask n ← shift mask n according to Δx
        mask n ← scale mask n according to d_1
        mask n + 1 ← get_final_shape(mask n, GMM_obj, GMM_bg) (see Sect. 4)
    end
end
```

kernel in range domain was set to $h_r = 4$, and the bandwidth of the one in spatial domain to $h_s = 5$. The minimal segment size was set to 5 pixels.

In order to evaluate the performance of GMM-SAMT the algorithm has been tested on several video sequences. We tested our algorithm on some self-recorded sequences, which are either showing a parking lot or the traffic on an airport apron, and we also used the sequence of *PETS 2000*.[1] In Fig. 7.6 the tracking results of GMM-SAMT are compared to the results of the traditional mean shift tracker combined with the $\pm 10\,\%$ method for tracking a car, which is backing out of a parking space. The visual evaluation of the tracking results already shows that GMM-SAMT clearly outperforms the traditional mean shift algorithm. While GMM-SAMT is able to adapt to the shape of the turning car, the traditional method even fails to fit the size of the ellipse indicating scale and position of the object being tracked to the size of the car.

More results of the standard mean shift tracker combined with the $\pm 10\,\%$ method and the proposed GMM-SAMT are shown in Figs. 7.7, 7.8 and 7.9. In the first row of Fig. 7.7 the results of the standard mean shift tracking for tracking a follow-me car, which turns right, are shown. Even though the ellipse is chosen such that no background color influences the target model, the mean shift tracker using

[1] ftp://ftp.pets.rdg.ac.uk/pub/PETS2000/

Fig. 7.6 Results of traditional mean shift tracking (*red* ellipse) compared to GMM-SAMT (*green* contour) for tracking a car on a parking lot

Fig. 7.7 Tracking a follow-me car using the standard mean shift tracker (*top row*) and using GMM-SAMT (*bottom row*)

the $\pm 10\%$ method is not able to track or to adapt to the size of the follow-me car. The mean shift tracker simply searches for the best match, which can be found inside the true object area, that's why the $\pm 10\%$ method is not working at all. Finally, the standard mean shift tracker even fails to track the object. Whereas GMM-SAMT is able to adapt to the changes of the follow-me car, although the contour of the car is changing drastically while the car moves through the sequence.

Figure 7.8 compares the tracking results of the standard mean shift tracker and GMM-SAMT for tracking a red car in the *PETS 2000* sequence. The results of the standard mean shift method are shown in the top row of Fig. 7.8, while the GMM-SAMT results can be seen in the bottom row of Fig. 7.8. Even though the standard

Fig. 7.8 Tracking a *red* car using the standard mean shift tracker (*top row*) and using GMM-SAMT (*bottom row*)

Fig. 7.9 Tracking an airplane using the standard mean shift tracker (*top row*) and using GMM-SAMT (*bottom row*)

mean shift tracker performs much better than in the case of the follow-me car, GMM-SAMT still provides the more precise tracking results.

By comparing the GMM-SAMT results of tracking an airplane, see bottom row of Fig. 7.9, to the results of the standard mean shift tracker given in the top row of Fig. 7.9, one can see that GMM-SAMT is also able to track the contour of objects of quite complex shape. Only if the color similarity between object color and background color is to high, GMM-SAMT misses some small parts of the airplane.

For a more objective evaluation the tracking results are compared to a manually labeled ground truth and the tracking error t_{err} in pixels is estimated by computing the averaged Euclidean distance of the tracked centroids to the ground truth centroids. The tracking error for the standard mean shift tracker are given in Table 7.1, while Table 7.2 lists the tracking error for GMM-SAMT. Since the standard mean shift method fails to track the follow-me car, the tracking error is quite high in that case and certainly does not represent the general performance of the mean shift tracker. However, GMM-SAMT outperforms the standard mean shift tracking in all other

Table 7.1 Tracking error as well as Recall, Precision and F_1 measure of standard mean shift tracking

Target	Ground truth frames	Standard mean shift			
		t_{err}	Recall	Precision	F_1 score
Parking car	30	9	0.96	0.52	0.68
Follow-me car	20	88	0.23	0.14	0.60
Airplane	15	32	0.75	0.25	0.37
Red car	15	8	0.80	0.79	0.80

Table 7.2 Tracking error as well as Recall, Precision and F_1 measure of GMM-SAMT

Target	Ground truth frames	GMM-SAMT			
		t_{err}	Recall	Precision	F_1 score
Parking car	30	3	0.98	0.86	0.92
Follow-me car	20	3	0.99	0.83	0.90
Airplane	15	4	0.84	0.76	0.80
Red car	15	1	0.97	0.91	0.94

Table 7.3 The tracking error t_{err} and F_1 measure of GMM-SAMT compared to standard mean shift tracking

Target	Parking car	Follow-me car	Airplane	Red car
Δt_{err}	−6	−85	−28	−7
ΔF_1	0.24	0.30	0.43	0.14

cases as well. For a better comparison the difference of the tracking error, estimated by subtracting the tracking error of the standard mean shift tracker from the tracking error of GMM-SAMT, is shown in Table 7.3.

To further evaluate the tracking performance the information retrieval measurements *Recall* and *Precision* were also computed by comparing the detection results to the ground truth as follows:

$$\text{Recall} = \frac{N_{cor}}{N_{gt}}, \tag{7.19}$$

$$\text{Precision} = \frac{N_{cor}}{N_{det}}, \tag{7.20}$$

where N_{cor} is the number of correctly detected object pixels, N_{det} is the number of all detected object pixels and N_{gt} represents the number of object pixels in the ground truth. The *Recall* and *Precision* scores given in Table 7.1 and in Table 7.2 confirm the impression of the visual inspection, since for all test sequences GMM-SAMT achieves better results as the standard mean shift method. In addition to the information retrieval measurements, we also calculated the even more significant F_1 measure:

$$F_1 = 2 \cdot \frac{\text{Recall} \cdot \text{Precision}}{\text{Recall} + \text{Precision}}. \tag{7.21}$$

The F_1 scores of the standard GMM method and of the Auto GMM-SAMT detection unit are compared in Table 7.3. Again the visual impression is confirmed.

7.6 Conclusions

GMM-SAMT extends the traditional mean shift algorithm to track the contour of objects with changing shape without the help of any predefined shape model. Since GMM-SAMT works with an object mask based kernel, the influence of background color on the target model is avoided. Thus, the tracking algorithm is much more robust than standard mean shift tracking. To adapt the kernel to the changing object shape, a GMM of the object and a GMM of the surrounding background are used to segment the object area from the background. The kernel is then adapted to the segmented object shape. Thus, the proposed algorithm is able to track the position and the contour of an object quite robust, even if the object is performing out-of-plane rotations.

However, in case of very similar object colors GMM-SAMT also has to deal with errors, since segmentation errors can occur. Hence, in future work we will focus on the color similarity problem. The investigation of other additional object features might provide a first solution to this problem.

Acknowledgments This work has been supported by the Gesellschaft für Informatik, Automatisierung und Datenverarbeitung (iAd) and the Bundesministerium für Wirtschaft und Technologie (BMWi), ID 20V0801I.

References

1. Comaniciu D, Ramesh V, Meer P (2000) Real-time tracking of non-rigid objects using mean shift. In: Proceedings of IEEE conference on computer vision and pattern recognition, IEEE Press, New York, pp 142–149
2. Comaniciu D, Meer P (2002) Mean shift: a robust approach toward feature space analysis. IEEE Trans Pattern Anal Mach Intell 24:603–619
3. Cheng Y (1995) Mean shift, mode seeking, and clustering. IEEE Trans Pattern Anal Mach Intell 17:790–799
4. Bradski GR (1998) Computer vision face tracking for use in a perceptual user interface. Intel Technol J 2:12–21
5. Comaniciu D, Ramesh V, Meer P (2003) Kernel-based object tracking. IEEE Trans Pattern Anal Mach Intell 25:564–575
6. Collins RT (2003) Mean-shift blob tracking through scale space. In: Proceedings of IEEE computer society conference on computer vision and pattern recognition, pp 234–240

7. Qifeng Q, Zhang D, Peng Y (2007) An adaptive selection of the scale and orientation in kernel based tracking. In: Proceedings of the third international IEEE conference on signal-image technologies and internet-based system, IEEE Press, New York, pp 659–664
8. Vilaplana V, Marques F (2008) Region-based mean shift tracking: application to face tracking. In: Proceedings of 15th IEEE international conference on image processing, IEEE Press, New York, pp 2712–2715
9. Yilmaz A (2007) Object tracking by asymmetric kernel mean shift with automatic scale and orientation selection. In: Proceedings of IEEE conference on computer vision and pattern recognition, IEEE Press, New York, pp 1–6
10. Quast K, Kaup A (2009) Scale and shape adaptive mean shift object tracking in video sequences. In: Proceedings 17th European signal processing conference, pp 1513–1517
11. Nowak A, Hörchens L, Röder J, Erdmann M (2006) Colourbased video segmentation for tv studio applications. In: Proceedings of the 51st international scientific colloquium, 2006
12. Stauffer C, Grimson WEL (2000) Learning patterns of activity using real-time tracking. IEEE Trans Pattern Anal Mach Intell 22(8):747–757
13. Quast K, Kaup A (2010) Real-time moving object detection in video sequences using spatio-temporal adaptive Gaussian mixture models. In: Proceedings of international conference on computer vision theory and applications (VISAPP '10), Angers, France, 2010
14. Dempster AP, Laird NM, Rubin DB et al (1977) Maximum likelihood from incomplete data via the EM algorithm. J Royal Stat Soc. Series B (Methodological) 39(1):1–38
15. Ihlow A, Heuberger A (2009) Sky detection in fisheye images for photogrammetric analysis of the land mobile satellite channel. In: Proceedings of the 10th workshop digital broadcasting, pp 56–60

Part III
High-Level Descriptors and Video Retrieval

Chapter 8
Forensic Reasoning upon Pre-Obtained Surveillance Metadata Using Uncertain Spatio-Temporal Rules and Subjective Logic

Seunghan Han, Bonjung Koo, Andreas Hutter and Walter Stechele

Abstract This chapter presents an approach to modeling uncertain contextual rules using subjective logic for forensic visual surveillance. Unlike traditional real-time visual surveillance, forensic analysis of visual surveillance data requires mating of high level contextual cues with observed evidential metadata where both the specification of the context and the metadata suffer from uncertainties. To address this aspect, there has been work on the use of declarative logic formalisms to represent and reason about contextual knowledge, and on the use of different uncertainty handling formalisms. In such approaches, uncertainty attachment to logical rules and facts are crucial. However, there are often cases that the truth value of rule itself is also uncertain thereby, uncertainty attachment to rule itself should be rather functional. *The more X then the more Y* type of knowledge is one of the examples. To enable such type of rule modeling, in this chapter, we propose a reputational subjective opinion function upon logic programming, which is similar to fuzzy membership function but can also take into account uncertainty of membership value itself. Then we further adopt subjective logic's fusion operator to accumulate the acquired opinions over time. To verify our approach, we present a preliminary experimental case study on reasoning likelihood of being a good witness that uses metadata extracted by a person tracker and evaluates the relationship between the tracked persons. The case study

S. Han (✉) · B. Koo · A. Hutter
Siemens AG, Corporate Technology, Otto-Hahn-Ring 6, Munich, Germany
e-mail: hanseunghan@googlemail.com

B. Koo
Department of Computer Science, Sogang University, Shinsu dong 1, Seoul, Korea
e-mail: kbj9090@gmail.com

A. Hutter
e-mail: andreas.hutter@siemens.com

S. Han · W. Stechele
Institute for Integrated Systems, Technische Universität München, Arcisstrasse 21, Munich, Germany
e-mail: Walter.Stechele@tum.de

N. Adami et al. (eds.), *Analysis, Retrieval and Delivery of Multimedia Content*, Lecture Notes in Electrical Engineering 158, DOI: 10.1007/978-1-4614-3831-1_8,

is further extended to demonstrate more complex forensic reasoning by considering additional contextual rules.

Keywords Visual surveillance · Forensic reasoning · Logic programming · Subjective logic

8.1 Introduction

As traditional computer vision technology further matures, higher level forensic semantic understanding of visual surveillance data has been gaining increasing attention. Such forensic semantic analysis deals with a propositional assumption to be investigated after an incident and the answer to the propositional assumption should be an epistemic reasoning result upon pre-observed evidential and contextual cues. Therefore, such forensic semantic analysis of visual surveillance data requires intelligent reuse of low level vision analytic results with additional visual, and non visual, contextual cues. However, unlike domains that can solely rely on deterministic knowledge model, in visual surveillance, contextual knowledge as well as low level vision analytic results are fraught with facets of *uncertainties*, *incompleteness* and *inconsistencies*. Therefore, the key challenges for such high level analysis approaches are the choice of an appropriate contextual knowledge representation and the proper reasoning mechanism under uncertainty. Depending on how such approaches handle uncertainty, they can be roughly categorized into *intensional* and *extensional* approaches [1]. In intensional approaches, also known as state based approaches, uncertainty is attached to 'subsets of possible states' and handle uncertainty taking into account relevance between the states. In extensional approaches, also known as rule-based systems treat uncertainty as a generalized truth value attached to formulas and compute the uncertainty of any formula as a function of the uncertainties of its sub formulas. There is trade-off between the two approaches. Intentional approaches assume completeness of the state model, therefore, semantically clear but computationally clumsy. Extensional approaches are computationally convenient but semantically sloppy. In forensic visual surveillance, however, considering the variety of possible semantics in scenes, extensional approaches have advantages in the flexibility and expressive power due to their ability to derive a new proposition based only on what is currently known (a) regardless of anything else in the knowledge base (*locality*) and (b) regardless of how the current knowledge was derived (*detachment*). *locality* and *detachment* are together referenced to as *modularity* [1]. Due to the advantage of extensional approaches, there has been some extent of work on the use of logic programming language with different uncertainty handling formalisms for visual surveillance and computer vision problems. In such approaches, intermediate metadata comes from vision analytics and additional visual or non visual contextual cues are encoded as either symbolized facts or rules. Then uncertainty comes with vision analytics are represented according to the chosen uncertainty formalism and attached to their symbolized facts. Similarly, uncertainty as general trustworthiness or priority among rules is also represented according to the chosen uncertainty formalism and attached to given contextual rules. Once such an uncertainty attachment is done, principled inference, which is often nonmonotonic, is conducted. The examples of such principled inferences are default reasoning [2] to handle inconsistent information, abduction [3] to find most probable hypothesis of given observation and belief revision over time upon the change of observation, etc. In this pipeline, therefore, appropriate uncertainty assignment as well as proper uncertainty formalism plays an important role. However, there are often cases that the trustworthiness of rule itself is also uncertain thereby, uncertainty attachment to rule itself should be rather functional. *the more X then the more Y* type of knowledge is one of the examples. To enable such type of rule modeling, in this chapter, we further explorer our previous work [4–6], where we proposed the use of subjective logic [7] with logic programming and demonstrated that the proposed approach can cover inconsistent information handling as default reasoning and bidirectional reasoning as can be typically done in intensional approaches. We first propose a reputational subjective opinion function that is similar to fuzzy membership function but also can take into account uncertainty of membership value itself. Then we further adopt subjective logic's fusion operator to accumulate the acquired opinions over time. To demonstrate reasoning under uncertain rules, we present a preliminary experimental case study by intentionally restricting the type of available metadata to the results from human detection and tracking algorithms. Automatic human detection and tracking is one of the common analytics and becoming more widely employed in automated visual surveillance systems. The typical types of meta-information that most human detection analytic modules generate comprise, for instance, localization information such as coordinate, width, height, time and (optionally) additional low-level visual feature vectors. We intend to use further such information for evaluating the relationship between two persons and, more specifically, for estimating whether one person could serve as a witness of another person in a public area scene. Examples for (linguistic) domain knowledge applicable to this scenario include: (1) (At least) two distinct people are required for building a relationship. (2) The closer the distance between two people is, the higher is the chance that they can identify each other. (3) If two persons approach each other directly (face-to-face) then there is a higher chance that they can identify each other. Such linguistic knowledge can be modeled and encoded as rules by the proposed approach. The case study is further extended to demonstrate more complex forensic reasoning by considering additional contextual rules together with the shown uncertain rules.

The rest of the chapter is organized as follows. In Sect. 8.2, we briefly review related work regarding intensional and extensional approaches with more focus on the latter one. In Sect. 8.3, we will first give a short introduction to subjective logic theory. In Sect. 8.4, we introduce our approach to modeling uncertain rules. Section 8.5 presents a case study scenario in a typical public area scene and deals with rule encoding and preliminary experimental demo results. Section 8.6 further extend the scenario with more complex situational rules. Finally, Sect. 8.7 concludes with discussions and future research directions.

Table 8.1 A comparison of previous extensional approaches

Approach	Akdemir et al. [9]	Jianbing et al. [10]	Shet et. al [11, 12]	Anderson et al. [13]	Han et al. [4–6]
Knowledge modeling	Ontology	Rule based	Rule based	Rule based	Rule based
Uncertainty formalism	–	Dempster Shafer	Bilattice	Fuzzy logic	Subjective logic
Traditional logic operators	–	–	√	√	√
Arithmetic operators	–	–	–	–	√
Info. fusion operators	–	√	√	–	√
Extra operators (MP, MT, reputation, etc.)	–	–	–	–	√
Default reasoning	–	–	√	–	√
Belief revision	–	–	√	–	√
Bidirectional inference	–	–	–	–	√
Uncertain rule modeling	–	–	–	√	√ (by this work)

8.2 Related Work

To address high level context modeling and reasoning in the visual surveillance domain, traditionally, whole model based approaches such as Bayesian networks have been used. Such approaches are called 'intensional'. Bremont et al. [8] employs a context representation scheme for surveillance systems. Hongeng et al. [14] considers an activity to be composed of action threads and recognizes activities by propagating constraints and likelihood of event threads in a temporal logic network. Other approaches use a qualitative representation of uncertainty [15], HMM to reason about human behaviors based on trajectory information [16], a use of bayesian network and AND/OR tree for the analysis of specific situations [17] or a GMM based scene representation for reasoning upon activities [18]. In such approaches, contextual knowledge is represented as a graph structure having nodes that are considered as symbolic facts. In the sense of logic, connected two nodes can be interpreted as a propositional logic rule that can consider only one relation, the causality implication. A piece of propositional knowledge segment should exist within the whole graph structure, thereby, once uncertainty propagation mechanism is learnt, adding additional pieces of knowledge will require restructuring causality influence relation of the whole graph structure. This aspect restricts expressive power and increases the modeling cost. Due to this complexity and lack of modularity, such approaches have been focusing on relatively narrow and specific semantics. However, as forensic sense of semantics in visual surveillance is gaining more attention, more flexible knowledge representation and uncertainty handling mechanism is required. For this reason, there has been some work on the use of logic programming languages to achieve better expressive power and on the use of different uncertainty handling formalisms to reason under uncertainty. The achievement of better expressive power is mainly due to the first-order predicate logic that logic programming provides. While propositional logic deals with simple declarative propositions, first-order logic additionally covers predicates and quantifiers. Akdemir et al. [9] proposed an ontology based approach for activity recognition, but without uncertainty handling mechanism (In ontology community, Description Logics (DLs) are often used as knowledge representation formalism and DLs are decidable fragments of first-oder-logic.). Shet et al. [11] proposed a system that adopts Prolog based logic programming for high-level reasoning. In [12] the same authors extended their system with the bilattice framework [19] to perform the task of detecting humans under partial occlusion based on the output of parts based detectors. Jianbing et al. [10] used rule-based reasoning with Dempster Shafer's Theory [20] for a bus surveillance scenario. Anderson et al. [13] used fuzzy logic [21] to model human activity for video based eldercare. Han et al. [4–6] proposed the use of logic programming and subjective logic [7] to encode contextual knowledge with uncertainty handling, then demonstrated bidirectional conditional inference and default reasoning. Such logic framework based uncertainty handling approaches can be categorized as 'extensional'. Table 8.1 shows a brief comparison of the previously proposed extensional approaches. the table shows that the coverage of the subjective logic based approach is most broad. For example, while some provides information fusion capability for fusing two contradictory information sources, such as Dempster Shafer's fusion operator, bilattice's operator and subjective logic's consensus operator, only some of them support default reasoning that handles such contradictory information to draw reasonable decision and belief revision. Indeed, bidirectional inference is only supported by subjective logic based approach. In this chapter, we further propose an approach to modeling uncertain propositional rules and inference under such uncertain rules for high level semantic analysis of visual surveillance data. In the sense of linguistic interpretation of the rules, the most similar previous approach to the proposed work would be [13]. In the work, quantitative low level features from human detection are linguistically symbolized into terms such as 'high', 'medium', 'low' and 'very low' according to their corresponding membership functions. Therefore, in such approach, defining membership function is critical. Then the linguistic symbols are used to form a conjunctive logical patterns of a human activities. This means, rules contain symbolized static facts. In our approach, rules allow to contain variable itself. Indeed, our approach even allows uncertainty on a membership-like function by the use of the reputation operator in subjective logic thereby, relieves the burden of defining exact form of membership-like function.

8.3 Subjective Logic Theory

Jøsang [22, 7] introduced subjective logic as a framework for artificial reasoning. Unlike traditional binary logic or probabilistic logic (the former can only consider

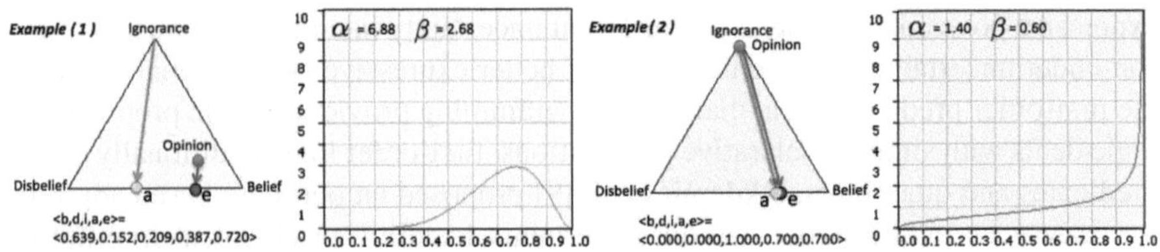

Fig. 8.1 Opinion triangle and beta distribution (Colour figure online)

true or false, and the latter can consider degrees of truth or falseness), subjective logic explicitly represents the amount of 'lack of information (ignorance) on the degree of truth about a proposition' in a model called *opinion* and comes with a rich set of operators for the manipulation of opinions [7]. The idea of explicit representation of ignorance is introduced from belief theory and the interpretation of an opinion in bayesian perspective is possible by mapping opinions to beta distributions. It is also different from fuzzy logic: while fuzzy logic maps quantitative measure to non-crisp premises called fuzzy sets (e.g. 'fast', 'slow', 'cold', 'hot' etc.), subjective logic deals with the uncertain belief itself on a crisp premise (e.g. 'intrusion happened', 'accident happened', etc.). However, in the sense of interpretation, mapping of an opinion into the linguistic certainty fuzzy set (i.e., 'very certainly true', 'less certainly true', etc) is also possible. In general, subjective logic is suitable for modeling real situations under partial ignorance on a proposition's being true or false. Known application areas are trust network modeling, decision supporting, etc. However, to the best of our knowledge, the application of subjective logic in computer vision related domains has been limited to [4–6] that demonstrated the capability of default reasoning and bidirectional interpretation of conditional rules. In this section, we will give a brief introduction to subjective logic theory.

Definition 8.1 (*Opinion*) [7] Let $\Theta = \{x, \overline{x}\}$ be a state space containing x and its complement $\overline{x}$. Let b_x, d_x, i_x represent the belief, disbelief and ignorance in the truth of x satisfying the equation: $b_x + d_x + i_x = 1$ and let a_x be the base rate of x in Θ. Then the opinion of an agent ag about x, denoted by w_x^{ag}, is the tuple $w_x^{ag} = (b_x^{ag}, d_x^{ag}, i_x^{ag}, a_x^{ag})$.

Definition 8.2 (*Probability expectation*) [7] Let $w_x^{ag} = \{b_x^{ag}, d_x^{ag}, i_x^{ag}, a_x^{ag}\}$ be an opinion about the truth of x, then the probability expectation of w_x^{ag} is defined by: $E(w_x^{ag}) = b_x^{ag} + a_x^{ag} i_x^{ag}$.

Opinions can be represented on an so called *opinion triangle* as shown in Fig. 8.1. A point inside the triangle represents a (b_x, d_x, i_x) triple. The corner points marked with Belief, Disbelief or Ignorance represent the extreme cases, i.e., no knowledge (0, 0, 1), full disbelief (0, 1, 0) and full belief (1, 0, 0). The base rate a_x represents the prior knowledge on the tendency of a given proposition to be true and can be indicated along the base line (the line connecting Belief and Disbelief). The probability expectation E is then formed by projecting the opinion onto the base line, parallel to the base rate projector line (see the blue line) that is built by connecting the a_x point with the Ignorance corner (see the red line). An interesting property of subjective opinions is their direct mapping to beta distributions. Beta distributions are normally denoted as $Beta(\alpha, \beta)$ where α and β are its two parameters (α represents the number of positive observations and β represents amount of negative observations about a crisp proposition respectively). The beta distribution of an opinion $w_x = (b_x, d_x, i_x, a_x)$ is the function $Beta(\alpha, \beta)$ where $\alpha = 2b_x/i_x + 2a_x$ and $\beta = 2d_x/i_x + 2(1 - a_x)$. In Fig. 8.1, Example (1) shows an opinion about a proposition of an agent, that can be interpreted as *seems likely and slightly uncertain true*, and Example (2) shows full ignorance (a.k.a. *vacous* opinion) at the time of judgement about a proposition. Assuming base rate to be 0.7 in the example we get expectation value also to be 0.7 and the beta distribution appears biased towards 'True' though the opinion represents full ignorance.

8.4 Modeling Uncertain Rule Using Subjective Logic

The proposed uncertain rule modeling approach mainly relies on rule-based system that enables logic programming. The traditional rule-based system, which can only handle binary logic, is extended to allow representation of uncertainty using subjective opinions and operators. For a given propositional knowledge, we assume a fuzzy-like membership function that grades degree of truth. Then we focus on that the interpretation of such membership function can be dogmatic, thereby, when the function is projected on the opinion space, it only lays on the bottom line of the opinion space. Indeed, in many cases, the exact shape of the function is hard to determine. To address this aspect, we introduce a reputational function that evaluates the trust worthiness of the fuzzy-like membership function. Then we introduce accumulation of the resulted opinions overtime. In this section, we will first give a brief overview how rules are expressed in logic programming. Thereafter, comes with further details of the uncertain rule modeling.

8.4.1 Logic Programming

Logic programming mainly consists of two types of logical formulae, rules and facts. Rules are of the form $A \leftarrow f_0, f_1, \ldots, f_m$ where A is rule head and the right hand side is called body. Each f_i is an atom and ',' represents logical conjunction. Each atom is of the form $p(t_1, t_2, \ldots, t_n)$, where t_i is a term and p is a predicate symbol that takes n terms (i.e. arity n). Terms could either be variables or constant symbols. Negation is represented with the symbol $\neg$ such that '$A = \neg\neg A$'. Both positive and negative atoms are referenced to as literals. Given a rule $head \leftarrow body$, we interpret the meaning as *IF body THEN head*. Traditionally, resolved facts that matches to a rule is called *extension*. In extensional approaches [11, 12, 10, 4–6] mentioned in

Sect. 8.2, rules have been used to define and reason about various contextual events or activities.

8.4.2 Logic Programming Extended Using Subjective Logic

To extend logic programming with subjective logic, the CLIPS [23] rule engine was used as a basis to provide flexibility for defining complex data structure as well as for providing a rule resolving mechanism. To extend this system, a data structure *opinion(agent,proposition,b,d,i,a)* was defined that can be interpreted as a fact of arity 6 with the following terms, *agent* (*opinion owner*), *proposition*, *belief*, *disbelief*, *ignorance*, and *atomicity*. To represent propositions, we extended the structure so that it can take arity n properties as well. Therefore, given a predicate p the proposition can be described as $p(a_1, a_2, \ldots, a_n)$. In our system, therefore, each fact is represented as the form of $w^{agent}_{p(a_1,a_2,\ldots,a_n)}$. Namely, rules are defined with the opinion and proposition structure. Additionally, functions of subjective logic operators taking opinions as parameters were defined. In this way, uncertainty in the form of opinion triangle is attached to rules and facts. This aspect is depicted as follows:

Definition 8.3 (*Opinion Assignment*) Given a knowledge base $\mathcal{K}$ in form of declarative language and Subjective Opinion Space O, an opinion assignment over sentences $k \in \mathcal{K}$ is a function $\phi : k \rightarrow O$. s.t.

1. $\phi_{fact} : Fact \rightarrow O$, e.g. $w^{a}_{p(a_1,a_2,\ldots,a_n)} = (b, d, u, i)$
2. $\phi_{Rule} : Rule \rightarrow O$, e.g. $(w^{a_c}_{p_c(a_{c1},\ldots,a_{cn})} \leftarrow w^{a1}_{p_1(a_{11},\ldots,a_{1n})}, \ldots, w^{ai}_{p_n(a_{i1},\ldots,a_{in})}) = (b, d, u, i)$
3. $\phi_{RuleEval} : Rule\ Head \rightarrow \left(\circledast_{w^{a_i}_{p_i} \in Rule\ Body} w^{a_i}_{p_i(a_{ai1},\ldots,a_{in})} = O \right)$, where $\circledast$ indicates one of subjective logic's operators.

Example for a given rule $w^{a_c}_{p_c(a_{c1},\ldots,a_{cn})} \leftarrow w^{a1}_{p_1(a_{11},\ldots,a_{1n})}, \ldots, w^{ai}_{p_n(a_{i1},\ldots,a_{in})}$,

$$w^{a_c}_{p_c(a_{c1},\ldots,a_{cn})} = w^{a1}_{p_1(a_{11},\ldots,a_{1n})} \circledast \cdots \circledast w^{ai}_{p_n(a_{i1},\ldots,a_{in})} = (b, d, u, i).$$

$\phi_{inference}$ denoted $cl(\phi) : q \rightarrow O$, where $\mathcal{K} \models q$ called Closure.

It is important to note that there are different ways of opinion assignment. While Definition 8.3—2 assigns an opinion to a whole rule sentence itself, Definition 8.3—3 assigns an opinion to the consequence part of the rule (rule head). The assigned opinion is functionally calculated out of opinions in the rule body using appropriate subjective logic operators. Definition 8.3—2 especially plays an important role for prioritizing or weighting rules for default reasoning [6]. Given the initial opinion assignment by Definition 8.3—1 and 2, the actual inference is performed by Definition 8.3—3 and 4, where Definition 8.3—4 is further defined as follows:

Definition 8.4 (*Closure*) Given a knowledge base $\mathcal{K}$ in form of declarative language and an opinion assignment ϕ, labeling every sentence $k \in \mathcal{K}$ into Subjective Opinion Space O, then the closure over $k \in \mathcal{K}$, is the opinion assignment function $cl(\phi)(q)$ that labels information q entailed by $\mathcal{K}$ (i.e. $\mathcal{K} \models q$).

For example, if ϕ labels sentences $\{a, b, c \leftarrow a, b\} \in \mathcal{K}$ as $\phi_{fact}(a)$, $\phi_{fact}(b)$ and $\phi_{Rule}(c \leftarrow a, b)$, then $cl(\phi)$ should also label c as it is information entailed by $\mathcal{K}$. The assignment can be principled by the definition of closure. For example, an opinion assignment to c, in a simple conjunctive sense can be $\phi_{fact}(a) \cdot \phi_{fact}(b) \cdot \phi_{Rule}(c \leftarrow a, b)$, where $\cdot$ represent conjunction in Subjective Logic. In our system, to support the rich set subjective logic operators, we made the specification of Definition 8.3—3 in rule description as follows (note that, most of rule based systems also support describing actions in the head part of a rule):

$$\begin{aligned} &ACTION : Assert\ new\ Opinion\ w^{a_c}_{p_c(a_{c1},\ldots,a_{cn})},\ \text{where}\ w^{a_c}_{p_c(a_{c1},\ldots,a_{cn})} = \\ &w^{a1}_{p_1(a_{11},\ldots,a_{1n})} \circledast \cdots \circledast w^{ai}_{p_n(a_{i1},\ldots,a_{in})} \leftarrow w^{a1}_{p_1(a_{11},\ldots,a_{1n})}, \ldots, w^{ai}_{p_n(a_{i1},\ldots,a_{in})}. \end{aligned} \tag{8.1}$$

Due to the redundancy that arises when describing rules at the opinion structure level, we will use abbreviated rule formulae as follows:

$$w^{a_c}_{p_c(a_{c1},\ldots,a_{cn})} \leftarrow w^{a1}_{p_1(a_{11},\ldots,a_{1n})} \circledast \cdots \circledast w^{ai}_{p_n(a_{i1},\ldots,a_{in})}. \tag{8.2}$$

where $\circledast$ indicates one of subjective logic's operators. This way of representing rules, we can build a propositional rules that comprise opinions about a predicate as facts, check logical conjunction based existence of involved opinions and finally define resulted predicate with opinion attached by calculating opinion values with subjective logic operators. To realize this concept, a prototype system integrating binary logic programming and subjective logic calculus has been implemented. For the logic programming part, the CLIPS [23] rule engine was used.

8.4.3 Uncertain Propositional Rules

In logic programming, a conditional proposition $y \leftarrow x$ is interpreted as *IF x THEN y*. However, there are often cases that we may want to interpret the meaning as *the more x then the more y* or *the more x then the less y*, etc. In this case, the opinion attached to the consequence of the rule should be rather functional in terms of the elements within the rule body. Therefore, the opinion assignment suit to this interpretation is Definition 8.3—3. In the sense of intrinsic linguistic uncertainty of the rule, it resembles fuzzy rules shown by Anderson et al. [13, 21]. In the work, quantitative low level features of human detection results such as 'centroid', 'eigen-based height' and 'ground plane normal similarity' are linguistically mapped into non-crisp premises (i.e. fuzzy sets) as '(H)igh', '(M)edium', '(L)ow' and '(V)ery Low'.

Then fuzzy rules defines the conjunctive combination of those linguistic symbols to draw higher semantics such as 'Upright', 'In Between' and 'On the ground' (e.g. $Upright(L) \leftarrow Centroid(H), EigenHeight(M), Similarity(H)$ [13]). Therefore, introducing appropriate fuzzy membership functions for each linguistic terms and proper handling of the membership functions is an important issue. In this view, Mizumoto et al. [24] showed comparison of sophisticated mathematical handling of ambiguous concepts such as 'more or less' having various shapes. One another thing worth to note concerning fuzzy logic is that, even if there are Zadeh's original logical operators, there are yet another ways of defining logical operators as well. For example, for given two quantitative variables x and y come with corresponding membership functions μ_a and μ_b, Zadeh's AND operator is defined as x AND $y = min(\mu_a(x), \mu_a(y))$. In so-called 't-norm fuzzy logic', any form of t-norms can be considered as AND operators. For example, in the case of using product t-norm, the AND operator can be defined as x AND $y = \mu_a(x) \cdot \mu_b(x)$ [25]. This aspect still remains controversial among most statisticians, who prefer Bayesian logic [26]. Contrary, as explained in the Sect. 8.3, subjective logic can be interpreted in the sense of bayesian and also the final quantitative opinion space can also be interpreted in the sense of fuzziness (i.e. 'very certainly true', 'less certainly true', etc). This way, we believe that subjective logic can better bridge the interpretation of fuzzy intuitive concepts with better bayesian sense. The basic idea of our approach is as follows:

1. For a given propositional rule 'the less (more) $y \leftarrow$ the more x' we could introduce a membership-like function $\mu_i : x \rightarrow y$.
2. It is clear that the function μ_i should be monotonically decreasing (increasing) but the shape is not quite clear.
3. Considering potentially possible multiple membership like functions μ_i, however the values of $\mu_i(x)$ at the two extreme point of $(min_x \leq x \leq max_x)$ tend to converge but the values in between are diverge therefore, the values of later cases are more uncertain.
4. Considering the aspect of 3. we introduce so-called reputational opinion function on the function μ_i and combine it with raw opinion obtained from μ_i using subjective logic's reputation operator.

This idea is depicted in Fig. 8.2, where the actual reputational operation is defined as follows:

Definition 8.5 (*Reputation*) [27] Let A and B be two agents where A's opinion about B's recommendations is expressed as $w_B^A = \{b_B^A, d_B^A, u_B^A, a_B^A\}$, and let x be a proposition where B's opinion about x is recommended to A with the opinion $w_x^B = \{b_x^B, d_x^B, u_x^B, a_x^B\}$. Let $w_x^{A:B} = \{b_x^{A:B}, d_x^{A:B}, u_x^{A:B}, a_x^{A:B}\}$ be the opinion such that:

$$\begin{cases} b_x^{A:B} = b_B^A b_x^B & d_x^{A:B} = d_B^A d_x^B \\ u_x^{A:B} = d_B^A + u_B^A + b_B^A u_x^B & a_x^{A:B} = a_x^B \end{cases}$$

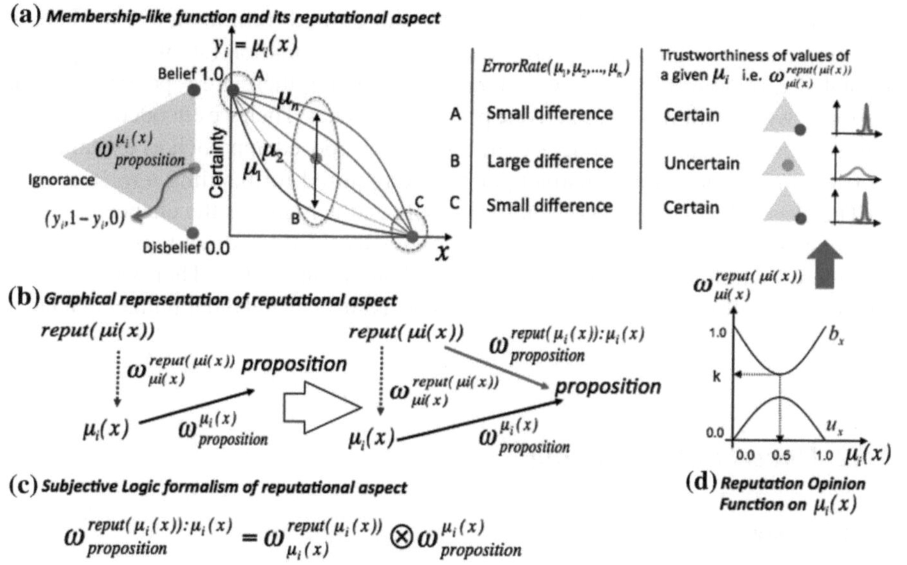

Fig. 8.2 Uncertain rule modeling using subjective logic's reputation operator

then $w_x^{A:B}$ is called the reputation opinion of A. By using the symbol $\otimes$ to designate this operation, we get $w_x^{A:B} = w_B^A \otimes w_x^B$.

For actual evaluation of a given function μ_i, an opinion assignment function on the given μ_i need to be defined. Although there could be also another ways of such function, in our approach, this is modeled as follows:

$$w_{\mu_i(x)}^{reput^{\mu_i(x)}} = \begin{cases} b_x = k + 4(1-k)(\mu_i(x) - \frac{1}{2})^2 \\ d_x = \frac{1-b_x}{Dratio} \\ u_x = 1 - b_x - d_x. \end{cases} \tag{8.3}$$

where k, represents the minimum boundary of belief about the value from $\mu_i(x)$, and the $Dratio$ indicates the ratio for assigning the residue of the value μ_i to disbelief and uncertainty. This is depicted as Fig. 8.2d.

8.5 Case Study I

8.5.1 Scenario Setting for Case Study

At this stage we focused on evaluating the modeling approach itself rather than the reliability of the person detection algorithm. Therefore, we manually annotated a test video from one of i-LIDS [28] data sample with ground truth metadata for human detection comprising bounding boxes and timing information (shown in Fig. 8.3). In total, 1 minute of test video was annotated in which there are 6 people. For our purposes, we intentionally marked one person as suspect. Then we encoded following linguistic contextual knowledge according to the proposed approach as explained in Sect. 8.4. (1) (At least) two distinct people are required for building a relationship. (2) The closer the distance between two people is, the higher is the chance that they can identify each other. (3) If two persons approach each other directly (face-to-face) then there is a higher chance that they can identify each other. Then we calculate subjective opinions between the person marked as suspect and other human instances over time.

8.5.2 Uncertainty Modeling

8.5.2.1 Distance

The distance between a pair of people would be one of the typical pieces of clue for reasoning whether one person could serve as a witness of another person. This relates to the general human knowledge that *The closer two people are in distance, the more chances of perceiving the other are*. Humans are very adapted to operating upon such type of uncertain and ambiguous knowledge. Exactly modeling such a relation is not trivial, but we can approximate it with a monotonic decreasing function about the possibility of perceiving each other. This aspect is depicted as three possible curves in the middle of Fig. 8.4a, where x represents the distance between the persons as calculated from the person detection metadata and μ_i represents the likelihood that two persons at this distance would perceive each other, $maxdist$ is the maximum possible (i.e. diagonal) distance in a frame and a_i is the estimated probability that two humans could have recognized each other at the $maxdist$ distance. However, the value derived from such function is not fully reliable due to the variety of real world and uncertainty in the correctness of the function and uncertainty in the distance value itself. Considering the aspect of distance, it is clear that both the extreme cases i.e. very close or very far are much more certain than in the middle of the range. Thus, to better model the real world situation, the reputational opinion function need to be applied to any chosen function μ_i. This is modeled as opinion on the reliability of $\mu_i(x)$ by applying Eq. (8.3). In order to evaluate the impact of choosing different functions in Fig. 8.4a, three different types of μ_i functions (a concave, convex and linear) have been applied. The derived reputational opinions showed similar aspects having peaks of certain belief at each extreme cases as shown in Fig. 8.5.

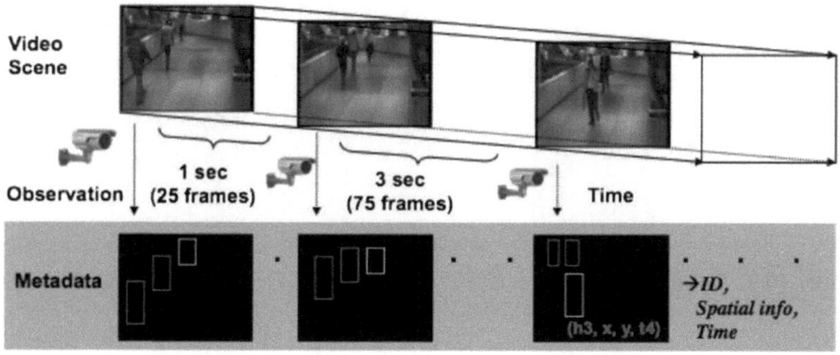

Fig. 8.3 Scenario setting for case study I

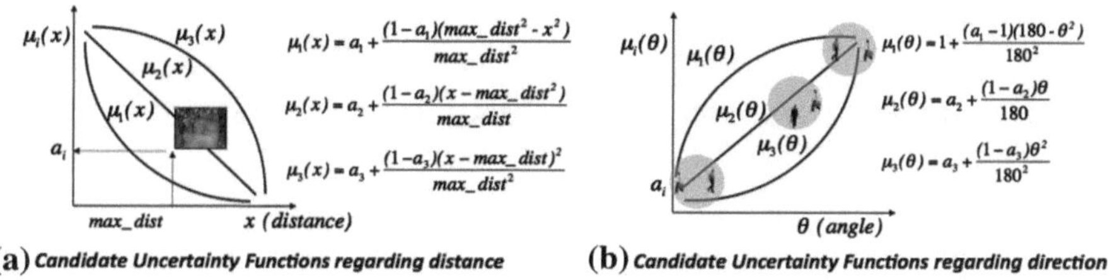

Fig. 8.4 Candidate uncertainty functions regarding distance and direction

8.5.2.2 Direction

Similarly, we also used direction information between two persons. The linguistic knowledge to be modeled is *if two persons approach each other directly (face-to-face) then there is a higher chances of perceiving each other*. The corresponding direction-based relevance function is shown in Fig. 8.4b, where Θ represents the angle between the persons heading directions as calculated from the person detection metadata and μ_i represents the likelihood that two persons at the angle would perceive each other and a_i is the expected minimum probability that two humans could have recognized each other at any angle. However, again the trustworthiness of the values from such functions μ_i is uncertain, especially in the middle range of the Θ. To roughly model such aspect, for a chosen function $\mu_i(\Theta)$, the same reputational function from Eq. (8.3) was used again. The impact of choosing different μ_i showed similar behavior as of direction based opinions as shown in Fig. 8.5.

8.5.3 Rule Encoding

In addition to the uncertainty modeling, logic programming is used to represent the given contextual rules as explained in Sect. 8.4.2. Encoded rules in form of Eq. (8.2) are as follows:

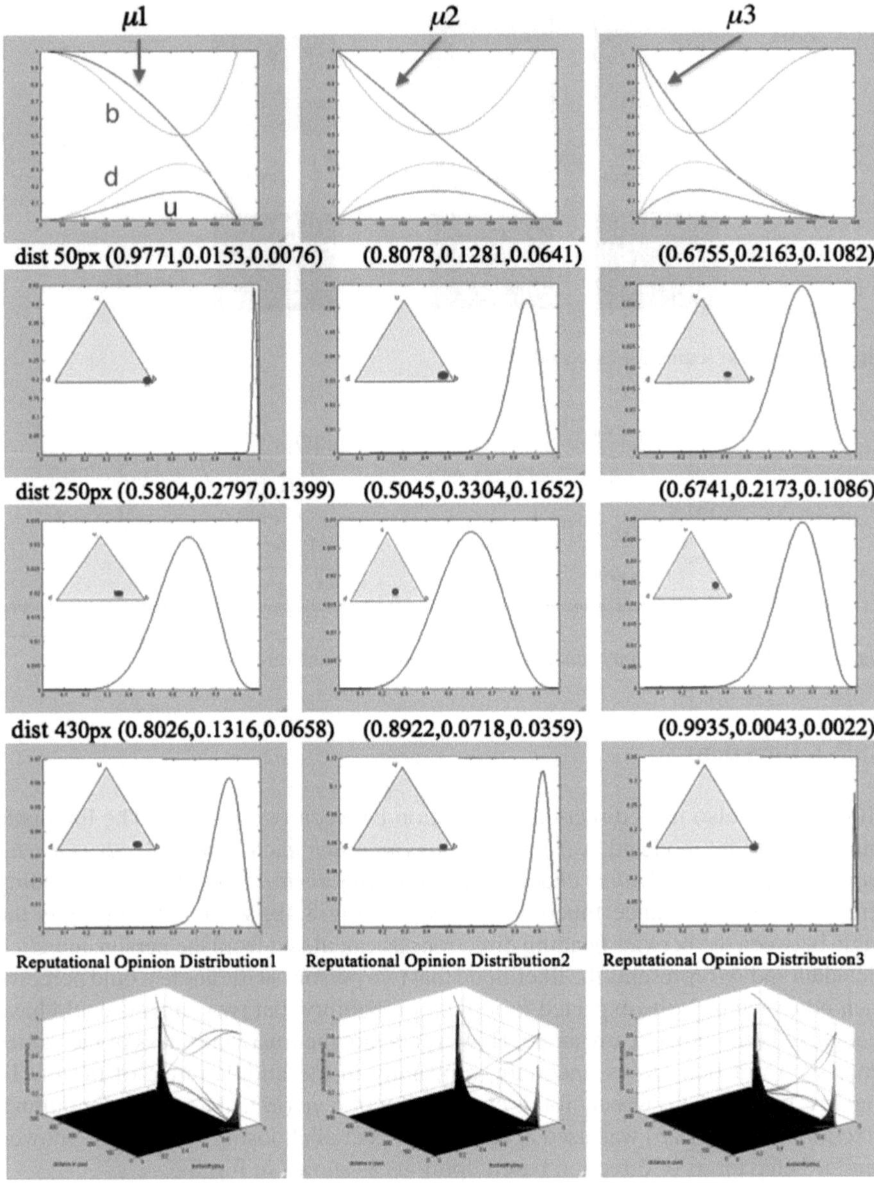

Fig. 8.5 Samples of reputational opinion according to distance and Eq. (8.3)

$$w^{Rule1}_{witness(H_1,H_2,T_1)} \leftarrow \left(w^{Human_{Detector}}_{human(H_1,T_1)} \wedge w^{Human_{Detector}}_{human(H_2,T_1)}\right) \otimes \left(w^{\mu_{dist}(d)}_{witness(H_1,H_2,T_1)} \otimes w^{reput^{\mu(d)}}_{\mu_{dist}(d)}\right). \tag{8.4}$$

$$w^{Rule2}_{witness(H_1,H_2,T_1)} \leftarrow \left(w^{HumanDetector}_{human(H_1,T_1)} \wedge w^{HumanDetector}_{human(H_2,T_1)}\right) \otimes \left(w^{\mu_{dir}(d)}_{witness(H_1,H_2,T_1)} \otimes w^{reput^{\mu(d)}}_{\mu_{dir}(d)}\right). \tag{8.5}$$

$$w^{Rule3}_{witness(H_1,H_2,T_1)} \leftarrow \left(w^{Rule1}_{witness(H_1,H_2,T_1)} \wedge w^{Rule2}_{witness(H_1,H_2,T_1)}\right). \tag{8.6}$$

$$w^{Rule4}_{witness(H_1,H_2,T_n)} \leftarrow \oplus_{i=1}^{n} w^{Rule3}_{witness(H_1,H_2,T_i)}. \tag{8.7}$$

The first rule (8.4) starts considering the necessary condition, meaning that there should be a distinct pair of two people. Therefore the conjunction operation $\wedge$ on two opinions [29] is used that is very similar to the operation $P(A) \cdot P(B)$ except that in subjective logic the opinion can additionally represent ignorance. Then, for the resulting set of frames the reputational opinion about the distance opinions is calculated as described in Sect. 8.5.2. Each result is assigned to a new opinion with the predicate of the appropriate arity and is assigned the name of agent with the final belief values. In this case, the final opinion value represents that there is an opinion about two persons being potential witnesses of each other from an agent named $Rule1$. The second rule (8.5) is almost same as rule (8.4). The only different part of this rule is that the reputational opinion is about direction. The third rule (8.6) combines the evidences coming from rule (8.4) and (8.5). The conjunction operator $\wedge$ is used to reflect that for reliable positive resulting opinions both evidences should have appeared with a certain amount of belief. The last rule (8.7) is about accumulating the belief over time using the consensus operator $\oplus$ that is defined as follows:

Definition 8.6 (*Consensus*) [30] Let $w_x^A = (b_x^A, d_x^A, i_x^A, a_x^A)$ and $w_x^B = (b_x^B, d_x^B, i_x^B, a_x^B)$ be opinions respectively held by agents A and B about the same state x, and let $k = i_x^A + i_x^B - i_x^A i_x^B$. When $i_x^A, i_x^B \to 0$, the relative dogmatism between w_x^A and w_x^B is defined by γ so that $\gamma = i_x^B / i_x^A$. Let $w_x^{A,B} = (b_x^{A,B}, d_x^{A,B}, i_x^{A,B}, a_x^{A,B})$ be the opinion such that:

$$k \neq 0: \begin{cases} b_x^{A,B} = (b_x^A i_x^B + b_x^B i_x^A)/k \\ d_x^{A,B} = (d_x^A i_x^B + d_x^B i_x^A)/k \\ i_x^{A,B} = (i_x^A i_x^B)/k \\ a_x^{A,B} = \frac{a_x^A i_x^A + a_x^B i_x^A - (a_x^A + a_x^B) i_x^A i_x^B}{i_x^A + i_x^B - 2 i_x^A i_x^B} \end{cases}$$

$$k = 0: \begin{cases} b_x^{A,B} = \frac{\gamma b_x^A + b_x^B}{\gamma + 1} \\ d_x^{A,B} = \frac{\gamma d_x^A + d_x^B}{\gamma + 1} \\ i_x^{A,B} = 0 \\ a_x^{A,B} = \frac{\gamma a_x^A + a_x^B}{\gamma + 1}. \end{cases}$$

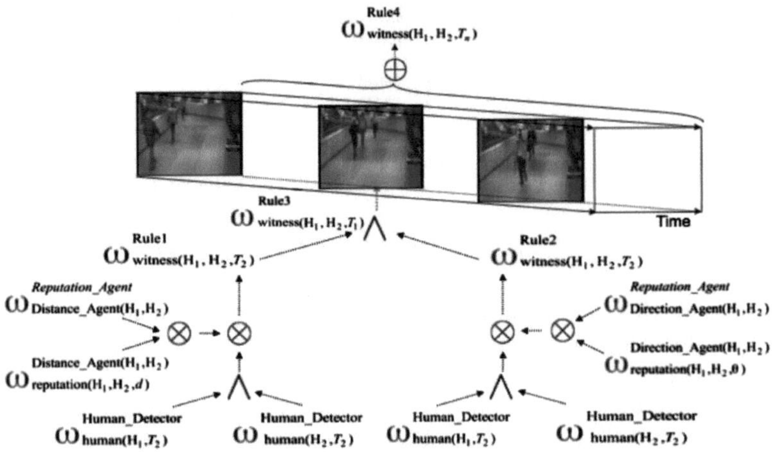

Fig. 8.6 Tree representation of rules

Then $w_x^{A,B}$ is called the consensus opinion between w_x^A and w_x^B, representing an imaginary agent $[A, B]$'s opinion about x, as if that agent represented both A and B. By using the symbol $\oplus$ to designate this operator, we define $w_x^{A,B} = w_x^A \oplus w_x^B$.

Figure 8.6 shows a graphical representation of the rules in a tree form.

8.5.4 Experimental Result

Using the rules described in Sect. 8.5.3, we calculated subjective opinions between a person marked as suspect and other human instances over time. Figure 8.7 shows a snapshot of the visualization in the prototype comprising a video player and an opinion visualizer. While the video is being played the corresponding metadata is transformed into the corresponding opinion representation. The translated opinions are fed into the rule-engine which automatically evaluates the rules. The right part of Fig. 8.7 shows the opinion about the proposition 'human 5 is a witness for the suspects marked red' and its corresponding mapping to beta distribution. For verification of these results, a questionnaire was prepared to collect scores about the witnessing chances for each of the 'pairs' in the scene (e.g. human1 and suspect, human2 and suspect , etc). Seven people from our lab took part in the questionnaire. Then changing the uncertainty functions on uncertain rules, we tested the behavior of the proposed approach to check whether it well models human intuition. Although there can be 9 possible combinations of uncertainty functions (i.e. 3 distance functions and 3 direction functions), to better contrast the impact of changing such uncertainty functions, we have fixed the direction function to the type of μ_3 defined in Fig. 8.4b and tested with 3 different direction functions shown in Fig. 8.4a. Then the mean

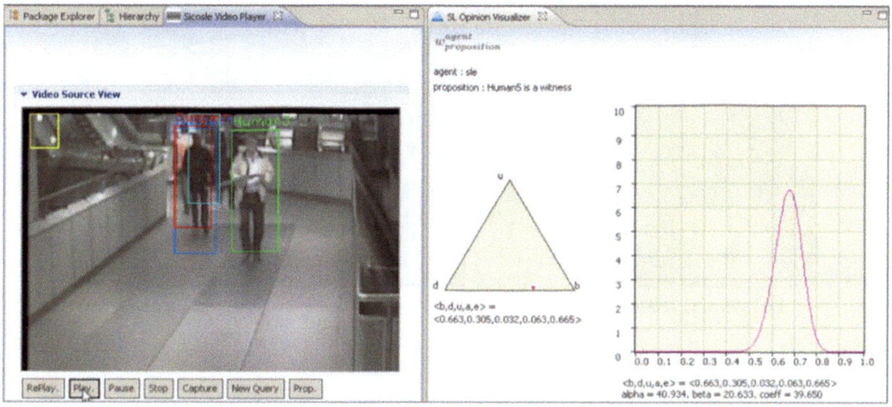

Fig. 8.7 Visualization of the experiment

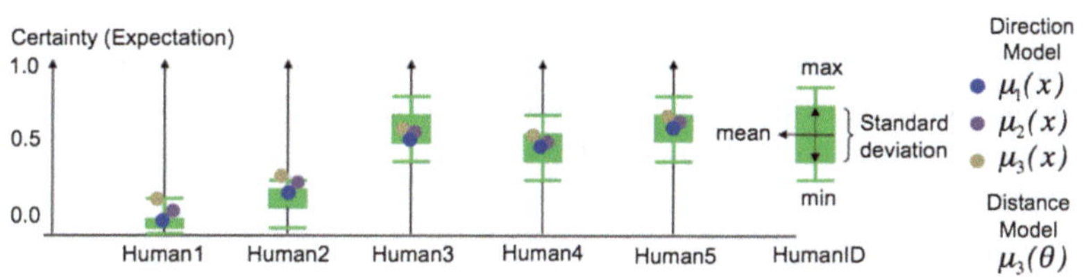

Fig. 8.8 Experimental result

and standard deviation, *min* and *max* of the 'human opinions' were calculated and compared to the computed results. According to [7], the following criteria should be applied to the computed results.

(1) *The opinion with the greatest probability expectation is the greatest opinion.*
(2) *The opinion with the least uncertainty is the greatest opinion.*
(3) *The opinion with the least relative atomicity is the greatest opinion.*

In the described experiment, due to the small size of possible pairs, only the first criterion was applied and the final expectation values of each opinion for candidate pairs were plotted jointly with the questionnaire based result as shown in Fig. 8.8. The final result turns out to be following the tendency of questionnaire based human 'opinions'. The change of uncertainty function seems not introducing that critical differences. However, there were more differences between the expected values, when the final expectation values were low, for instance, though it was a slight differences, μ_3 tend to yield larger expectation value then μ_2 and μ_1. The differences ware smaller when the final expectation values were getting higher. However, in any cases, the order on the ranking of witnesses show the same results. Therefore, in the sense of human like reasoning, it seems that the proposed approach well models human intuition.

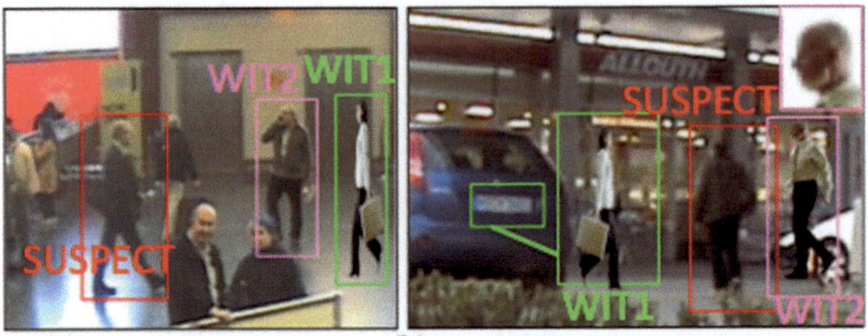

(a) *Witness Talking on Phone* (b) *Witness with License Plate and Face Detection*

Fig. 8.9 Scenario setting for case study 2

8.6 Case Study II

In this section, we further explorer the proposed case study scenario for more complex contextual forensic reasoning. Especially, we will consider the situation that is needed to be modeled in the sense of so-called default reasoning [2].

8.6.1 Scenario Setting for Case Study II

Consider a conceptual scenario that a security personnel wants to get suggestions of most probable witnesses of a selected suspect in a scene. Given an assumption that automatic vision analytics are running and extracting basic semantics, we will also assume two virtual situations as shown in Fig. 8.9, where, witnesses are reasoned according to the uncertain spatio-temporal rules as demonstrated in Sect. 8.5. In all situation we will assume that 'witness2' has higher opinion then 'witness1'. In addition to this, we will assume optional cases that additional evidential cues are detected. In Fig. 8.9a, 'witness2' is talking on the phone. In Fig. 8.9b, the optional case is the detection of a license plate of the car seems to belong to the 'witness1' and 'witness2' comes with face detection.

8.6.2 Reasoning Examples

Given the scenario with optional cases, we will also assume that (1) people usually do not recognize well when they are talking on the phone, (2) identifiable witness is a good witness. (3) License plate is better identifiable source than face detection because we can even fetch personal information of the owner easily. Therefore, under optional assumption, for example, in Fig. 8.9a, 'witness1' should be better witness, and in Fig. 8.9b, 'witness1' should be suggested as a better witness. This kind of non monotonic reasoning under inconsistent information is called default reasoning and defined as follows:

Definition 8.7 (*Default theory*) [2] Let $\Delta = (D, W)$ be a default theory, where W is a set of logical formulae (rules and facts) also known as the definite rules and D is a set of default rules of the form $\frac{\alpha:\beta}{\gamma}$, where α is known as the precondition, β is known as the justification and γ is known as the conclusion.

Han et al. [6] showed that this aspect can be modeled using subjective logic as well under the opinion assignment defined in Definition 8.3 in Sect. 8.4.2. Here, it is important to note that unlike the case of uncertain rule modeling, the type of opinion assignment to prioritize belong to Definition 8.3—2. and the default inference scheme belongs to Definition 8.3—4. As shown in [6], we set $T \simeq (1, 0, 0)$ (full truth), $DT_1 \simeq (0.5, 0, 0.5)$ (weak default true), $DT_2 \simeq (0.8, 0, 0.2)$ (strong default true), $F \simeq (0, 1, 0)$ (full false), $DF_1 \simeq (0, 0.5, 0.5)$ (weak default false), $DF_2 \simeq (0, 0.8.0.2)$ (strong default false), $* \simeq (0.33, 0.33, 0, 34)$ (contradiction), $U \simeq (0, 0, 1)$ (full uncertainty) and $\perp \simeq (0.5, 0.5, 0)$ (full contradiction). For the rest of truth values we will use opinion triple representation (b,d,i). The default inference scheme using subjective logic is as follows:

Definition 8.8 ($Default\ inference_{sl}$) [6] Given a query sentence q and given S and S' that are sets of sentences such that $S \models q$ and $S' \models \neg q$, then the default inference is the truth value assignment closure $cl_{sl_{di}}(\phi)(q)$ given by:

$$cl_{sl_{di}}(\phi)(q) = \bigoplus_{S \models q} u \sqcup \left[\prod_{p \in S} cl_{sl}(\phi)(p) \right] \oplus \neg \bigoplus_{S' \models \neg q} u \sqcup \left[\prod_{p \in S'} cl_{sl}(\phi)(p) \right]. \quad (8.8)$$

Example 1 (Witness talking on the phone) Assume the following set of rules about determining good witness including the uncertain spatio-temporal relation based witness reasoning rule described in Sect. 8.5.3. Then also assume the following opinion assignment that $witness2$ (denoted as wit_2) has higher opinion being the witness than $witness1$ (denoted as wit_1).

$$\phi_{Rule}\left[w^{Rule4}_{witness(H_1)} \leftarrow \oplus^{n}_{i=1} w^{Rule3}_{witness(H_1,H_2,T_i)}\right] = DT_1.$$

$$\phi_{Rule}\left[\neg w_{witness(H_1)} \leftarrow w_{talking\_on\_phone(H_1)}\right] = DT_2.$$

$$\phi_{RuleEval}\left[w^{Rule4}_{witness(wit\_1)}\right] = (0.6, 0.15, 0.25).$$

$$\phi_{RuleEval}\left[w^{Rule4}_{witness(wit\_2)}\right] = (0.7, 0.10, 0.20).$$

Given two default true and default false rules and facts that can be seen as definite true, the inference for reasoning better witness using default logic with subjective logic is as follows.

$$cl_{sl_{di}}(\phi)(w_{witness(wit\_1)}) = [U \sqcup ((0.6, 0.15, 0.25) \cdot DT_1)].$$
$$= [U \sqcup (0.44, 0.15, 0.41)] = (0.44, 0.15, 0.41) \sim (Expectation = 0.54).$$
$$cl_{sl_{di}}(\phi)(w_{witness(wit\_2)}) = [U \sqcup ((0.7, 0.10, 0.20) \cdot DT_1)].$$
$$= [U \sqcup (0.50, 0.10, 0.40)] = (0.50, 0.10, 0.40) \sim (Expectation = 0.60).$$

Above result shows that given the weak rules, 'witness2' is more probable witness candidate than 'witness1'. Then, let us consider the weak opinion assignment to the additional contextual cue that witness2 is using the phone. This semantics can be interpreted as 'the witness seems to using a phone but not quite sure'.

$$\phi_{fact}[w_{talking\_on\_phone(wit\_2)}] = (0.6, 0.15, 0.25).$$

Given the additional information, the inference on witness2 is being witness is as follows.

$$\begin{aligned} & cl_{sl_{di}}(\phi)(w_{witness(wit\_2)}) \\ &= [U \sqcup ((0.7, 0.10, 0.20) \cdot DT_1)] \oplus \neg[U \sqcup ((0.6, 0.15, 0.25) \cdot DT_2)] \\ &= [U \sqcup (0.50, 0.10, 0.40)] \oplus \neg[U \sqcup (0.59, 0.15, 0.26)] \\ &= (0.50, 0.10, 0.40) \oplus \neg(0.59, 0.15, 0.26) \\ &= (0.50, 0.10, 0.40) \oplus (0.15, 0.59, 0.26) \\ &= (0.34, 0.47, 0.19) \sim (Expectation = 0.39). \end{aligned}$$

The resulting opinion (0.34,0.47,0.19) on witness2's being a good witness now weaker than (0.44,0.15,0.41) which is for the case of witness1's being a good witness. The expectation values also captures this aspect. Thus, this result shows that the inference scheme well models human intuition.

Example 2 (Witness with face detection vs. license plate detection) Consider the following set of rules about determining good witness and the following opinion assignment to capture the scenario described in Sect. 8.6.1 and depicted in Fig. 8.9b.

$$\phi_{Rule}\left[w^{Rule4}_{witness(H_1)} \leftarrow \oplus^{n}_{i=1} w^{Rule3}_{witness(H_1,H_2,T_i)}\right] = DT_1.$$
$$\phi_{Rule}\left[w_{witness(H_1)} \leftarrow w^{Rule4}_{witness(H_1)} \cdot w_{hasFaceDetectInfo(H_1)}\right] = DT_1.$$
$$\phi_{Rule}\left[w_{witness(H_1)} \leftarrow w^{Rule4}_{witness(H_1)} \cdot w_{hasLicenseDetectInfo(H_1)}\right] = DT_2.$$
$$\phi_{RuleEval}\left[w^{Rule4}_{witness(wit\_1)}\right] = (0.6, 0.15, 0.25).$$
$$\phi_{RuleEval}\left[w^{Rule4}_{witness(wit\_2)}\right] = (0.7, 0.10, 0.20).$$
$$\phi_{fact}\left[w_{hasLicenseDetectInfo(wit\_1)}\right] = (0.6, 0.15, 0.25).$$
$$\phi_{fact}\left[w_{hasFaceDetectInfo(wit\_2)}\right] = (0.6, 0.15, 0.25).$$

Given two default true and default false rules and facts that can be seen as definite true, the inference for reasoning better witness using default logic with subjective logic is as follows.

$$
\begin{aligned}
& cl_{sl_{di}}(\phi)(w_{witness(wit\_1)}) \\
& \quad = [U \sqcup ((0.6, 0.15, 0.25) \cdot DT_1 \cdot (0.6, 0.15, 0.25) \cdot DT_2)] \\
& \quad = [U \sqcup ((0.44, 0.15, 0.41) \cdot (0.59, 0.15, 0.26))] \\
& \quad = (0.33, 0.28, 0.39) \sim (Expectation = 0.36).
\end{aligned}
$$

$$
\begin{aligned}
& cl_{sl_{di}}(\phi)(w_{witness(wit\_2)}) \\
& \quad = [U \sqcup ((0.7, 0.10, 0.20) \cdot DT_1 \cdot (0.6, 0.15, 0.25) \cdot DT_1)] \\
& \quad = [U \sqcup ((0.5, 0.1, 0.4) \cdot (0.44, 0.15, 0.41))] \\
& \quad = (0.3, 0.24, 0.47) \sim (Expectation = 0.33).
\end{aligned}
$$

Above result shows that given the evidences, 'witness2' is slightly more probable witness candidate than 'witness1' because license plate info is more informative thereby strongly considered than face related information by the opinion assignment. However, due to the opinion on the fact level is not certain, the values were not strongly forced the belief but rather increased the uncertainty in the final opinion. The expectation values also captures this aspect. Thus, this result shows that the inference scheme well models human intuition.

8.7 Discussions and Conclusion

Intelligent forensic reasoning upon metadata acquired from automated vision analytic modules is an important aspect of surveillance systems with high usage potential. The knowledge expressive power of the reasoning framework and the ability of uncertainty handling are critical issues in such systems. In this chapter, based on our previous work on the use of logic programming with subjective logic, we extended the framework so that it can also handle uncertain propositional rules. The approach is mainly based on the fuzzy-like membership function and the reputational operation on it. Although we still need to extend this concept to large scale data, we advocate that this work showed the potential of the proposed approach. The main advantage of the proposed approach is that it offers more choices to model complex contextual human knowledge by enriching the expressive power of the framework. The other advantage of the proposed approach is that the modeled uncertain rules can be used with another principled reasoning scheme. In this chapter, especially, we have demonstrated how the reasoning results from uncertain spatio-temporal rules could be used with default reasoning. Another interesting property of the system is that, unlike traditional probability based conditional reasoning, this approach allows for representing lack of information about a proposition. We could also roughly assign our subjective priors with lack of information, and observations can also be represented with any degree of ignorance, therefore we believe this better reflects human intuition and real world situations. Another beneficial property is the flexibility of assigning opinions to formulae. Especially, rule can embed its own opinion calculation scheme thereby, allows for sophisticated propagation of opinions through the inference pipeline. There are, however, still several open issues such as how to better model the reputational function, how to automatically assign proper prior opinions to rules, etc. Our future research will cover further extending and applying the shown approach to more complicated scenarios using automatically generated large scale data.

Acknowledgments The work presented here was partially funded by the German Federal Ministry of Economy and Technology (BMWi) under the THESEUS project.

References

1. Pearl J (1988) Probabilistic reasoning in intelligent systems: networks of plausible inference. Morgan Kaufmann, San Mateo
2. Reiter R (1980) A logic for default reasoning. Artif Intell 13:68–93
3. Denecker M, Kakas A (2002) Abduction in logic programming. In: Computational logic: logic programming and beyond. Lecture notes in artificial intelligence, vol 2407, pp 402–437. Springer, Berlin
4. Han S, Hutter A, Stechele W (2009) Toward contextual forensic retrieval for visual surveillance: challenges and an architectural approach. In: 10th international workshop on image analysis for multimedia interactive services (WIAMIS'09), London, 6–8 May 2009
5. Han S, Koo B, Hutter A, Shet V, Stechele W (2010) Subjective logic based hybrid approach to conditional evidence fusion for forensic visual surveillance. In: 7th IEEE international conference on advanced video and signal based surveillance (AVSS'10), Boston, 29 Aug–1 Sept 2010
6. Han S, Koo B, Stechele W (2010) Subjective logic based approach to modeling default reasoning for visual surveillance. In: 4th IEEE international conference on semantic computing (ICSC'10), Pittsburgh, 22–24 Sept 2010
7. Jøsang A (2001) A logic for uncertain probabilities. Int J Uncertain Fuzz Knowl-Based Syst 9:279–311
8. Bremond F, Thonnat M (1996) A context representation for surveillance systems. In: ECCV worshop on conceptual descriptions from images, Cambridge, Apr 1996
9. Akdemir U, Turaga P, Chellappa R (2008) An ontology based approach for activity recognition from video. In: ACM conference on multimedia (ACM-MM'08), Oct 2008
10. Jianbing M, Weiru L, Paul M, Weiqi Y (2009) Event composition with imperfect information for bus surveillance. In: 6th IEEE international conference on advanced video and signal based surveillance (AVSS'09), Genoa, 2–4 Sept 2009
11. Shet V, Harwood D, Davis L (2005) VidMAP: video monitoring of activity with prolog. In: 2nd IEEE international conference on advanced video and signal based surveillance (AVSS'05), Como, 15–16 Sept 2005
12. Shet V, Neumann J, Ramesh V, Davis L (2007) Bilattice-based logical reasoning for human detection. In: IEEE international conference on computer vision and pattern recognition (CVPR'07), Minneapolis, 18–23 June 2007

13. Anderson D, Luke RH, Keller JM, Skubic M (2007) Modeling human activity from voxel person using fuzzy logic. IEEE Trans Fuzzy Syst 17(1):39–49
14. Hongeng S, Nevatia R, Bremond F (2004) Video-based event recognition: activity representation and probabilistic recognition methods. Comput Vis Image Understand 96(2):129–162
15. Feryhough J, Cohn AG, Hogg DC (1998) Building qualitative event models automatically from visual input. In: 6th IEEE international conference on computer vision (ICCV'98), pp 350–355
16. Makris D, Ellis T (2002) Spatial and probabilistic modelling of pedestrian behaviour. In: 13th British machine vision conference (BMVC'02), Cardiff, 2–5 Sept 2002
17. Cupillard F, Bremond F, Thonnat M (2003) Behavior recognition for individuals, groups of people, and crowd. In: IEEE seminar intelligent distributed surveillance systems, London, Mar 2003
18. Makris D, Ellis T (2005) Learning semantic scene models from observing activity in visual surveillance. IEEE Trans Syst Man Cybernet Part B 35(3):397–408
19. Ginsberg ML (1988) Multivalued logics: a uniform approach to inference in artificial intelligence. Comput Intell 4(3):256–316
20. Shafer G (1976) A mathematical theory of evidence. Princeton University Press, Princeton
21. Zadeh LA (1973) Outline of a new approach to the analysis of complex systems and decision processes. IEEE Trans Syst Man Cybernet (SMC)(3):28–44
22. Jøsang A (1997) Artificial reasoning with subjective logic. In: 2nd Australian workshop on commonsense reasoning, Perth
23. NASA (1993) Basic programming guide. NASA JSC-25012
24. Mizumoto M, Zimmermann HJ (1982) Comparison of fuzzy reasoning methods. Fuzzy Sets Syst 8:253–283
25. Gottwald S, Hajek P (2005) Triangular norm based mathematical fuzzy logic. In: Klement EP, Mesiar R (eds) Logical, algebraic, analytic and probabilistic aspects of triangular norms, pp 275–300. Elsevier, Amsterdam
26. Zadeh LA (2008) Is there a need for fuzzy logic? Inf Sci 178:2751–2779
27. Jøsang A, Marsh S, Pope S (2006) Exploring different types of trust propagation. In: 4th international conference on trust management
28. i-lids dataset.: http://www.ilids.co.uk
29. Jøsang A, McAnally D (2004) Multiplication and comultiplication of beliefs. Int J Approx Reason 142:19–51
30. Jøsang A (2006) The consensus operator for combining beliefs. Artif Intell J 38(1):157–170
31. Hakeem A, Shah M (2004) Ontology and taxonomy collaborated framework for meeting classification. In: 17th international conference on pattern recognition (ICPR'04), Washington
32. Jøsang A (2008) Conditional reasoning with subjective logic. J Multiple Valued Logic Soft Comput 15(1):5–38

Chapter 9
AIR: Architecture for Interoperable Retrieval on Distributed and Heterogeneous Multimedia Repositories

Florian Stegmaier, Mario Döller, Harald Kosch, Andreas Hutter and Thomas Riegel

Abstract Nowadays, multimedia data is produced and consumed at an ever increasing rate. Similarly to this trend, diverse storage approaches for multimedia data have been introduced. These observations lead to the fact that distributed and heterogeneous multimedia repositories exist, whereas an easy and unified access to the stored multimedia data is not given. This chapter presents an architecture, named AIR, that offers the aforementioned retrieval possibilities. To ensure interoperability, AIR makes use of recently issued standards, namely the MPEG Query Format (multimedia query language) and the JPSearch transformation rules (metadata interoperability).

Keywords External metasearch · Heterogeneous databases · Interoperability · Standardization

9.1 Introduction

Multimedia data is produced in an immense rate and speed. By investigating solutions and approaches for storing and archiving the produced data, one rapidly ends up in a highly heterogeneous environment of data stores. In series, the involved domains feature individual sets of metadata formats for describing content, technical or structural information of multimedia data [17]. Furthermore, depending on the management and retrieval requirements, these data sets are accessible in different systems supporting a multiple set of retrieval models and query languages. By

F. Stegmaier (✉) · M. Döller · H. Kosch
Chair of Distributed Information Systems University of Passau, Passau, Germany
e-mail: stegmai@dimis.fim.uni-passau.de

A. Hutter · T. Riegel
Corporate Technology Siemens AG, 81739 Munich, Germany

N. Adami et al. (eds.), *Analysis, Retrieval and Delivery of Multimedia Content*, Lecture Notes in Electrical Engineering 158, DOI: 10.1007/978-1-4614-3831-1_9,

summing up all these obstacles, an easy and efficient access and retrieval across those system borders is a very cumbersome task [16].

Standards are one way to introduce interoperability among different peers. Recent developments and achievements in the domain of multimedia retrieval concentrated on the establishment of a multimedia query language (MPEG) [6], standardized image retrieval (JPEG) and the heterogeneity problem between metadata formats (JPEG [5] & W3C [17]).

In this context, the chapter introduces an architecture for a middleware component abstracting the heterogeneous environment of multimedia data stores.

The development of our framework pursues the following main requirements: modular architectural design, implemented as an external metasearch engine, a broad scope of multimedia retrieval paradigms (e.g. query by example), unified multimedia requests and cross system multimedia retrieval (cross metadata as well as cross query language). Furthermore AIR supports multiple query processing strategies (autonomous and federated). Concepts and modules/placeholders for intelligent query segmentation and distribution as well as result set aggregation are the highlights of the current implementation. However, the actual retrieval process of the multimedia data is performed inside the connected backends.

Constitutively on our article [19], this chapter deepens these findings and is organized as follows: Sect. 9.2 introduces two examples for heterogeneous retrieval whereas Sect. 9.3 identifies possible search concepts in this domain. Section 9.4 highlights the overall system architecture. A description of two projects, that utilize AIR takes place in Sect. 9.5. Related work will be introduced in Sect. 9.6. The chapter will be concluded in Sect. 9.7.

9.2 Heterogeneous Retrieval Scenarios

In the following, let us consider some heterogeneous retrieval scenarios demonstrating the complexity for an unified access.

(i) Scenario 1—Video surveillance: Airports, train stations and other popular public places are already well equipped with surveillance systems (e.g. security cameras). Today, the huge amount of available data is hardly annotated with metadata and thus, an analysis ends in inspecting the footage manually. Such tasks are highly time consuming, defective and inefficient. Therefore, let us assume a future surveillance network consisting of several autonomous systems of different vendors (e.g. security systems of a bank and a grocery store). These systems store the captured multimedia data in different databases that use diverse query languages (e.g. MOQL [9] or SQL/MM [11]) and extract metadata information using diverse metadata formats (e.g. MPEG-7 [10] or Dublin Core [7]). A possible query for this scenario could be:

"Find video segments where an identified person (e.g. selected by a bounding rectangle) is running through the scene!"

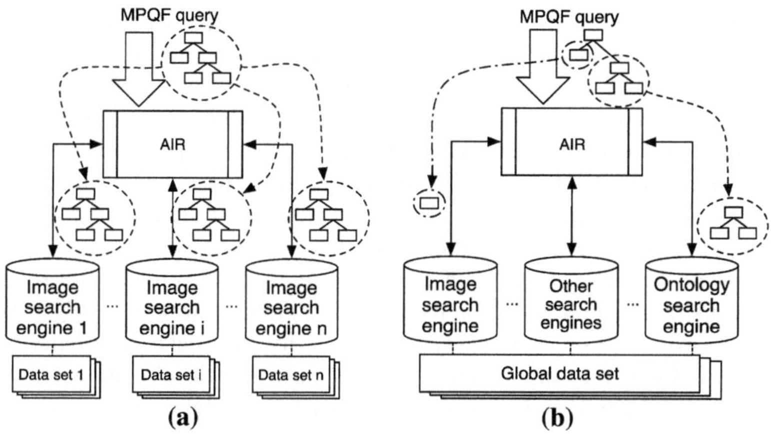

Fig. 9.1 AIR query processing strategies. **a** Local processing. **b** Distributed processing

(ii) Scenario 2—Medical examination: Today's advances in medical imaging and the digitalization of patient data leads to the development of intelligent medical image search engines. The main purpose is to improve the current workflow of a physician (e.g. radiologist). In a medical system, there are different forms of a patient studies, such as textual descriptions (e.g. findings or scientific papers) or digital images (e.g. CT scan). This diversity leads to heterogeneous systems, where the distributed knowledge base is semantically linked among each isolated peer, for example by the patients name and his birthdate. Let us assume a scenario, where the data about patients is subdivided into an image retrieval system dealing with visual representations of CT scans, DICOM information about the patient data and an RDF triple store containing semantic annotations about their diseases and anatomy. An example query could be:

"Select CT scan, where disease is lesion, anatomy near liver and the CT scan is similar to aGivenCTScan.jpg!"

9.3 Query Processing Strategies

The AIR framework can be operated in many different facets within a distributed and heterogeneous multimedia search and retrieval framework. In general, the tasks of every internal component of AIR highly depends on the registered retrieval services or knowledge bases. In this context, two main query processing strategies have to be distinguished, as illustrated in Fig. 9.1.

The first paradigm deals with registered and participating retrieval systems that are able to process the whole query locally, see Fig. 9.1a. In this sense, those heterogeneous systems may provide their local metadata format (e.g. Dublin Core or MPEG-7) and a local/autonomous data set as described in Scenario 1. A query transmitted to such a environment is understood as a whole by each peer and the items of the result set are the outcome of an execution of the query. Of course, transformation of the used metadata format (e.g. from Dublin Core to MPEG-7) may be needed for some systems. In addition, depending on the degree of overlap among the data sets (e.g. the same image is annotated in all databases), the individual result sets may contain duplicates. However, a result aggregation process needs to perform an overall ranking of the result items of the involved retrieval systems. Here, duplication elimination algorithms may be applied as well.

The second paradigm deals with registered and participating retrieval systems that allow distributed processing on the basis of a global data set, see Fig. 9.1b. The involved heterogeneous systems may depend on different data representation (e.g. ontology based semantic annotations and XML based low level features) and query interfaces (e.g. SPARQL and XQuery), but describe a global data schema as shown in Scenario 2. In this context, a query transmitted to AIR needs to be evaluated and optimized. This results into a specific query execution plan. In series, segments of the query are forwarded to the respective engines and executed. Now, the result aggregation has to deal with a correct consolidation of the partial result sets. In this context, AIR behaves like a federated multimedia database management system.

9.4 AIR Architecture

Figure 9.2 illustrates an end-to-end workflow scenario in a distributed multimedia retrieval scenario by the use of AIR. At its core, AIR uses the MPEG Query Format (MPQF)[1], which is part 12 of the MPEG-7 standard and especially designed for the retrieval of multimedia data. MPQF is an XML-based multimedia query language which defines the format of queries and replies to be interchanged between clients and servers in a multimedia information search and retrieval environment. The normative parts of the MPEG Query Format define three main components: The Input Query Format provides means for describing query requests from a client to a multimedia information retrieval system (MMRS). The Output Query Format specifies a message container for MMRS responses and finally the Query Management Tools provide means for functionalities such as service discovery, service aggregation and service capability description (e.g. which query types or multimedia formats are supported). Note, the term service refers to all MMRS including single databases as well as service providers administrating a set of MMRSs. A query request can be composed of three different parts. A declaration part points to resources (e.g. image file or its metadata description, etc.) that are reused within the query condition or output description part. The output description part allows, by using the respective MMRS metadata description, the definition of the structure as well as the content of the expected result set. Finally, the query condition part denotes the search criteria by providing a set of different query types (e.g. QueryByMedia) and expressions

[1] http://www.mpegqueryformat.org

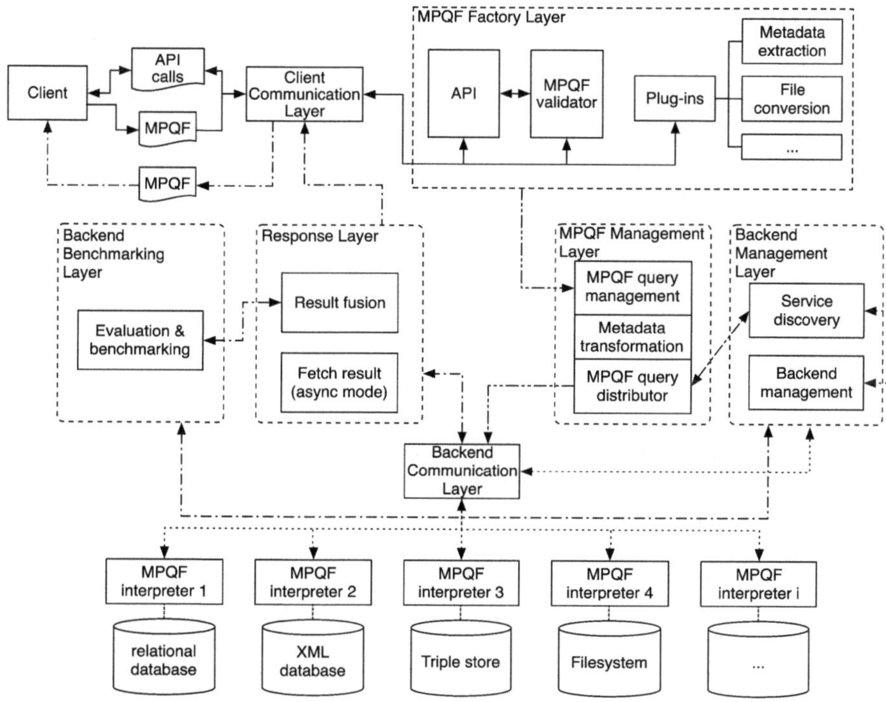

Fig. 9.2 Overview of the AIR components

(e.g. GreaterThan) which can be combined by Boolean operators (e.g. AND). In order to respond to MPQF query requests, the Output Query Format provides the ResultItem element and attributes signaling paging and expiration dates. A detailed description about MPQF can be found in [6].

9.4.1 Backend Management Layer

The main functionalities of the Backend Management Layer are the (de-)registration of backends with their capability descriptions and the service discovery for the distribution of incoming MPQF queries. As already mentioned, these capability descriptions are standardized in MPQF, allowing the specification of the retrieval characteristics of registered backends. Such characteristics consider for instance the supported query types or metadata formats. In series, depending on those capabilities, this component is able to filter registered backends during the search process (service discovery). For a multimedia retrieval system, it is very likely that not all functions specified in MPQF are supported. In such an environment, one of the important tasks for a MPQF client is to identify the backends which provide the desired query functions or support the desired media formats identified by an MIME type using the service discovery.

9.4.2 MPQF Factory Layer

The main purpose of the MPQF Factory Layer is the generation and validation of MPQF queries. This layer also encapsulates interfaces for inserting preprocessing plug-ins. These could for example expose methods to generate media specific metadata (e.g. MPEG-7 low level features) or to perform file conversion (e.g. smaller Bit rate for better bandwidth usage). Two possible modes of operations are implemented in order to pass a query request to AIR.

The first mode assumes that an instance of the MPQF query was already created in another place. Then, the query is directly transmitted to AIR. In a next step, the query runs through the validation process and is mapped into the internal MPQF object structure.

As not every client is able to deliver complete MPQF queries to AIR, the second mode addresses the creation of a MPQF query through an API. In general, a MPQF query consists of two main parts. First, the QueryCondition element holds the filter criteria in an arbitrary complex condition tree. In this context, the condition tree has to be build bottom up. Second, the OutputDescription element defines the structure of the result set. In this object, the needed information about required result items, grouping or sorting is stored. After finalizing the query creation, the generated MPQF query will be transmitted to AIR. A set of query templates at the client side can be established to simplify the query creation process using the API approach.

9.4.3 MPQF Management Layer

The MPQF Management Layer organizes the registration of MPQF queries and their distribution to the applicable retrieval services. After the registration with a unique identifier of the entire query, the distribution of the query depends on the underlying search concept. For the local processing scenario, the whole query is transmitted to the backends. In contrast to that, in a distributed processing scenario, the query will be automatically divided in segments by analyzing the query types used in the condition tree. Here, well known tree algorithms like depth-first search can be used. The key intention of this segmentation is that every backend only gets a query segment, which it can process as a whole. Besides query splitting and distribution, the transformation between metadata formats is another crucial task of the management layer. In this context, the JPSearch transformation rules [5] have been integrated for antagonizing metadata heterogeneity. They define XML-based syntactical mappings between the JPSearch core schema and an arbitrary (XML based) metadata format. The present implementation is able to transform metadata informations inside an isolated, metadata specific query type (without touching the overall query semantic). Currently, a second approach for improving the metadata interoperability issues will be integrated. This approach is developed by the *W3C Media Annotation Working Group* [21], which aims to improve interoperability between multimedia metadata formats on the Web by providing an interlingua ontology and an API designed to facilitate cross-community data integration of information related to media resources on the Web. To do so, syntactic as well as semantic mappings between the so-called media ontology defined by the group and a large number of metadata formats have been identified in a group report (available on the working group page).

In order to highlight the progress of the received queries, AIR introduces the following MPQF query lifecycle: *pending* (query registered, process not started), *retrieval* (search started, some results missing), *processing* (all results available, aggregation in progress), *finished* (result can be fetched) and *closed* (result fetched or query lifetime expired). These states are also valid for query segments, since they are also valid MPQF queries.

9.4.4 MPQF Interpreter

The MPQF Interpreter is located at the backend and is supposed to act as a mediator between AIR and a particular retrieval service. Its main purpose is to receive a MPQF query and to transform it into native calls of the underlying query language (e.g. SQL/MM in a Oracle database). After a successful retrieval, the MPQF Interpreter forwards the result set (converted in a MPQF response) to the AIR framework. in case of an error, a meaningful system message will be generated an inserted into a special section of the result set.

9.4.5 Backend Benchmarking Layer

In order to describe the main advantage of the Backend Benchmarking Layer (BBL), let us assume a scenario as shown in Fig. 9.1a. There, for instance image retrieval is realized by a query by example search. A MPQF query is send directly to AIR and the query is distributed to the applicable backends. The major issue in this case is not the distribution, but the result aggregation of the different result sets. AIR has to aggregate the results on the one side by eliminating all duplicates and on the other side by performing an *ideal* ranking of the individual result items. A first implementation uses the round robin approach [1], which guarantees efficient processing of result sets of autonomous retrieval systems. However, it is supposable that different backends use different implementations and quality measures for processing the fuzzy retrieval that leads to quality discrepancies between the result sets. Therefore, similar to approaches such as [2] where statistics about sources are collected, the BBL will provide information about the *search quality* of a backend which leads to a more intelligent re-ranking and aggregation of the result set. The idea is to calculate a specific benchmarking score for a system on the basis of the produced/derivable quality measures. This helps to compare different retrieval systems, potentially evaluated by different benchmarks. Here, we assume, that the benchmarks share a certain degree of overlap in the extracted measures (e.g. normalized discounted cumulative gain). As already said, the most important information for an external metasearch engine (for reranking) is, whether a relevant item is well ranked in the result set or not. This information and a respective MPQF query classification model will be realized by a new benchmarking environment that allows to rate the search quality of registered backends. This topic is currently under development and a close collaboration with related organizations (JPEG [3] and ImageCLEF [14]) is initiated. First results have been proposed in [18].

9.4.6 Response Layer

The MPQF Response Layer performs in general the result aggregation and returns the aggregated result set. This follows the definition of an external metasearch engine given by Montague and Aslam in [13]. External metasearch treats existing search engines, which are potentially operating on diverse document sets, as black boxes and consolidates their output. Recent work [4] already highlighted useful result aggregation algorithms for MPQF. These algorithms could build the basis for a new approach, which also takes the advantages of the BBL into account.

9.5 Projects Utilizing AIR

The following two real world projects are using the current prototype of the AIR framework. These two have been selected, because they differ (a) in their covered domain and (b) in the way, the query is being processed (cf. Sect. 9.2). This diversity clearly shows the applicability of AIR in a wide range of usage scenarios.

9.5.1 THESEUS (Application Scenario MEDICO)

The THESEUS project[2] is funded by the German Federal Ministry of Economics and Technology. Its challenge is to find ways of providing users with simple and efficient access to this enormous amount of knowledge available on the Web. The applications of this project should develop new mechanisms for automatic annotation of data, rapid processing of multimedia documents or innovative ontology management. The main project is subdivided in six sub-projects, that are settled in a variety of domains (e.g. digital libraries). The mission of the MEDICO application scenario is to establish an intelligent and scalable search engine for the medical domain by combining medical image processing and semantically rich image annotation vocabularies.

Figure 9.3 sketches an end-to-end workflow inside the MEDICO system. It provides the user with an easy-to-use web-based form to describe his/her search query.

[2] http://www.theseus-programm.de

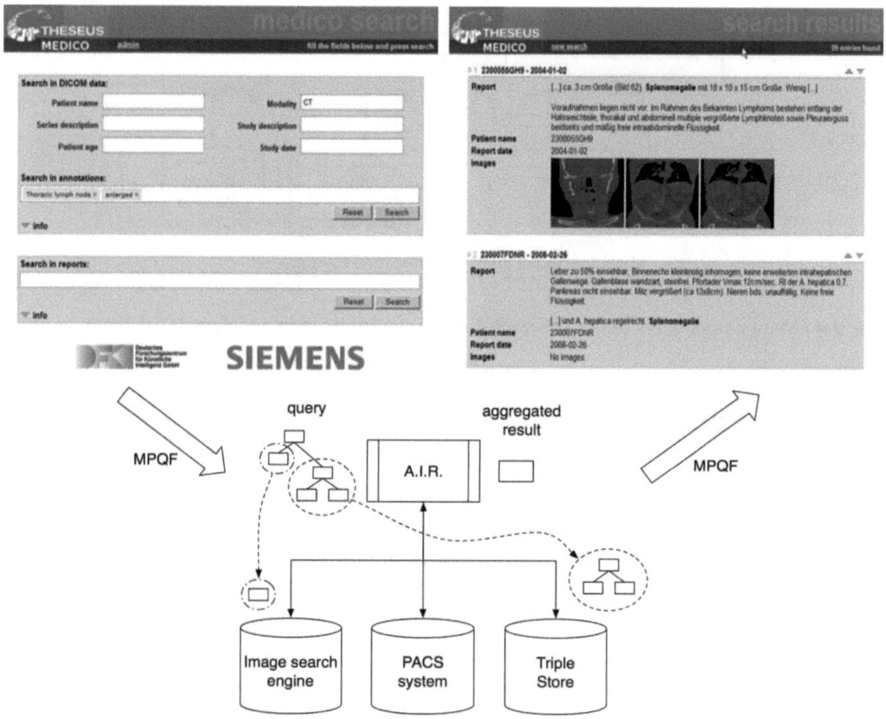

Fig. 9.3 Search infrastructure: end-to-end workflow between the Medico web interface and AIR

Currently, this user interface utilizes a semantically rich data set composed of DICOM tags, image annotations, text annotations and gray-value based 3D CT images. This leads to a heterogeneous multimedia retrieval environment with multiple query languages: DICOM tags are stored in a PACS system, image/text annotations are saved in a triple store and the CT scans are accessible by a image search engine performing a similarity search. Apparently, all these retrieval services are using their own query languages for retrieval (e.g. SPARQL) as well as the actual data representation for annotation storage (e.g. RDF/OWL). Beside all differences, these different data sources describe a common (semantically linked) global data set. To fulfill a meaningful semantic search, the present interoperability issues have to be solved. Furthermore, it is essential to formulate queries that take the aforementioned diverse retrieval paradigms into account. For this purpose, MEDICO integrates the AIR multimedia middleware framework, following the federated query processing strategy as described in Sect. 9.2. An evaluation regarding the retrieval abilities of the MEDICO system can be found in [15].

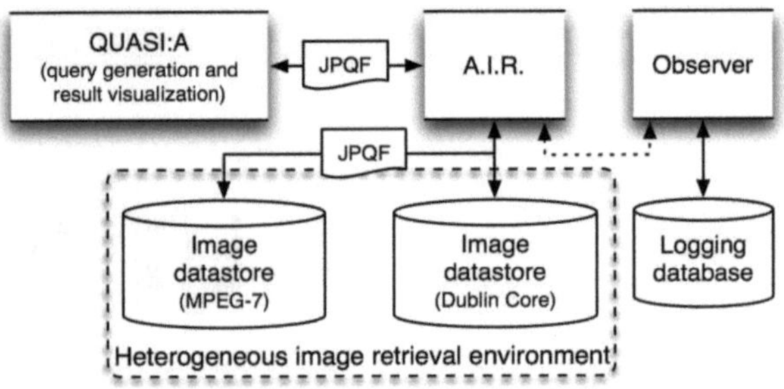

Fig. 9.4 Main components of the JPSearch based Interoperable Image Search

9.5.2 Interoperable Image Search

Figure 9.4 shows the image retrieval system consisting of three independent parts that are connected by the AIR middleware framework. All components are implemented in Java, using the SOAP protocol[3] for communication. The retrieval process is based on the JPSearch standard[4] [8], issued by ISO/IEC SC29 WG1 (commonly known as JPEG). The goal of JPSearch is to standardize interfaces for an abstract image retrieval system. Within this standard, a specific query language—JPEG Query Format (JPQF)—has been defined using a subset (tailored to image retrieval) of MPQF. In the following, the different parts will be highlighted from a functional point of view.

Note: It is no restriction for AIR, that this scenario is based on JPSearch, esp. JPQF. AIR is able to process MPQF as well as JPQF due to the subset relation. Further, a few components, e.g. implementation of JPSearch Transformation Rules and QUASI:A, will serve as an official reference implementation for the standard.

9.5.3 Heterogeneous Image Retrieval Environment

The data source is a heterogeneous image retrieval environment, whereas the engaged data stores act autonomous. In this context, autonomous means that the engaged data stores have no direct correlation/connection in the first place. The following assumptions are made: the data stores feature retrieval services in order to process the incoming JPQF query as a whole (no segmentation of queries needed). Furthermore the image data sets are not overlapping, but are annotated with diverse metadata formats, here MPEG-7 and Dublin Core. Therefore, duplicate elimination plays only a minor role in the aggregation process. The main challenge is to manage heterogeneity that is expressed by (i) different metadata formats for annotation and (ii) different query languages for retrieval following the local processing strategy of Sect. 9.2.

[3] http://www.w3.org/TR/soap12-part1/

[4] http://www.jpsearch.org/

9.5.4 User Interfaces

The *query and search for images application (QUASI:A)* is JavaFX based and supposed to offer JPQF query generation (cf. Fig. 9.5a) as well as result presentation functionalities. As a proof of concept, only a subset of JPSearch functionalities has been implemented, focusing on the specified interoperability issues. Therefore, it is restricted to the three JPQF query types: QueryByMedia, QueryByDescription and QueryByRelevanceFeedback. The first query type is an implementation of the well-known Query–By–Example paradigm. Here, a user is able to specify a picture (e.g. accessible on the internet or via file upload) that serves as an input for a similarity search. This picture can also be modified (e.g. crop or resize), as shown in Fig. 9.5b, where a special region of interest has been selected. The second query offers the possibility to define a metadata based search. Here, a user may fill out a form containing elements of the JPSearch core schema to perform an exact metadata search. These query types and the comparison types can be linked by the use of Boolean operators (e.g. AND) in a tree based manner, as illustrated in Fig. 9.5a. This visualization technique ensures clarity and usability. The images stored in the aggregated result set will be presented in a gallery fashioned way. Here, a single image of the gallery can be directly used as an input for a further similarity search (browsing) or a subset (positive as well as negative examples) of the result set defining a relevance feedback query.

The *Query Observer* is a subproject of AIR. Its main intention is to visualize the process units of a JPQF query inside AIR. In other words it acts like a query tracing system. This system is divided into two parts: tracing service and visualization interface. The tracing service is implemented as a SOAP web service and receives query events directly from AIR. A single event consists of the event name, a generic comment, the processing time, the complete XML code of the JPQF query and the assigned JPQF ID. These events are stored in a relational database. Figure 9.6 shows the visualization interface presenting the stored events of a query. After a preprocessing by the global JPQF ID, a query to evaluate can be selected. The system distinguishes different process units of a JPQF query and presents the stored event informations. For example, moving the cursor over an event symbol the corresponding comment will be displayed as a tooltip. For every single event, the complete XML code of a JPQF query can be displayed. This feature is especially useful to show the metadata transformation process. Finally, the visualization can be run in two modes of operation, namely automatic and manual playback using playback functionalities (e.g. pause or play).

9.6 Related Work

Recently, several approaches for accessing multimedia data in a possibly distributed and heterogeneous environment have been proposed:

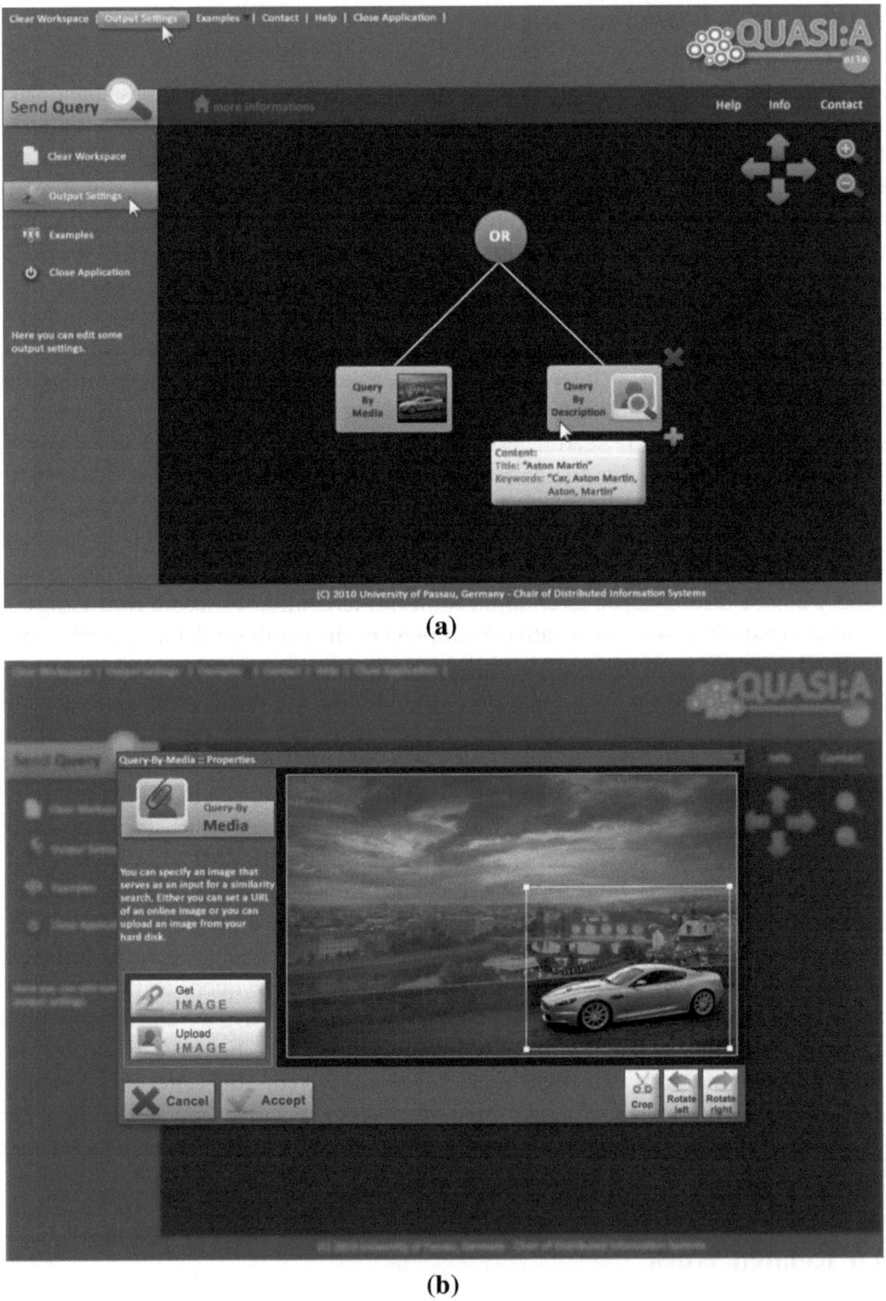

Fig. 9.5 JavaFX based user interface QUASI:A. **a** Query generation and result presentation **b** QueryByMedia functionalities

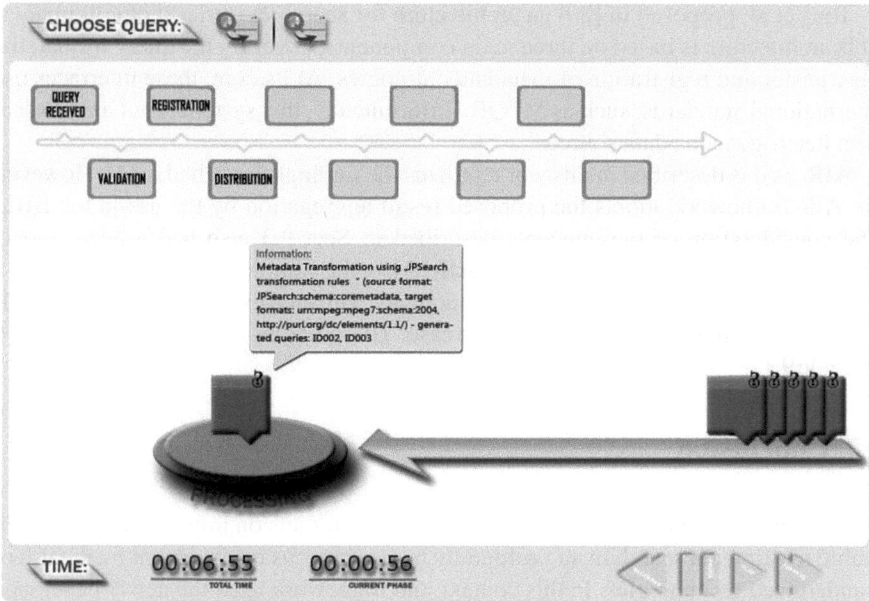

Fig. 9.6 Query Observer—JPQF visualization and tracing

Table 9.1 Comparison of the three systems referring to the requirements defined in Sect. 16.1

	Generic framework	LEGO-like architecture	AIR
Modularity	+	–	++
Multimedia specific query types	4 q.t. (proprietary)	9 q.t. (standardized)	9 q.t. (standardized)
Cross system retrieval	Unified access	Not given	Federated access
Metadata interoperability	Limited to DICOM	Metadata ontologies	Transformation rules & Metadata ontologies

Möller et al. issued in [12] a generic framework for medical search and retrieval. The application consists of a graphical metadata extractor, an annotation interface and a search interface. Here, the search interface is rather limited regarding the multimedia search capabilities and the metadata extractor is closed to the DICOM standard, but it is able to address heterogeneous data sources.

Tous et al. proposed in [20] an architecture for search & retrieval of still images. This architecture is based on three main components, covering the query format, the file transfer and registration of metadata ontologies. At its core, these interfaces use international standards, such as MPQF. Unfortunately, this system is not able to deal with heterogeneous data sources.

AIR, as it is described in this work, utilizes the findings described in [4]. However, the AIR framework adopts the proposed result aggregation by the use in the BBL. The consideration of the concepts described in Sect. 9.3 also had a deep impact regarding the presented architecture of AIR. Also, the complete architecture as shown in Sect. 9.4 is composed of modular components. This makes it possible to tailor AIR specifically to the needs of a specific use case. These findings have been summarized in Table 9.1.

9.7 Conclusion

This chapter introduced the AIR framework which targets on implementing interoperable multimedia search in an profoundly heterogeneous environment by the use of standardized technologies. In this context, the framework used the newly developed MPEG Query Format for unifying multimedia search requests. Besides, metadata heterogeneity is antagonized by the established metadata transformation approach of JPSearch/JPEG. These features are completed by means for query management and distribution as well as service discovery and result set aggregation techniques. The comparison in Table 9.1 clearly show the advances of the proposed architecture. While the generic framework provides an unified access to heterogeneous data sources, it lacks in the expressivness of proprietary multimedia queries and metadata interoperability. In contrast to that, the LEGO-like architecture sets the focus on metadata interoperability, but the heterogeneity problem remains unstudied. Further developments will concentrate on incorporating the quality of involved retrieval systems by applying benchmarking results. In addition, work in the direction of a federated multimedia database management system will be applied.

Acknowledgments This work has been partially supported by the THESEUS Program, which is funded by the German Federal Ministry of Economics and Technology. We would like to thank the following students for contributing: Sebastian Freilinger–Huber, Sebastian Unverricht, Wolfgang Hans, Ludwig Bachmeier, Udo Gröbner and Atanas Yakimov.

References

1. Berretti S, Bimbo AD, Pala P (2003) Merging results of distributed image libraries. In: Proceedings of the international conference on multimedia and expo. Baltimore, Maryland, USA, pp 33–36
2. Craswell N, Hawking D, Thistlewaite PB (1999) Merging results from isolated search engines. In: Proceedings of the Australasian database conference. Auckland, New Zealand, pp 189–200

3. Döller M (2010) Draft of call for JPSearch benchmark requirements. ISO/IEC JTC 1/SC 29/WG1 N5399 (JPEG)
4. Döller M, Bauer K, Kosch H, Gruhne M (2008) Standardized multimedia retrieval based on web service technologies and the MPEG query format. J Digit Inf 6(4):315–331
5. Döller M, Stegmaier F, Kosch H, Tous R, Delgado J (2010) Standardized interoperable image retrieval. In: Proceedings of the ACM symposium on applied computing, track on advances in spatial and image-based information systems. Sierre, Switzerland, pp 881–887
6. Döller M, Tous R, Gruhne M, Yoon K, Sano M, Burnett IS (2008) The MPEG query format: on the way to unify the access to multimedia retrieval systems. IEEE Multimedia 15(4):82–95
7. Dublin core metadata initiative: dublin core metadata element set—version 1.1: reference description (2008). http://dublincore.org/documents/dces/
8. Dufaux F, Ansorge M, Ebrahimi T (2007) Overview of JPSearch: a standard for image search and retrieval. In: Proceedings of the 5th international workshop on content-based multimedia indexing. Bordeaux, France
9. Li JZ, özsu MT, Szafron D, Oria V (1997) MOQL: a multimedia object query language. In: Proceedings of the 3rd international workshop on multimedia information systems. Como Italy, pp 19–28
10. Martinez JM, Koenen R, Pereira F (2002) MPEG-7. IEEE Multimedia 9(2):78–87
11. Melton J, Eisenberg A (2001) SQL multimedia and application packages (SQL/MM). SIGMOD Rec. 30(4):97–102
12. Möller M, Sintek M (2007) A generic framework for semantic medical image retrieval. In: Proceedings of the 1st knowledge acquisition from multimedia content workshop in conjunction with SAMT, vol 253. Genova, Italy, pp 18–32
13. Montague M, Aslam J (2002) Condorcet fusion for improved retrieval. In: Proceedings of the 11th international conference on information and knowledge management. McLean, Virginia, USA, pp 538–548
14. Müller H, Geissbuhler A (2004) Benchmarking image retrieval applications. In: Proceedings of the 7th international conference on visual information systems. San Francisco, California, USA
15. Seifert S, Thoma M, Stegmaier F, Hammon M, Döller M, Kriegel HP, Cavallaro A, Huber M, Comaniciu D (2011) Combined semantic and similarity search in medical image databases. In: Proceedings of the SPIE medical imaging. Lake Buena Vista, Florida, USA
16. Smith JR (2008) The search for interoperability. IEEE Multimedia 15(3):84–87
17. Stegmaier F, Bailer W, Bürger T, Döller M, Höffernig M, Lee W, Malaisé V, Poppe C, Troncy R, Kosch H, de Walle RV (2009) How to align media metadata schemas? design and implementation of the media ontology. In: Proceedings of the 10th international workshop of the multimedia metadata community on semantic multimedia database technologies in conjunction with SAMT, vol 539, Graz, Austria, pp 56–69
18. Stegmaier F, Bürger T, Döller M, Kosch H (2010) Knowledge based multimodal result fusion for distributed and heterogeneous multimedia environments: concept and ideas. In: Proceedings of the 8th international workshop on adaptive multimedia retrievalm. Linz, Austria
19. Stegmaier F, Döller M, Kosch H, Hutter A, Riegel T (2010) AIR: architecture for interoperable retrieval on distributed and heterogeneous multimedia repositories. In: Proceedings of the 11th international workshop on image analysis for multimedia interactive services. Desenzano del Garda, Italy, pp 1–4
20. Tous R, Delgado J (2009) A LEGO-like metadata architecture for image search & retrieval. In: Proceedings of the 3rd workshop on multimedia data mining and management in conjunction with 20th international conference on DEXA. Linz, Austria, pp 246–250
21. World wide web consortium (W3C): media annotations working group. Video in the Web activity. (2008). http://www.w3.org/2008/WebVideo/Annotations/

Chapter 10
Local Invariant Feature Tracks for High-Level Video Feature Extraction

Vasileios Mezaris, Anastasios Dimou and Ioannis Kompatsiaris

Abstract In this work the use of feature tracks for the detection of high-level features (concepts) in video is proposed. Extending previous work on local interest point detection and description in images, feature tracks are defined as sets of local interest points that are found in different frames of a video shot and exhibit spatio-temporal and visual continuity, thus defining a trajectory in the 2D+Time space. These tracks jointly capture the spatial attributes of 2D local regions and their corresponding long-term motion. The extraction of feature tracks and the selection and representation of an appropriate subset of them allow the generation of a Bag-of-Spatiotemporal-Words model for the shot, which facilitates capturing the dynamics of video content. Experimental evaluation of the proposed approach on two challenging datasets (TRECVID 2007, TRECVID 2010) highlights how the selection, representation and use of such feature tracks enhances the results of traditional keyframe-based concept detection techniques.

Keywords Feature tracks · Video concept detection · Trajectory · LIFT descriptor · Bag-of-Spatiotemporal-Words

V. Mezaris (✉) · A. Dimou · I. Kompatsiaris
Information Technologies Institute, Centre for Research and Technology Hellas,
6th Km Charilaou-Thermi Road, 57001 Thermi, Greece
e-mail: bmezaris@iti.gr

A. Dimou
e-mail: dimou@iti.gr

I. Kompatsiaris
e-mail: ikom@iti.gr

N. Adami et al. (eds.), *Analysis, Retrieval and Delivery of Multimedia Content*, Lecture Notes in Electrical Engineering 158, DOI: 10.1007/978-1-4614-3831-1_10,

10.1 Introduction

The development of algorithms for the automatic understanding of the semantics of multimedia and in particular of video content, and the semantic indexing by means of high-level features (concepts) corresponding to semantic classes (objects, events) is currently one of the major challenges in multimedia research. This is motivated by the ever-increasing pace at which video content is generated, rendering any annotation scheme that requires human labor unrealistically expensive and unpractical for use on anything but a very restricted subset of the generated content, which may be of unusually high value or importance (e.g. cinema productions, medical content).

Research efforts towards the goal of high-level video feature extraction have followed in the last decade or so several different directions that have the potential to contribute to this goal, ranging from temporal or spatio-temporal segmentation [1, 2] to key-frame extraction, video content representation using global shot or image features, local interest point detection and description [3], creation of visual lexicons for video representation (Bag-of-Words [4]), machine learning for associating low-level and high-level features, etc. Typically, techniques belonging to several of the aforementioned categories need to be carefully combined for extracting high-level video features. The latter are useful in a wide variety of media organization and analysis tasks, including interactive retrieval and the detection of scenes and high-level events in video [5, 6].

This work focuses on video content representation, and in particular builds upon previous work on local interest point detection and description to propose the extraction, selection and representation of feature tracks. These features compactly describe the appearance and the long-term motion of local regions and are invariant, among others, to camera motion, in contrast to both 2D interest point descriptors and their known extensions to spatio-temporal interest points. The proposed feature tracks are shown to be suitable for the generation of a Bag-of-Spatiotemporal-Words (BoSW) model that facilitates capturing the dynamics of video content, allowing the more reliable detection of high-level features that have a strong temporal dimension (e.g. "people-dancing").

The rest of the chapter is organized as follows: in Sect. 10.2, previous work on local interest point detection and description is discussed. In Sect. 10.3, feature track extraction and selection are presented, while the representation of feature tracks using the LIFT descriptor and the use of such descriptors for building a BoSW model are discussed in Sect. 10.4. Experimental results are reported in Sect. 10.5 and finally conclusions are drawn in Sect. 10.6.

10.2 Related Work

Several approaches to scale-invariant interest point detection and description in still images have been proposed and are widely used in still image understanding tasks (image classification, object detection, etc.), as well as in other applications. SIFT

[3] is probably the most widely adopted method; SIFT-based descriptors are shown in [7] to outperform several previously proposed techniques for local region description. More recent work on this topic includes SURF [8], which focuses mostly on speeding-up the interest point detection and description process, and [9], which examines the introduction of color information to the original grey-value SIFT. For the application of high-level feature extraction in generic image collections, the above descriptors are typically used to build a Bag-of-Words (BoW) model [4], which involves the definition of a "vocabulary" of visual words (typically, created by clustering the interest point descriptors coming from a large number of images and then selecting the resulting centroids as words) and the subsequent representation of each image as the histogram of the visual words (i.e., corresponding interest points) found in it.

Large-scale video analysis for the purpose of high-level feature extraction, using local features, is in most cases performed at the key-frame level [10]. Thus, the video analysis task reduces to still image analysis. This has obvious advantages in terms of computational complexity, but on the other hand completely disregards the temporal dimension of video and the wealth of information that is embodied in the evolution of the video frames along time. The temporal evolution of the video signal, i.e. motion, is generally considered to convey very important information in video, being a key element of several video understanding and manipulation tasks, e.g. retrieval [11]. Long-term region trajectories in particular, rather than the motion at the frame level, have been shown to be very useful for video segmentation, indexing and retrieval in several works (e.g. [1]). Similarly to other analysis tasks, the use of video data in excess of one single key-frame (e.g. using multiple key-frames per shot [12], or treating all frames as key-frames and also considering their temporal succession [13]) for high-level feature extraction has been shown to lead to improved results.

In order to introduce temporal information in the interest-point-based representation of video shots, in [14] spatial interest points are detected using the SIFT methodology and additional motion constraints; the detected points are described using both visual and motion information. In [15], the use of spatio-temporal (as opposed to spatial-only) interest point detectors is proposed. Spatio-temporal interest points are defined as locations in the video where intensity values present significant variations both in space and in time. In [16] and other works, such points are used for human action categorization, since the abrupt changes in motion that trigger the detection of spatio-temporal interest points can be useful in discriminating between different classes of human activity (walking, jumping, etc.). However, spatio-temporal interest points define 3D volumes in the video data that typically neither account for possible camera motion nor capture long-term local region trajectories. To alleviate these drawbacks, the tracking of spatial interest points across successive frames has been proposed for applications such as object tracking [17] and the visualization of pedestrian traffic flow in surveillance video [18]. In [19], the problem of object mining in video is addressed by tracking SIFT features and subsequently clustering them, to identify differently moving objects within a shot. In [20, 21], interest points are tracked and either the motion information alone [20] or appearance and motion information in separate BoW models [21] are used for action recognition in video.

However, neither one of the previous works on tracking spatial interest points uses the outcome of tracking for defining a BoSW model of the shot, as in the present work.

10.3 Feature Tracks

10.3.1 Feature Track Extraction

Let S be a shot comprising T frames, $S = \{I_t\}_{t=0}^{T-1}$, coming from the temporal subsampling of the original video shot $S^0 = \{I_\tau\}_{\tau=0}^{T^0-1}$ by a factor of a; $T = \lceil T^0/a \rceil$.

Application of one of the available combinations of interest point detection and description techniques (e.g. [3, 8, 9]) on any frame I_t of S results in the extraction of a set of interest point descriptions $\Phi_t = \{\phi_m\}_{m=1}^{M_t}$, where M_t is the total number of interest points detected in the frame, and interest point ϕ_m is defined as $\phi_m = [\phi_m^x, \phi_m^y, \phi_m^d]$. ϕ_m^x, ϕ_m^y denote the coordinates of the corresponding local region's centroid on the image grid and ϕ_m^d is the local descriptor vector, e.g. an 128-element SIFT vector. In this work, the SIFT method was used for interest point detection and description, due to its well-documented [3, 7] invariance properties.

Having detected and described interest points in all frames of S, a temporal correspondence between an interest point $\phi_m \in \Phi_t$ and one interest point of the previous frame can be established by local search in a square spatial window of dimension $2 \cdot \sigma + 1$ of frame I_{t-1}, i.e., by examining if one or more $\phi_n \in \Phi_{t-1}$ exist that satisfy the following conditions:

$$|\phi_m^x - \phi_n^x| \leq \sigma \ , \tag{10.1}$$

$$|\phi_m^y - \phi_n^y| \leq \sigma \ , \tag{10.2}$$

$$d(\phi_m^d, \phi_n^d) \leq d_{sim} \ , \tag{10.3}$$

where σ is a constant whose value is chosen such that a reasonably-sized square spatial window is considered during local search, and d(.,.) is the Euclidean distance. The latter was also used in [3] for keypoint matching across different images, and is chosen in this work for consistency with the K-Means clustering that is used at a later stage for assigning the extracted tracks to words of the BoSW model (Sect. 10.4.3). If multiple interest points satisfying (10.1)–(10.3) exist, the one for which quantity $d(\phi_m^d, \phi_n^d)$ is minimized is retained. When such an interest point ϕ_n exists, the interest point $\phi_m \in \Phi_t$ is appended to the feature track where the former belongs, while otherwise (as well as when processing the first frame of the shot) the interest point ϕ_m is considered to be the first element of a new feature track.

Repeating the temporal correspondence evaluation for all interest points and all pairs of consecutive frames in S results in the extraction of a set Ψ of feature tracks, $\Psi = \{\psi_k\}_{k=1}^{K}$, where $\psi_k = [\psi_k^x, \psi_k^y, \psi_k^d]$. ψ_k^d is the average descriptor vector of a feature track, estimated by element-wise averaging of all interest point descriptor vectors ϕ_m^d of the feature track, as in [19], while ψ_k^x is the corresponding time-series of camera-motion-compensated interest point displacement in the x-axis between successive frames of S in which the feature track is present. ψ_k^y is defined similarly for the y-axis. Thus, $\xi_k = [\psi_k^x, \psi_k^y]$ is the long-term trajectory of the interest point that generates the feature track: $\psi_k^x = [\psi_k^{x,t_{k1}}, \psi_k^{x,t_{k1}+1}, \dots \psi_k^{x,t_{k2}}]$ where $t_{k2} > t_{k1}$ (and similarly for ψ_k^y). The values $\psi_k^{x,t}$ are estimated for any given t by initially using the differences $\phi_m^x - \phi_n^x$, $\phi_m^y - \phi_n^y$ for all identified valid pairs of interest points between frames I_t, I_{t-1} to form a sparse, non-regular motion field for the corresponding pair of frames; subsequently, the 8 parameters of the bilinear motion model, representing the camera motion, are estimated from this field using least-squares estimation and an iterative rejection scheme, as in [1]. Then $\psi_k^{x,t}$ and $\psi_k^{y,t}$ are eventually calculated as the differences between the initial displacement of the corresponding interest point's centroid between times $t-1$ and t, and the estimated camera motion at the location of the centroid.

The simple interest point matching between successive frames of S, which is used as part of the proposed feature track extraction process, was chosen primarily for its simplicity; more elaborate techniques for tracking across frames have been proposed (e.g. [18]) and can be used instead, for producing more accurate feature tracks, if the added computational complexity is not a limiting factor. An example of the feature tracks that are extracted by the proposed procedure is shown in Fig. 10.1.

10.3.2 Feature Track Selection

The feature track extraction process, described in the previous section, typically results in the extraction of a large number of feature tracks (e.g. in the order of tens of thousands) for every shot. These exhibit significant differences in their temporal duration, with the track length $t_{k2} - t_{k1}$ ranging from 0 to $T-1$, T being the number of frames in the shot (Fig. 10.2). Besides the practical problems associated with storing and using such a large number of descriptors for every shot, the possible presence of noisy or otherwise erroneous tracks among those originally extracted may adversely affect concept detection. Therefore, selecting a suitable subset of these feature tracks is proposed.

One possible criterion for selecting a subset of feature tracks is their repeatability under variations (e.g. perspective, scale, and illumination variations). Repeatability is among the main requirements for any descriptor. In this work, it is hypothesized that the repeatability of a track can be approximated by examining the temporal duration of it. More specifically, let us assume that R denotes the real-world scene that is depicted in shot S. Under constant illumination conditions and assuming no local (object) motion, the result of capturing scene R with an ideal static camera would be an ideal image I_r. Then, every image $I_t \in S$ can be seen as a different noisy observation of I_r, affected by image acquisition noise and possible global and

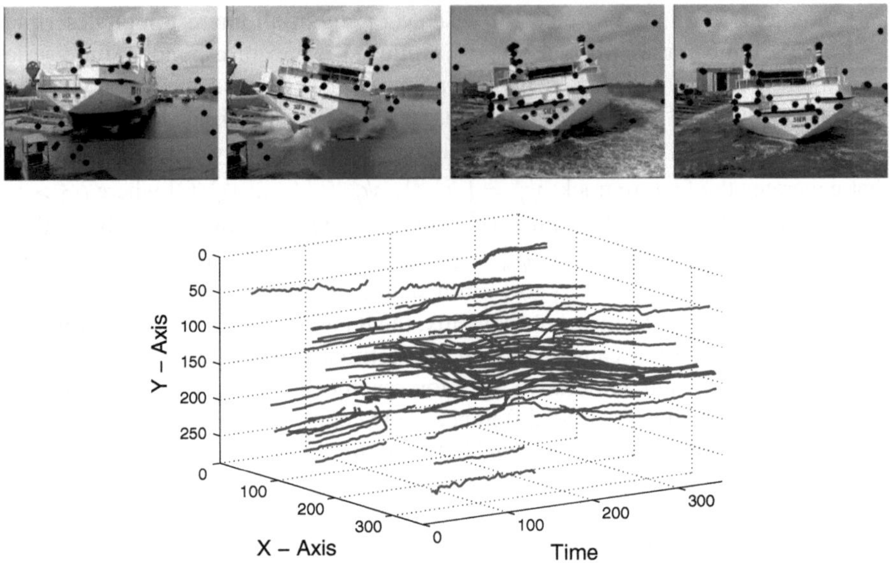

Fig. 10.1 Example of a few interest points that belong to extracted feature tracks (marked on *four indicative frames* of a shot), and an overview of the corresponding feature tracks in the 2D+Time space for the whole shot

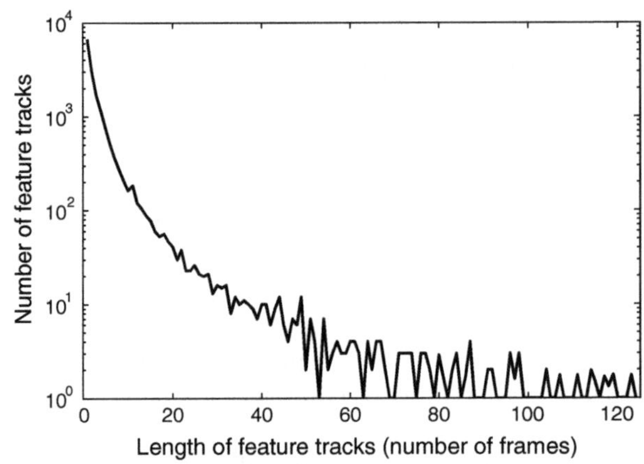

Fig. 10.2 Example of the distribution of feature tracks extracted for a shot, according to their temporal duration

local motion, as well as perspective, scale, and illumination variations. Similarly, every interest point in image I_t that is part of an extracted feature track ψ_k can be perceived as the result of detecting the corresponding ideal interest point of I_r under the specific variations affecting image I_t. Of course, the assumption made here is that the correspondences established with the use of (10.1)–(10.3) are not erroneous. Consequently, the probability of a specific feature track being present in one frame of S can be used as a measure of the repeatability of the interest point that defines this feature track, thus also as a measure of the relevant repeatability of the feature track itself, in comparison to other feature tracks of the shot.

Following this discussion, in this work the probability of a specific feature track being present in one frame of S is calculated as the number of frames in which the track extends, divided by the total number of frames of the shot,

$$p(\psi_k) = \frac{t_{k2} - t_{k1}}{T - 1}, \tag{10.4}$$

and is used as a measure of the feature track's repeatability. Consequently, the feature tracks of set Ψ generated for shot S are ordered according to $p(\psi_k)$ (equivalently, in practice, according to $t_{k2} - t_{k1}$) in descending order, and the N first tracks are selected for generating the BoSW model of the shot.

It should be emphasized that repeatability is just one possible criterion for selecting feature tracks, and the most repeatable features are not necessarily the most informative ones as well; thus, jointly considering repeatability and additional criteria may be beneficial. Furthermore, note that the temporal duration of a track being a good approximation of its repeatability is only a hypothesis that we make; this needs to be experimentally verified. To this end, the track selection strategy described above, which is based on this hypothesis, is evaluated against two other possible such strategies in the experimental results section.

10.4 Bag-of-Spatiotemporal-Words

10.4.1 Feature Track Representation

The selected feature tracks are variable-length feature vectors, since the number of elements comprising ψ_k^x and ψ_k^y is proportional to the number of frames that the feature was tracked in. This fact, together with other possible track artefacts (e.g. the extraction of partial tracks, due to failure in interest point matching between consecutive frames, occlusions, etc.) make the matching of feature tracks non-trivial and render their current representation unsuitable for direct use in a BoW-type approach. For this reason, each motion trajectory is transformed to a fixed-length descriptor vector that attempts to capture the most important characteristics of the motion.

To capture motion at different time-scales, ψ_k^x and ψ_k^y are initially subject to low-pass filtering using a filter bank shown in Fig. 10.3, based on the lowpass Haar filter $H(z) = \frac{1}{2}(1 + z^{-1})$. This results in the generation of a family of trajectories, $\xi_{k,q} = [\psi_{k,q}^x, \psi_{k,q}^y]$, $q = 0, \ldots, Q - 1$, as shown in Fig. 10.3, which due to the simplicity of the Haar filter are conveniently calculated as follows:

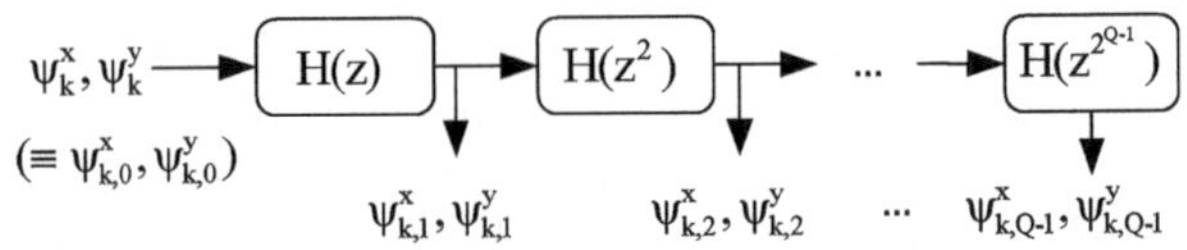

Fig. 10.3 Filter bank used for capturing motion at different time-scales

$$\psi^x_{k,q} = [\psi^{x,t_{k1}+2^q-1}_{k,q}, \ \psi^{x,t_{k1}+2^q}_{k,q}, \ \ldots \ \psi^{x,t_{k2}}_{k,q}] \ , \tag{10.5}$$

$$\psi^{x,t}_{k,q} = \frac{1}{2^q} \sum_{i=0}^{2^q-1} \psi^{x,t-i}_{k} \ . \tag{10.6}$$

The y-axis elements of the trajectory are calculated similarly.

For any trajectory $\xi_{k,q}$, the histogram of motion directions at granularity level θ is defined as a histogram of $\frac{\pi}{\theta}$ bins: $[0, \theta)$, $[\theta, 2\cdot\theta)$,..., $[\pi-\theta, \pi)$. When $\pi \leq \theta < 2\cdot\pi$, $\theta' = \theta - \pi$ is used instead of θ for assigning the corresponding elementary motion to the appropriate bin of the histogram. The value of each bin is defined as the number of elementary motions $[\psi^{x,t}_{k,q}, \ \psi^{y,t}_{k,q}]$ of the trajectory that fall into it, normalized by division with the overall number of such elementary motions that belong to the examined trajectory. $\lambda(\xi_{k,q}, \theta)$ is defined as the vector of all bin values for a given $\xi_{k,q}$ and a constant θ.

Then, the initial trajectory ξ_k can be represented across different time-scales and at various granularity levels as a fixed length vector μ_k:

$$\begin{aligned}\mu_k = \Big[& \lambda\left(\xi_{k,0}, \frac{\pi}{2}\right), \ \lambda\left(\xi_{k,1}, \frac{\pi}{2}\right), \ \ldots \lambda\left(\xi_{k,Q-1}, \frac{\pi}{2}\right) \\ & \lambda\left(\xi_{k,0}, \frac{\pi}{4}\right), \ \lambda\left(\xi_{k,1}, \frac{\pi}{4}\right), \ \ldots \lambda\left(\xi_{k,Q-1}, \frac{\pi}{4}\right), \ \ldots \\ & \lambda\left(\xi_{k,0}, \frac{\pi}{2^J}\right), \ \lambda\left(\xi_{k,1}, \frac{\pi}{2^J}\right), \ \ldots \lambda\left(\xi_{k,Q-1}, \frac{\pi}{2^J}\right) \Big].\end{aligned} \tag{10.7}$$

The corresponding Local Invariant Feature Track (LIFT) descriptor is defined as:

$$LIFT(\psi_k) = [\psi^d_k, \ \mu_k] \ . \tag{10.8}$$

The LIFT descriptor is a fixed-length vector that compactly captures both the 2D appearance of a local image region and its long-term motion.

10.4.2 Invariance Concerns

The definition of the LIFT representation was guided by the need to introduce, to the extent possible, some invariance with respect to the scale and direction of the extracted tracks. Starting with the interest point detection and description in the 2D, the SIFT method was used, due to its well-documented [3, 7] and desirable invariance properties; other similar methods [8, 9] could also be used instead. Concerning the feature track extraction, camera-motion-compensated trajectories were estimated and employed to ensure that the final LIFT representation will not be affected by camera motion. Camera motion could also be useful for representing the shots, but should in any case be separated from the local motion of the different local features within the shot, rather than being allowed to corrupt the latter.

In the subsequent representation of the tracks by histograms, only the direction of each elementary motion of the track was employed, rather than the direction and magnitude of it. This was done for introducing some degree of invariance to image scale, since the same motion (e.g. a person picking up the phone) will result in different motion vector magnitudes depending on the focal length of the camera and its distance from the plane of the motion; on the contrary, the direction of motion is not affected by these parameters.

Histograms at various time-scales were selected for representing the tracks, instead of e.g. comparing the overall displacement of the interest point along the track, to allow for partial matches when considering partial tracks (i.e., when the beginning and end of the different extracted tracks that correspond to the same class of actions do not coincide with each other and with the actual beginning and end of the depicted action). Although the adopted solution may be non-optimal, the reliable matching of partial tracks would otherwise require the use of a computationally expensive optimization-based technique for evaluating the similarity of them, in place of the Euclidean distance typically used in K-Means when creating the "words" used in the Bag-of-Words approach.

The use of motion direction histograms at different granularity levels θ (instead of using a single histogram with a high number of bins) aims at allowing again for partial matches between tracks using a simple metric (i.e., L1/L2 rather than e.g. the Earth Mover's Distance), in the case of small variations in the direction of motion. When considering only a very fine granularity level θ, significant such variations between similar shots could be caused by even small differences in camera angle/viewpoint. The combined use of multiple (from coarse to fine) granularity levels can alleviate this effect to some degree. Alternatively, the weighted assignment of every elementary motion to more than one neighboring bins, when constructing each motion direction histogram, could be employed.

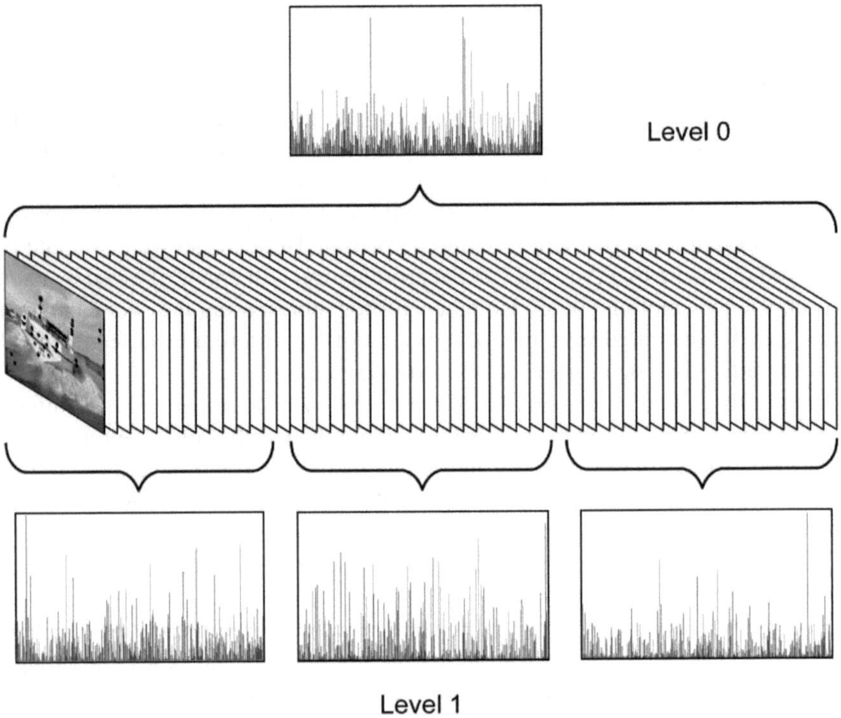

Fig. 10.4 Illustration of the temporal pyramidal methodology

10.4.3 Shot Representation

The LIFT descriptors of the feature tracks extracted and selected for a video shot, according to the processes of Sect. 10.3, can be used for generating a BoSW model. This will essentially describe the shot in terms of classes of "similarly-moving, visually-similar local regions", rather than simply "visually-similar local regions" (detected by either spatial or spatio-temporal interest point detectors), as in the current state-of-the-art, e.g. [12, 16]. The BoSW model is expected to allow for the improved detection of dynamic concepts in video, in contrast to the traditional keyframe-based BoW that by definition targets the detection of static concepts. Furthermore, since the shot features used in the BoW and BoSW models are different and, to some degree, complementary, it is expected that combining the two models can result in further improvement of the detection rates for both dynamic and static concepts.

For the generation of the BoSW model, the typical process of generating BoW descriptions from any set of local descriptors is followed. Thus, K-Means clustering, using a fixed number of clusters, is performed on a large collection of LIFT descriptors for initially identifying a set of words (i.e., the centroids of the clusters). Hard- or soft-assignment of each one of the LIFTs of a given shot to these words can then be performed for estimating the histogram that represents a given shot on the basis of the defined spatio-temporal words. Furthermore, techniques such as spatial pyramids [22] or temporal extensions of them (Fig. 10.4) can be used in combination with the BoSW model, similarly to the way spatial pyramids are combined with the BoW one.

10.5 Experimental Results

In the experimental evaluation of the proposed techniques, two datasets were used. The first one is the TRECVID[1] 2007 dataset, which is made of professionally-created videos (Dutch TV documentaries). The training and testing portions of it comprise 50 hours of video each, and 18120 and 18142 shots respectively; all these shots are annotated with 20 concepts that were defined for the TRECVID 2009 contest. This dataset was employed for evaluating different design choices of the proposed BoSW (e.g. the feature track selection strategy) and for comparing them with alternate approaches, as well as for comparing the overall proposed technique with the traditional SIFT-based BoW one. The second dataset is the TRECVID 2010 one, which is made of heterogeneous internet videos. The training and test portions of it comprise approximately 200 h of video each, and 118536 and 144971 shots respectively; the training portion is annotated with 130 concepts that were defined for the TRECVID 2010 contest. This dataset was used for further comparing the overall proposed technique with the traditional SIFT-based BoW one, on the basis of the 30 concepts (out of the overall 130 ones) that were evaluated for each run that was submitted to TRECVID 2010.

In the process of extracting the proposed LIFT features of the video shots, the temporal sub-sampling parameter a was set equal to 3. This represents a good compromise between the need for accurately establishing the SIFT point correspondences from frame to frame (which calls for a low value of a, ideally 1) and the need for speeding up the feature extraction process. For each frame of the temporally sub-sampled sequence, the method of [3] was used for interest point detection and description, resulting in a 128-element vector for the local region of each interest point. Parameter σ, defining the local window where correspondences between SIFT descriptors are evaluated, was set to 20, and parameter d_{sim}, used for evaluating the similarity of SIFT descriptors in different frames, was set to 40,000. Using four different timescales ($Q = 4$) and three granularity levels θ (i.e., $J = 3$ in (10.7)) for representing the trajectory information of the extracted feature tracks resulted in the LIFT descriptor of each feature track being a 184-element vector, while setting $J = 5$ in selected experiments (indicated below) resulted in a 376-element vector instead.

A first series of experiments was carried out on the TRECVID 2007 dataset, in order to evaluate the appropriate number of feature tracks that should be used for representing each shot, given the above feature track extraction and representation

[1] http://www-nlpir.nist.gov/projects/trecvid/

parameter choices. A BoSW model, using hard assignment and 500 words, was used to this end, together with Support Vector Machine classifiers. The latter produced a fuzzy class membership degree in the range [0,1] when used for evaluating the relevance of each shot of the TRECVID 2007 test dataset with every one of the considered high-level features, exploiting the BoSW model. Prior to this, the SVM classifiers were trained using the TRECVID 2007 training dataset and the common annotation; for each high-level feature, a single SVM was trained independently of all others. It should be noted that this is only a baseline configuration; it is used for efficiently evaluating certain characteristics of the proposed BoSW, and is neither optimal nor in par with SoA works such as [12], where 4000 words, soft assignment, multiple color SIFT variants, and additional techniques such as pyramidal decomposition are combined, increasing the dimension of the vector representing each shot from 500 (as in our baseline configuration) to about 100,000. The results (mean average precision, calculated for a maximum of 2000 returned samples per concept [10]) are shown in Fig. 10.5a, where it can be seen that using 2500 feature tracks per shot leads to the best results overall.

A second series of experiments was carried out to evaluate the soundness of the feature track selection process of Sect. 10.3.2 and of the hypothesis that this process has been based on. Specifically, the selection of the 2500 tracks with the highest probability $p(\psi_k)$, as proposed in Sect. 10.3.2 (denoted as selection criterion "BB" in the sequel) was compared with a) the selection of the 2500 tracks with the highest probability $p(\psi_k)$ after removing from set Ψ those feature tracks used by selection criterion "BB" (denoted as "SB" in the sequel), and b) the random selection of 2,500 feature tracks from set Ψ (selection criterion "RR"). The LIFT descriptor was used in all the above cases for representing the selected tracks and for forming a 500-word BoSW model. Experimentation with the 500-word keyframe-based BoW model that uses SIFT descriptors was also carried out, for comparing BoSW and BoW when used in isolation. For creating the BoW model of each shot, the median frame of the shot was selected as a key-frame and SIFT descriptors were extracted from it. The results (Fig. 10.5b) show that selection criterion "BB" significantly outperforms criteria "SB" and "RR". The BoSW model using selection criterion "BB" by itself performs comparably to the keyframe-based BoW model overall, but considerably better than the latter when considering only dynamic concepts (i.e., a subset of the 20 defined high-level features, which is discussed in more detail below).

In a third series of experiments, the merit of combining the BoSW and BoW models was evaluated. The combination of the two was performed by concatenating the shot descriptions produced by each of them, similarly to how different BoW models based on different color SIFT variants are combined in [12]. In Table 10.1, BoW and the combination of BoW and BoSW (using selection criterion "BB") are compared using a) the baseline configuration used in the previous experiments: 500 words and hard assignment, and b) 500 words, soft assignment, a spatial pyramid of 2 levels for BoW and, in a similar fashion, the temporal pyramid of Fig. 10.4 for BoSW. Additionally, in the latter case 5 granularity levels θ (i.e., $J = 5$ in (10.7)), instead of 3, are used. The results of Table 10.1 document the contribution of the proposed BoSW model to improved performance when combined with the BoW

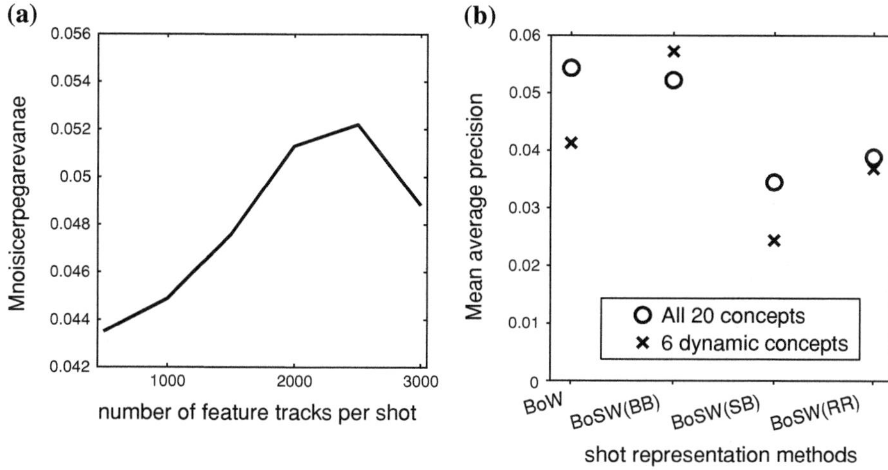

Fig. 10.5 Evaluation of (**a**) the impact of the number of feature tracks used for representing each shot, and (**b**) the impact of different shot representation techniques, on concept detection performance

Table 10.1 Comparison between BoW, combination of BoW and BoSW on the TRECVID 2007 dataset (mean average precision for all 20/6 dynamic concepts)

	BoW		BoW+BoSW(BB)	
Number of considered concepts:	20	6	20	6
500 words, hard assignment	0.054	0.041	0.068	0.056
500 words, soft assignment, spatial/temporal pyramidal decomposition	0.084	0.088	0.102	0.113

model, compared to the latter alone, as well as the applicability of techniques such as soft assignment and pyramidal decomposition (particularly temporal pyramids) to BoSW. Overall, considering the second of the the two tested configurations (500 words, soft assignment, spatial/temporal pyramidal decomposition), the SIFT-based BoW resulted in a mean average precision (MAP) of 0.084, whereas the combination of BoW and BoSW in a MAP of 0.102, representing an increase of the former by approximately 21 %. Considering only high-level features that have a strong temporal dimension ("people-dancing", "person-playing-soccer", etc.), i.e. features 5, 6, 7, 9, 11 and 13 of Fig. 10.6, the use of the proposed BoW and BoSW combination leads to an increase of MAP by approximately 28 % over using the SIFT-based BoW alone. The significance of taking into account motion information, as done by BoSW, for detecting such dynamic concepts can also be seen in Fig. 10.6, where the per-concept results (average precision) corresponding to the last row of Table 10.1 are shown.

Finally, the SIFT-based BoW and the combination of BoW and BoSW (using again 500 words, soft assignment, spatial/temporal pyramidal decomposition, and 5 granularity levels θ) were compared on the TRECVID 2010 dataset, by participating with the two corresponding runs to the TRECVID 2010 contest [23]. The results for

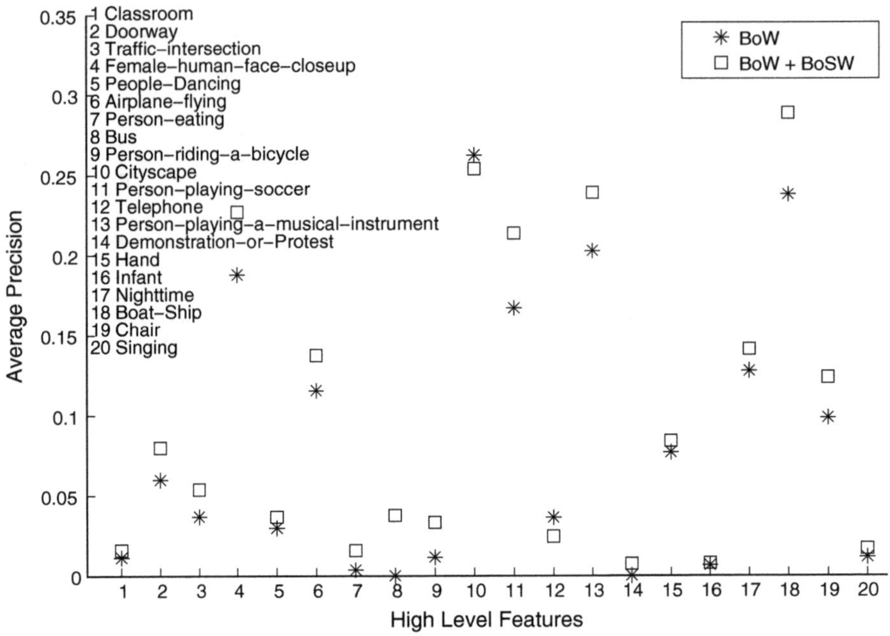

Fig. 10.6 Individual concept detection results on the TRECVID 2007 dataset for BoW alone and for the combination of BoW and BoSW, using 500 words, soft assignment, and spatial/temporal pyramidal decomposition

Table 10.2 Comparison between BoW, combination of BoW and BoSW on the TRECVID 2010 dataset (mean extended inferred average precision for all 30/8 dynamic concepts)

	BoW		BoW+BoSW(BB)	
Number of considered concepts:	30	8	30	8
500 words, soft assignment, spatial/ temporal pyramidal decomposition	0.030	0.020	0.038	0.039

the 30 concepts that were evaluated in this contest are reported in Table 10.2 and Fig. 10.7 (overall and per-concept results, respectively). Extended inferred average precision (xinfAP) and mean extended inferred average precision (MxinfAP) [24], calculated for a maximum of 2000 returned samples per concept, were used for quantifying the results, in order to account for the test portion of this dataset being annotated only in part. It can be seen that the SIFT-based BoW resulted in a MinfAP of 0.030, whereas the combination of BoW and BoSW in a MinfAP of 0.038, representing an increase of the former by approximately 26.7 %. Considering only high-level features that have a strong temporal dimension, i.e. features 1, 4, 7, 11, 23, 26, 28, and 30 of Fig. 10.7, the use of the proposed BoW and BoSW combination leads to an increase of MinfAP by approximately 95 % over using the SIFT-based BoW alone.

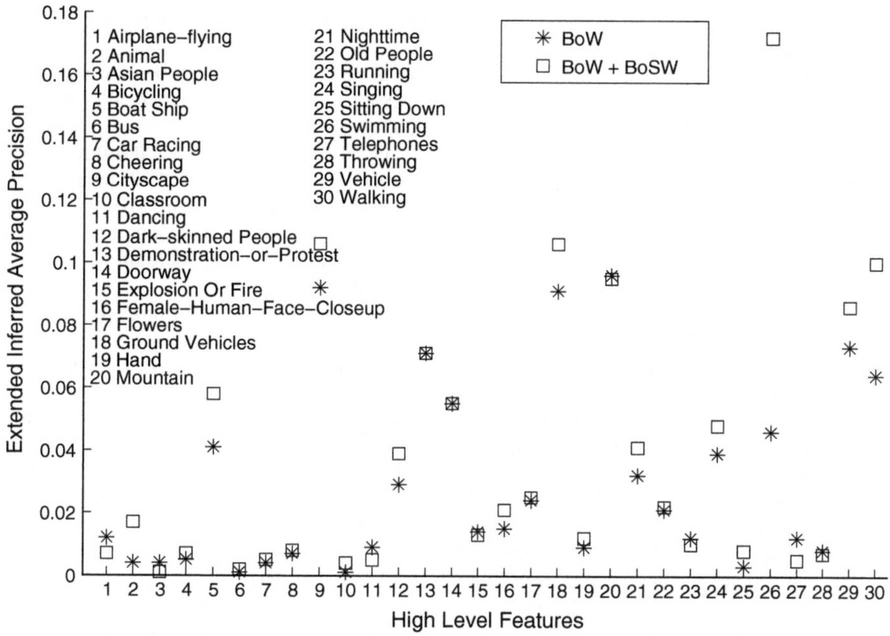

Fig. 10.7 Individual concept detection results on the TRECVID 2010 dataset for BoW alone and for the combination of BoW and BoSW, using 500 words, soft assignment, and spatial/temporal pyramidal decomposition

10.6 Conclusions

In this work the use of feature tracks was proposed for jointly capturing the spatial attributes and the long-term motion of local regions in video. In particular, techniques for the extraction, selection, representation and use of feature tracks for the purpose of constructing a BoSW model for the video shots were presented. Experimental evaluation of the proposed approach on two challenging test corpora (TRECVID 2007, TRECVID 2010) revealed its potential for concept detection in video, particularly when considering dynamic rather than static concepts.

Acknowledgments This work was supported by the European Commission under contract FP7-248984 GLOCAL.

References

1. Mezaris V, Kompatsiaris I, Boulgouris N, Strintzis M (2004) Real-time compressed-domain spatiotemporal segmentation and ontologies for video indexing and retrieval. IEEE Trans on Circuits Syst Video Technol 14(5):606–621
2. Mezaris V, Kompatsiaris I, Strintzis M (2004) Video object segmentation using Bayes-based temporal tracking and trajectory-based region merging. IEEE Trans Circuits Syst Video Technol 14(6):782–795

3. Lowe DG (2004) Distinctive image features from scale-invariant keypoints. Int J Comput Vision 60:91–110
4. Dance C, Willamowski J, Fan L, Bray C, Csurka G (2004) Visual categorization with bags of keypoints. In: Proceedings of the ECCV international workshop on statistical learning in computer vision, Prague, Czech Republic, May 2004
5. Mezaris V, Sidiropoulos P, Dimou A, Kompatsiaris I (2010) On the use of visual soft semantics for video temporal decomposition to scenes. In: Proceedings of the fourth IEEE international conference on semantic computing (ICSC 2010), Pittsburgh, PA, USA, Sept 2010
6. Gkalelis N, Mezaris V, Kompatsiaris I (2010) Automatic event-based indexing of multimedia content using a joint content-event model. In: Proceedings of the ACM multimedia 2010, Events in Multimedia workshop (EiMM10), Firenze, Italy, Oct 2010
7. Mikolajczyk K, Schmid C (2005) A performance evaluation of local descriptors. IEEE Trans Pattern Anal Mach Intell 27(10):1615–1630
8. Bay H, Ess A, Tuytelaars T, Gool LV (2008) Surf: speeded up robust features. Comput Vis Image Underst 110(3):346–359
9. Burghouts GJ, Geusebroek JM (2009) Performance evaluation of local colour invariants. Comput Vis Image Underst 113:48–62
10. Smeaton AF, Over P, Kraaij W (2009) High-level feature detection from video in TRECVid: a 5-Year retrospective of achievements. In: Divakaran A (ed) Multimedia content analysis, signals and communication technology. Springer, Berlin, pp 151–174
11. Piro P, Anthoine S, Debreuve E, Barlaud M (2010) Combining spatial and temporal patches for scalable video indexing. Multimedia Tools Appl 48(1):89–104
12. Snoek C, van de Sande K, de Rooij O et al (2008) The MediaMill TRECVID 2008 semantic video search engine. In: Proceedings of the TRECVID 2008 workshop, USA, Nov 2008
13. Ballan L, Bertini M, Bimbo AD, Serra G (2010) Video event classification using String Kernels. Multimedia Tools Appl 48(1):69–87
14. Chen M, Hauptmann A (2009) Mo SIFT: recognizing human actions in surveillance videos. Technical Report CMU-CS-09-161, Carnegie Mellon University
15. Laptev I (2005) On space-time interest points. Int J Comput Vision 64(2/3):107–123
16. Niebles JC, Wang H, Fei-Fei L (2008) Unsupervised learning of Human Action Categories using spatial-temporal words. Int J Comput Vision 79(3):299–318
17. Zhou H, Yuan Y, Shi C (2009) Object tracking using SIFT features and mean shift. Comput Vision Image Underst 113(3):345–352
18. Tsuduki Y, Fujiyoshi H (2009) A method for visualizing pedestrian traffic flow using SIFT feature point tracking. In: Proceedings of the 3rd Pacific-Rim symposium on image and video technology, Tokyo, Japan, Jan 2009
19. Anjulan A, Canagarajah N (2009) A unified framework for object retrieval and mining. IEEE IEEE Trans Circuits Syst Video Technol 19(1):63–76
20. Moenne-Loccoz N, Bruno E, Marchand-Maillet S (2006) Local feature trajectories for efficient event-based indexing of video sequences. In: Proceedings of the international conference on image and video retrieval (CIVR), Tempe, USA, July 2006
21. Sun J, Wu X, Yan S, Cheong L, Chua TS, Li J (2009) Hierarchical spatio-temporal context modeling for action recognition. In: Proceedings international conference on computer vision and pattern recognition (CVPR), Miami, USA, June 2009
22. Lazebnik S, Schmid C, Ponce J (2009) Spatial pyramid matching. In: Dickinson S, Leonardis A, Schiele B, Tarr M (eds) Object categorization: computer and human vision perspectives. Cambridge University Press, Cambridge
23. Moumtzidou A, Dimou A, Gkalelis N, Vrochidis S, Mezaris V, Kompatsiaris I (2010) ITI-CERTH participation to TRECVID 2010. In: Proceedings of the TRECVID 2010 workshop, USA, Nov 2010
24. Yilmaz E, Kanoulas E, Aslam J (2008) A simple and efficient sampling method for estimating AP and NDCG. In: Proceedings of the 31st annual international ACM SIGIR conference on research and development in, information retrieval (SIGIR), pp 603–610

Part IV
3D and Multi-View

Chapter 11
A New Evaluation Criterion for Point Correspondences in Stereo Images

Aleksandar Stojanovic and Michael Unger

Abstract In this chapter, we present a new criterion to evaluate point correspondences within a stereo setup. Many applications such as stereo matching, triangulation, lens distortion correction, and camera calibration require an evaluation criterion for point correspondences. The common criterion here is the epipolar distance. The uncertainty of the epipolar geometry provides additional information, and our method uses this information for a new distance measure. The basic idea behind our criterion is to determine the most probable epipolar geometry that explains the point correspondence in the two views. This criterion considers the fact that the uncertainty increases for point correspondences induced by world points that are located at a different depth-level compared to those that were used for the fundamental matrix computation. Furthermore, we show that by using Lagrange multipliers, this constrained minimization problem can be reduced to solving a set of three linear equations with a computational complexity practically equal to the complexity of the epipolar distance.

Keywords Fundamental matrix · Robust matching · Probabilistic epipolar geometry · Outlier elimination

11.1 Introduction

Over the past few decades, significant advances in the field of object recognition have been made. In the case of matching points in a stereo image pair, the epipolar geometry, once it has been estimated, can be used for the evaluation of further point matches. For a given point in one image, the epipolar geometry describes where the corresponding point in the second image may be located. In fact, a so-called epipolar

---

A. Stojanovic (✉) · M. Unger
RWTH Aachen University, Aachen, Germany
e-mail: stojanovic@ient.rwth-aachen.de

N. Adami et al. (eds.), *Analysis, Retrieval and Delivery of Multimedia Content*,
Lecture Notes in Electrical Engineering 158, DOI: 10.1007/978-1-4614-3831-1_11,

line in the second image can be derived, and the corresponding point must be located on this line. Stereo correspondence algorithms benefit from this in the sense that a correspondence search can be restricted to the epipolar line.

In practice, due to noise and outliers, only an approximation of the epipolar geometry can be estimated. By consequence the distance of a correspondence from the epipolar line in stereo matching is a useful criterion for evaluation. Many different distance measures, summarized in [6], evaluate how well a pair of points satisfies the epipolar geometry. However, none of these measures take into account the uncertainty of the epipolar geometry.

While for planar 3-D data a homographic model is sufficient to explain point correspondences, for general data the epipolar geometry is the only constraint on corresponding image points. For real images, some point correspondences tend to accumulate on image planes, while others are spread over different depth levels. In [12], a unified correspondence framework was introduced, that covers the homographic and epipolar extremes and limits the search region for point correspondences. For that purpose, a probabilistic framework was developed, based on the covariance matrix of the fundamental matrix.

Recently, in [2] a method to compute a probability density function for point correspondences was introduced. A drawback of this method is that this function is depending on the depth of points in the scene. This is in contradiction with the basic idea of epipolar geometry, which provides a relation between views without any dependency on the depth of points in the scene.

In this chapter, we present a novel distance measure that takes into account the uncertainty of the epipolar geometry in a sound way. We show that, using Lagrange multipliers, the constrained minimization problem can be reduced to solving a set of three linear equations with a computational complexity practically equal to the complexity of calculating the epipolar distance, as defined in [6]. We show the benefits of our criterion for the fundamental matrix computation.

11.2 Estimation of the Fundamental Matrix and its Uncertainty

If two cameras observe several 3-D points, the locations of the accordingly mapped 2-D points into each camera plane are not arbitrary, but related to each other. Having only one camera, one knows that a 3-D point must be located somewhere on the ray defined by its image point and the camera center. A second camera may recognize this ray. This means that the projection of the 3-D point into the second camera plane must be located on the mapped ray. In fact, for each image point in one image, a line can be determined in the other image. This relationship between two cameras is described by the epipolar geometry. Stereo correspondence algorithms may benefit from this geometric property in the sense that a correspondence search may be restricted to a small area around the epipolar line, or that outliers may be detected. In [12], a unified correspondence framework was introduced that covers the homographic and epipolar extremes and limits the search region for point correspondences.

11.2.1 Basic Equations

The complete epipolar geometry is algebraically expressed by a 3×3 matrix, the so called fundamental matrix F.

A given image point $x = [x, y, w]^T$ in homogeneous coordinates is mapped onto a line $l' = [a, b, c]^T$ by using F:

$$l' = Fx. \tag{11.1}$$

Since the corresponding point $x' = [x', y', w']^T$ must be located on the line l', the following equation must hold:

$$l'x' = 0. \tag{11.2}$$

In order to calculate F, we may combine (11.1) and (11.2), obtaining a linear equation on the entries of F.

$$x'^T Fx = 0. \tag{11.3}$$

Due to the fact that the fundamental matrix is a homogeneous matrix, it is scale invariant. In addition, the determinant must be zero because the mapping of a point onto a line forces rank two.

At the first glance, it seems that the problem of calculating the fundamental matrix is straightforward. By stacking up at least seven equations of the form of (11.3), the linear equation system can be solved and F is derived. The problem becomes much more challenging if the point correspondences are perturbed by noise or if they are outliers.

11.2.2 Noisy Point Correspondences

Besides methods like Gaussian elimination, the most common method to solve over-determined linear equation systems is the least squares technique (LS). The main drawback of this method is its vulnerability to noise if the underlying problem is not homoscedastic, so that the variance of the residuals is independent from the data. However, the basic equations for estimating the parameters of the fundamental matrix show a heteroscedastic behavior. Common estimation approaches like the normalized eight-point algorithm [7] (which is actually a modified least squares approach), probabilistic methods like the RANSAC [5] or LMedS [4] and complex iterative approaches like [10], which take into account the heteroscedasticity of the underlaying problem, have been developed. A good overview on recent methods for computing the fundamental matrix is given in [11]. In the context that point correspondences are only perturbed by Gaussian noise, these methods can be regarded as being nearly optimal. As the point correspondences are also corrupted by outliers, these methods have a reduced accuracy and a decision whether it is an outlier has to be made for each correspondence, see [16].

11.2.3 Epipolar Distance

Many different distance measures summarized in [6] evaluate how well a point correspondence satisfies the epipolar geometry. The most simple distance measure is the algebraic error which is simply the residuum of the epipolar geometry constraint:

$$r = x'^T Fx. \tag{11.4}$$

One problem with this measure is that it depends on the scaling of F as well as the norms of the points x and x'. The scaling of F is constant for a set of correspondences and therefore does not directly influence comparisons between them. Points near the origin are treated differently compared to ones with larger coordinates, which is problematic when it comes to answering the question which one better fits the epipolar constraint.

To overcome this difficulty, the geometric error can be determined. Since $l = Fx$ defines a line, the normal vector of that line has to be normalized, so that the Euclidean distance of the point x' to that line can be computed. This distance measure is much more accurate especially for points, which have very large vector norms. The measure is not linear anymore in the entries of F. This makes it much more difficult to estimate F, since noise behaves in a way such that a least squares solver cannot find an optimal solution (Fig. 11.1).

All of these measures treat the epipolar distance equally throughout the image plane. In other words, it is irrelevant, where in the image plane the epipolar distance is computed and how well the fundamental matrix is situated at this spot. But none of these measures take into account the uncertainty of the epipolar geometry. In estimation theory it is well known that predictions of an estimator near the mass center of the data set, which was originally used to compute the parameters of the estimator, are more accurate compared to predictions at the border of the data set [9]. Furthermore, since the covariance matrix of an epipolar line depends on the position of the related point correspondence, the reliability of the epipolar geometry is not homogeneously distributed, see Figs. 11.2 and 11.3. In [2] a method for computing a probability density function for point correspondences was introduced, where for each pixel the summarized probability of all possible epipolar lines going through this pixel is computed. This implies that point correspondences, having a disparity similar to the mean disparity of the data set, become more likely than uncommon disparities. However, in this work, a novel distance measure is presented, which takes into account the uncertainty of the epipolar geometry in a sound way. Instead of computing the accumulated probability at a certain point, the probability of one epipolar line is computed, going through a certain point, while best explaining the present epipolar geometry at the same time. It is shown that, using Lagrange multipliers, this constrained minimization problem can be reduced to solving a set of three linear equations with a computational complexity practically equal to the complexity of calculating the epipolar distance, as defined in [6]. The benefits of

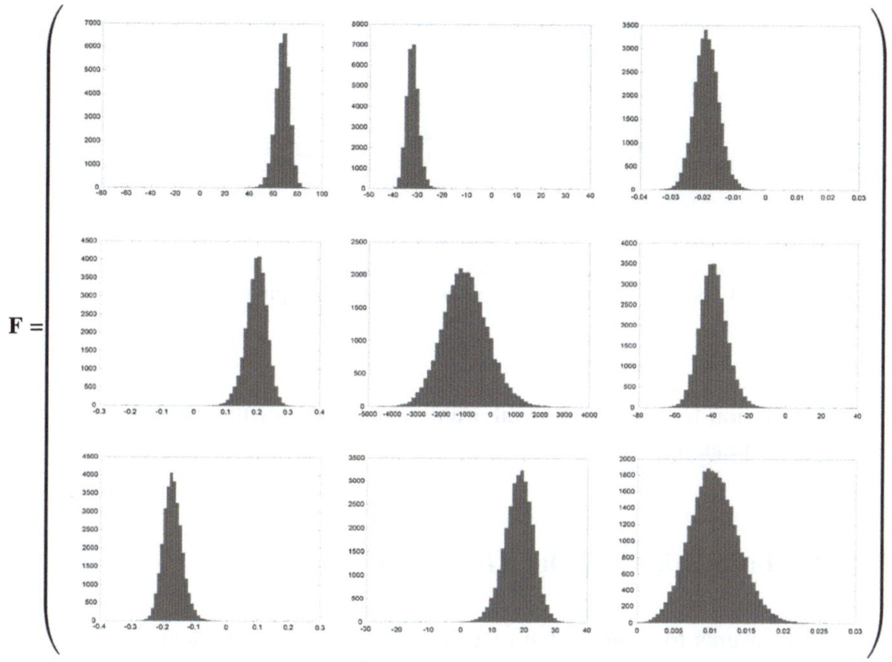

Fig. 11.1 Uncertainty of the fundamental matrix: for each entry of the fundamental matrix, a histogram was computed from several sets of noisy point correspondences, showing the impact of the noise

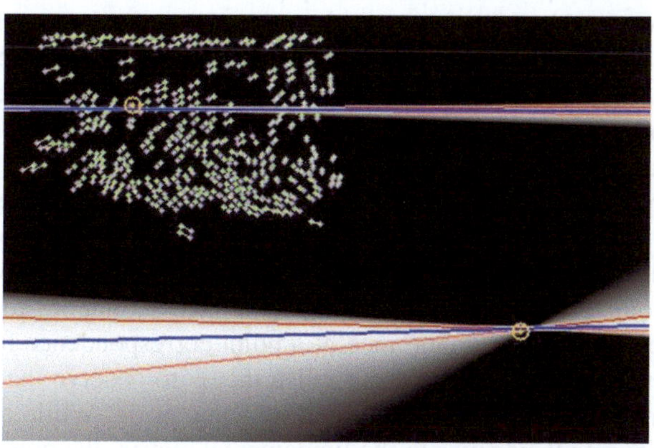

Fig. 11.2 Uncertainty of the epipolar geometry: point correspondences composing the data set, used to estimate the epipolar geometry and its uncertainty (*Upper left corner*). Epipolar band (k^2 values in *gray scale*), epipolar envelope (*green*, $\alpha = 0.95$ interval), and most probable *epipolar line* (*blue*) for a point near the mass center of the data set and outside of it (*yellow*). The narrowness of the epipolar band indicates the accuracy of the *epipolar line* estimation. Hence, *epipolar lines* for point correspondences located near the mass center of the data set are estimated more accurately compared to correspondences at the border

Fig. 11.3 Envelope of *epipolar lines*: point in the first image used for computation (*green*, $k^2 = 5.9915$, $\alpha = 0.95$ interval). Most likely *epipolar line* (*red*) and epipolar envelope (*green*) in the second image

the new criterion for outlier removal are demonstrated for the example of camera parameter estimation.

11.3 Probabilistic Epipolar Geometry

Existing measures to evaluate the quality of a point correspondence use the only available geometric constraint in a stereo setup, the epipolar geometry and its algebraic representation, the fundamental matrix. In practice, due to noise and outliers, only an approximation of the epipolar geometry can be estimated. By consequence, the distance of a correspondence from the epipolar line is a useful criterion for evaluation in stereo matching. In order to compute a robust fundamental matrix, point correspondences originating from uncommon depth levels are necessary. Having only coplanar 3-D points and hence correspondences, which can be explained by a homography, an epipolar geometry can not be found. If some additional correspondences, originating from 3-D points that are not coplanar with the other points, are available the quality of the resulting fundamental matrix will strongly depend on the quality of those additional point correspondences. If the ratio between coplanar points and other points is high, the epipolar geometry will be well suited for the coplanar points, but much less for the others. Hence, one has to derive a measure which incorporates the importance of a point correspondence for the epipolar geometry estimation. One way to accomplish this is to observe the impact on the fundamental matrix, if points are perturbed by some noise. While coplanar points marginally propagate their noise to the fundamental matrix, the noise from uncommon points has a significantly higher impact. Hence, point correspondences have to be treated differently when it comes to an evaluation of their quality. Therefore, a new distance measure has to be developed [13]. The basic idea for this is to invert the problem of finding a confidence interval with a certain probability α that an epipolar line is explained by the estimated epipolar geometry. Hence, for a given point x in the first view and a corresponding point x' in the second view, one line u going through x' and having at the same time the minimal Mahalanobis distance to the estimated epipolar line l has to be determined. For the Mahalanobis distance in this context, the inverse covariance matrix of the epipolar line is used, since it encapsulates the particular uncertainty of the epipolar geometry with respect to the correspondence $x \leftrightarrow x'$.

11.3.1 Determining the Covariance Matrix of the Fundamental Matrix

To capture the uncertainty of the epipolar geometry, it is not practical to just add noise to any single point, while evaluating the impact. Aside from computation complexity, some methods and especially the robust ones do not treat all point correspondences equally. Instead, the noise is added to all points at the same time, such that after a Monte Carlo simulation, the impact of the noise aggregates into the covariance matrix of the fundamental matrix. To compute the covariance matrix Σ_F of F, it is assumed that the noise on the correspondences follows a Gaussian distribution. After point correspondences are computed, several sets of point correspondences are generated by adding Gaussian noise to the original points. For each set, the fundamental matrix is computed using the normalized eight-point algorithm. Hence, a number of different fundamental matrices are obtained, showing how the epipolar geometry varies under slight changes in the point correspondences locations. In [15] a method for directly estimating the covariance matrix of the fundamental matrix is described.

11.3.2 Epipolar Lines and Epipolar Envelopes

As zero mean Gaussian noise on the point correspondences is assumed, for a given point x in the first image the estimated and most likely epipolar line l is given in homogeneous coordinates by

$$l = F\left(x + \begin{pmatrix} \mathcal{N}(0,\sigma) \\ \mathcal{N}(0,\sigma) \\ 0 \end{pmatrix}\right). \tag{11.5}$$

From the covariance matrix of the fundamental matrix Σ_F, the covariance matrix of the epipolar line is determined by

$$\Sigma_l = J \Sigma_F J^\mathsf{T}, \tag{11.6}$$

where J is the Jacobian of the mapping

$$l = \frac{(\mathsf{F}x)}{|\mathsf{F}x|}. \tag{11.7}$$

The squared Mahalanobis distance k^2 between an arbitrary line u and the estimated epipolar line l is then given by

$$k^2 = (l - u)^T \, \Sigma_l^+ \, (l - u)\,, \tag{11.8}$$

where Σ_l^+ is the pseudo-inverse of Σ_l, as the covariance matrix of the line might be rank deficient.

Thus, an approximation of how well an arbitrary line l matches with the knowledge acquired so far about the epipolar geometry and its uncertainty is acquired. For a given value of k^2, an envelope of epipolar lines containing all possible epipolar lines having a value less or equal to k^2 may be derived [6]. With the assumption that the elements of l have Gaussian distribution, k^2 has a cumulative χ_2^2 distribution and a probability that the true epipolar line is located within this envelope can be associated to the k^2-value. The region can be described by a conic C defined in homogeneous coordinates by

$$C = mm^{\mathsf{T}} - k^2 \Sigma_l, \tag{11.9}$$

where m is defined by

$$m = \mathsf{F}x. \tag{11.10}$$

In [3] a detailed description of the conic, describing a contour of equal likelihood, is given. The epipolar envelope is also used for guided matching [6], i.e., searching correspondences within the epipolar band after a first estimation of the fundamental matrix. It can be observed that if the fundamental matrix is computed from correspondences located only in the foreground, the uncertainty in the background becomes very high, see Fig. 11.2. This is the reason why correct matches lying in the corners of the image are often eliminated using a conventional criterion. To prevent this, we have to develop a new criterion being less stringent for matches in regions of high uncertainty.

11.3.3 New Distance Measure for Point Correspondence Evaluation

The basic idea for a new criterion is to invert the problem of finding an epipolar band for a given likelihood (i.e., a given k^2, respectively probability α). For a given point x' in the second view corresponding to a point x in the first view, the conic with minimal k^2 comprising the point x' has to be found. In other terms, the value k^2 in (11.9) that provides a hyperbola passing through x' is retrieved.

11.3.4 General Problem Statement

The point x' belongs to the conic C if the following equation holds

$$x'^T C x' = 0, \tag{11.11}$$

where C is given by (11.9). $F_2(k^2) = \alpha$ is the probability to find x' within C. It is not possible to retrieve the corresponding value k^2 directly from a point x' using the equations above. One possibility would consist in computing a multitude of different conics using a range of different values for k^2, and localize x' in between the conics. An approximation for k^2 would be obtained by interpolation, but for the problem there is a closed-form solution available, providing an exact result.

11.3.5 Closed-Form Solution

Assuming that the confidence, that any point x' in the second image, is corresponding to a point x in the first view is in relation with the probability of the epipolar geometry that would explain the correspondence pair $x \leftrightarrow x'$ and having at the same time the highest probability. In other terms, for a potential point correspondence $x \leftrightarrow x'$, the epipolar line l passing through x' having maximal probability is retrieved, i.e., minimal k^2 regarding (11.8). This assumption differs from the assumption made in [2], and we obtain a different criterion, which is more suitable for our purpose. If we denote the unknown epipolar line by u and the estimated line in the second image by l, the constrained minimization problem can be stated as follows:

$$\begin{cases} \min f(a,b,c) = (l-u)^T \,\Sigma_l^+\, (l-u) \\ \qquad g(a,b,c) = x'^T u = 0 \end{cases}, \tag{11.12}$$

with $x' = (x, y, 1)$, $m = (l_1, l_2, l_3)^\mathsf{T}$, $l = (a, b, c)^\mathsf{T}$, and the inverse covariance matrix Σ_l^+ obtained by an SVD, which is still a symmetric matrix of the form

$$\Sigma_l^+ = \begin{pmatrix} \sigma_{11} & \sigma_{12} & \sigma_{13} \\ \sigma_{12} & \sigma_{22} & \sigma_{23} \\ \sigma_{13} & \sigma_{23} & \sigma_{33} \end{pmatrix}. \tag{11.13}$$

In order to determine an exact solution, using Lagrange multipliers, this constrained minimization problem can be reduced to solving a set of three linear equations.

$$\begin{aligned} \nabla f(a,b,c) &= \lambda \nabla g(a,b,c) \\ g(a,b,c) &= 0 \end{aligned}. \tag{11.14}$$

After expanding $f(a,b,c)$ and computing its derivatives, it follows:

$$\begin{aligned}
\frac{\partial f}{\partial a} &= 2\,(a\sigma_{11} + b\sigma_{12} + c\sigma_{13} - l_1\sigma_{11} - l_2\sigma_{12} - l_3\sigma_{13}) \\
\frac{\partial f}{\partial b} &= 2\,(a\sigma_{12} + b\sigma_{22} + c\sigma_{23} - l_1\sigma_{12} - l_2\sigma_{22} - l_3\sigma_{23}) \\
\frac{\partial f}{\partial c} &= 2\,(a\sigma_{13} + b\sigma_{23} + c\sigma_{33} - l_1\sigma_{13} - l_2\sigma_{23} - l_3\sigma_{33})\,.
\end{aligned} \tag{11.15}$$

For $g(a,b,c)$ the following derivatives are determined:

$$\begin{aligned}
\partial g/\partial a &= x \\
\partial g/\partial b &= y \\
\partial g/\partial c &= 1
\end{aligned} \tag{11.16}$$

This results into

$$\begin{aligned}
\lambda \partial g/\partial a = \partial f/\partial a &\Leftrightarrow \lambda = x^{-1}\partial f/\partial a \\
\lambda \partial g/\partial b = \partial f/\partial b &\Leftrightarrow \lambda = y^{-1}\partial f/\partial b \\
\lambda \partial g/\partial c = \partial f/\partial c &\Leftrightarrow \lambda = z^{-1}\partial f/\partial c
\end{aligned} \tag{11.17}$$

As a problem with three unknown variables has to be solved, three equations are sufficient. Using the relations from (11.17) and the constraint that the point is located on the line l, the set of equations becomes

$$\begin{aligned}
\partial f/\partial a &= x\partial f/\partial c \\
\partial f/\partial b &= y\partial f/\partial c \\
ax + by + c &= 0
\end{aligned} \tag{11.18}$$

Expanding (11.18) the following linear equation system is obtained

$$\begin{pmatrix} & A & \\ x & y & 1 \end{pmatrix} \cdot \begin{pmatrix} a \\ b \\ c \end{pmatrix} = \begin{pmatrix} A \\ 0 \end{pmatrix} \cdot u, \tag{11.19}$$

where A is a 2×3 matrix:

$$A = \begin{pmatrix} \sigma_{11} - x\sigma_{13} & \sigma_{12} - x\sigma_{23} & \sigma_{13} - x\sigma_{33} \\ \sigma_{12} - y\sigma_{13} & \sigma_{22} - y\sigma_{23} & \sigma_{23} - y\sigma_{33} \end{pmatrix}. \tag{11.20}$$

This set of equations can be easily solved and the solution vector $(a,b,c)^{\mathsf{T}}$ is the line l passing through x' and having minimal k^2 in terms of (11.12). Finally, the corresponding value k^2 from (11.8) that induces an epipolar band delimited by a hyperbola passing through x' is obtained. The value k^2 is of special interest and this is the new distance measure, which can be used as an evaluation criterion for point correspondences in applications.

11.3.6 Application to Outlier Removal

The aim of computing the Mahalanobis distance k^2 for a point correspondence is to acquire a measurement for its compliance to the epipolar geometry. As stated before, 3-D points may agglomerate at similar depth levels or on planes. While point correspondences which originate from such configurations may be explained by a homography and the epipolar constraint may be well suited for them, the epipolar geometry will not fit well for arbitrary points and hence can not describe the camera parameters accurately. More often, the quality of a fundamental matrix is measured by the mean or the root mean square (RMS) of the epipolar distances of all available point correspondences. Nevertheless, computing a fundamental matrix with a low mean epipolar distance does not directly mean that it represents a stereo setup well. A fundamental matrix, computed from a set of point correspondences of coplanar 3-D points will show a low mean epipolar distance, but it will have a very poor performance when it comes to extracting camera parameters. Since the fundamental matrix encapsulates camera parameters, a decomposition into the extrinsic camera matrices should fit to the true rotation and translation of the cameras rather than just allowing many point correspondences to satisfy the epipolar constraint.

11.4 Results

In this section, a method for computing the fundamental matrix using the new distance measure derived above within an outlier removal step, is compared to one using the epipolar distance instead. In addition we compare it to RANSAC, as implemented in [1] and to the HEIV [10] method as a representative of the methods taking into account the heteroscedasticity of the fundamental matrix estimation problem.

11.4.1 Algorithm

To compare the new distance measure with the common epipolar distance, a modified iterative version of the normalized eight-point algorithm is used, including outlier removal. In each iteration, the minimal k^2 distance or respectively in the other methods, the epipolar distance of the point correspondences to their appropriate epipolar lines is determined. If the distance of a correspondence exceeds a threshold based on the mean distance value of all point correspondences, it is classified as an outlier and removed from further usage. In addition, the k^2 value or in the competing version, the epipolar distance, is used for weighting the linear equations of the least squares problem. After some iterations or if no significant outliers were detected, the process stops and a final fundamental matrix is returned, which has the minimal mean distance to all point correspondences, where in one case the most probable epipolar line is used and in the competitive method the common epipolar line.

In order to distinguish between the two methods, the one which uses the new distance measure is referred to as “Statistic”, while the method comprising the Euclidean epipolar distance is called ‘Euclidean’. Both methods are identical except for the distance measures used to eliminate outliers and weighting the linear equations. The resulting fundamental matrices have to be compared in a sound way. Since we replaced the epipolar distance, the “Statistic” method does not optimize the Sampson distance [6] any more. Hence, common quality measurements like RMS, or the mean epipolar distance are not practicable, even though the method is well recognized by them, as shown later. Instead, a simulation environment where a true fundamental matrix and the intrinsic camera parameters are known is used. This allows to evaluate the similarity between the estimated fundamental matrices and the ground truth. This is accomplished by separating the F-matrices into a rotation matrix and translation components, so that the angles between the rotations and translations can be compared, see Sect. 11.4.2. Nevertheless, results for the RMS-error and mean epipolar distance values are shown.

11.4.2 Essential Matrix Decomposition

In contrast to the fundamental matrix F, the essential matrix only contains the five extrinsic camera parameters. It can be regarded as a fundamental matrix, valid in the metric of the scene rather than the pixel domain. In order to compute the essential matrix, the intrinsic camera parameters have to be known. In conjunction with an intrinsic camera matrix

$$K = \begin{pmatrix} f_l & 0 & x \\ 0 & f_l & y \\ 0 & 0 & 1 \end{pmatrix}, \tag{11.21}$$

the essential matrix E can be determined by

$$E = K^T F K. \tag{11.22}$$

On the other hand, E can be computed from the rotation matrix R and translation vector t

$$E = R[t]_x. \tag{11.23}$$

In order to obtain R and t, the essential matrix may be further decomposed.

Using a singular value decomposition, matrices U, D, V may be computed such that

$$E = UDV^T \tag{11.24}$$

Ignoring the sign of the translation, t is equal to the last column vector of U. Also the rotation is ambiguous. The two possible rotation matrices R_1 and R_2 can be obtained by

$$\begin{aligned} R_1 &= UW_1V \\ R_1 &= UW_2V, \end{aligned} \tag{11.25}$$

with

$$W_1 = \begin{pmatrix} 0 & -1 & 0 \\ 1 & 0 & 0 \\ 0 & 0 & 1 \end{pmatrix} W_2 = \begin{pmatrix} 0 & 1 & 0 \\ -1 & 0 & 0 \\ 0 & 0 & 1 \end{pmatrix}. \tag{11.26}$$

The ambiguities of the translation as well as the rotation indicate that several camera setups exist, which lead to the same essential matrix. According to [6], there are four of such cases. They arise from the fact that the pinhole camera model does not reject points behind the camera plane. However, in a real camera, points behind a camera are not visible. Hence, only one of the four possible camera setups has a physical representation. It is simply the camera setup for which the 3-D reconstructed points are located before both camera planes. However, the rotation matrices may also be transposed, so that eight possible camera setups may be obtained. In [14] the decomposition of the essential matrix with eight solutions for two perspective cameras using an SVD is discussed.

11.4.3 Simulation Environment

A simulation environment with known camera matrices and 3-D world points is used for comparison. A 3-D point cloud is randomly created and mapped into the camera planes of a stereo camera system according to the rules of a pinhole camera. The distribution of the 3-D world points follows a 3-D Gaussian distribution. The mapped points in the first and second camera plane are interpreted as point correspondences. Those point correspondences are perturbed by zero mean Gaussian noise and a certain percentage of outliers is added. Using these point correspondences, the fundamental matrix is computed by several approaches. The two resulting fundamental matrices F_1 and F_2, which correspond to the "Statistic" and 'Euclidean' method are compared with additional fundamental matrices F_3 and F_4, computed from the HEIV and RANSAC approach. Despite measuring how well the three matrices fulfill the epipolar constraint, the fundamental matrices are transformed into essential matrices E_1, E_2, E_3 and E_4 using the known intrinsic camera parameter matrix K of the camera setup. The essential matrices are decomposed into rotation matrices R_1, R_2, R_3 and R_4 and normalized translation vectors t_1, t_2, t_3, t_4, such that $||t_i||_2 = 1$.

This comes from the fact that any 3-D reconstruction contains an arbitrary scaling so that only directions instead of distances can be determined. All these quantities are compared with the true rotation and translation between the cameras of the simulation.

Since the calculation of the rotation and translation takes place within the scale invariant projective space, the angles between the rotation R_i and the true rotation expressed by R_t and especially the angle between the translation vectors t_i and the true translation direction vector t_t are evaluated.

In order to compare the rotation matrices $R_i = [r_{x,i}, r_{y,i}, r_{z,i}]$ with the true rotation matrix $R_t = [r_{x,t}, r_{y,t}, r_{z,t}]$, the column vectors of the matrices are used.

After the rotation, the resulting vectors r_x, r_y, r_z are still orthogonal to each other. The similarity between the rotation matrices can be determined by computing the angles between the mappings of the unit vectors.

$$\begin{aligned} \alpha_i &= \arccos\left(r_{x,i} r_{x,t}\right) \\ \beta_i &= \arccos\left(r_{x,i} r_{x,t}\right) \\ \gamma_i &= \arccos\left(r_{x,i} r_{x,t}\right) \end{aligned} \tag{11.27}$$

The angles δ_i between t_t and the t_i are expressed by

$$\delta_i = \arccos\left(|t_t t_i|\right) \tag{11.28}$$

Depending on the direction of the translation vectors and hence on the sign of the argument, the δ_i can reach values between $[0..\pi]$. Since the t_i vectors are scale invariant, $\delta_i > \frac{\pi}{2}$ have to be adjusted, so that the direction of the t_i are inverted. This may be accomplished by taking the absolute value of the argument $t_t t_i$.

11.4.4 Gaussian Point Cloud

In a first experiment, the four methods are tested on data coming from a point cloud in 3-D, which was created at random. In order to capture the influence of the number of point correspondences, their number has been varied through several experiments. Since all methods comprise a kind of random generator, each experiment was carried out 100 times and the results were averaged. The 3-D points are mapped into the camera planes, where Gaussian noise with $\sigma = 0.5$ and 10 percent outliers were added (by adding noise with $\sigma = 5$).

In Fig. 11.4, for all methods the accuracy increases with the number of point correspondences. As expected, this comes from the fact that an estimation problem can be solved more accurately if more sample data is available. The "Statistic" method shows excellent results in all rotation and translation components compared to the other methods. Since a simple 3-D world setup, based on a Gaussian distributed point cloud is used, the statistical modeling and the fact that the uncertainty of the

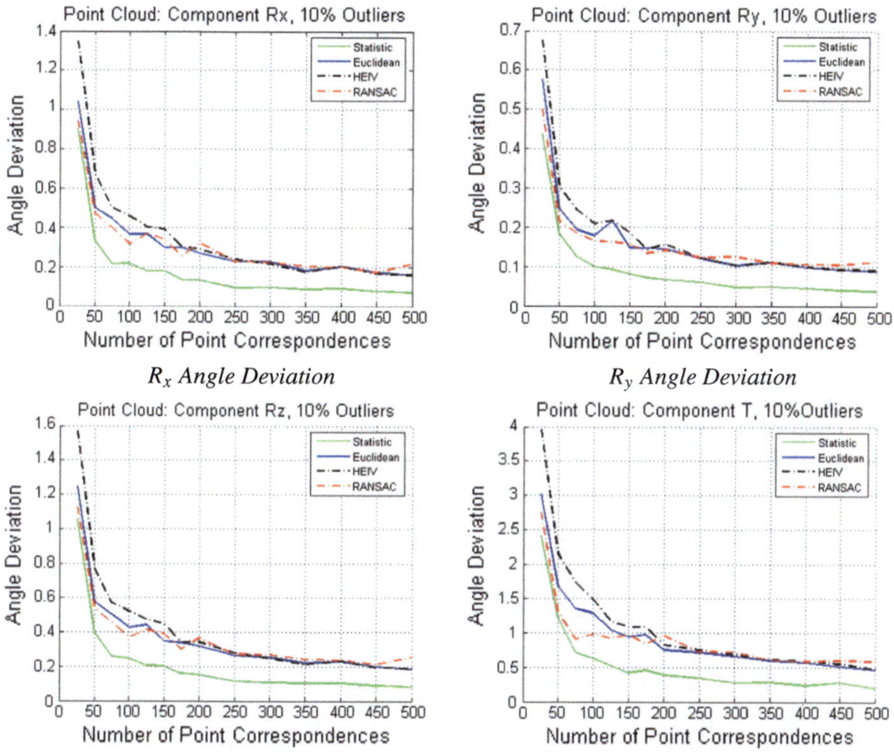

R_x *Angle Deviation* — R_y *Angle Deviation*

R_z *Angle Deviation* — *T Angle Deviation*

Fig. 11.4 Fundamental matrix decomposition results for different numbers of point correspondences: angle deviation in degree from ground truth for the R_x, R_y, R_z rotation and translation components for a Gaussian distributed point cloud in 3-D with 10 % outliers

fundamental matrix was made available payed off. In experiments with a low number of point correspondences, the RANSAC method performs also very well, while the other two methods outperform the RANSAC in accuracy when the number of point correspondences is increased. However, all three methods remain very close to each other, while the “Statistic” method is significantly better.

In Fig. 11.5 the results for common error measurements, namely the root mean square error (RMS) and the mean epipolar distance are shown. Again, the “Statistic” method performs best among all methods. While RANSAC is rated as the second best method by the mean epipolar distance, it is the worst one using the RMS error measurement. As there is a probability that RANSAC does not find any subset containing only inliers, there are cases where no appropriate fundamental matrix is found. Since the RMS error measurement is very sensitive to such events, the graph of RANSAC bounces. Since bounces are missing in the graphs of the other methods, it can be assumed that an appropriate fundamental matrix is always found by them.

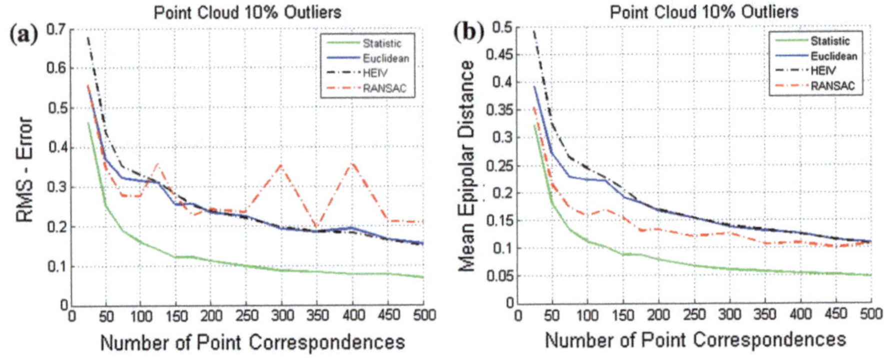

Fig. 11.5 Fundamental matrix evaluation varying the number of point correspondences: **a** RMS-error and **b** mean epipolar distance computed for each method on a Gaussian distributed point cloud in 3-D with 10 % outliers

11.4.5 More Scene Structure

In order to create a more reasonable scene model, several 3-D world setups are created at random, such that a percentage of the point correspondences are located on a wall-like structure while another percentage is located on a floor-like plane. Again, the 3-D points are mapped into the camera planes, where Gaussian noise with $\sigma = 0.5$ and 10 % outliers were added (by adding noise with $\sigma = 5$). The percentage was chosen according to the experience with matching algorithms in natural images, where a correspondence deviates more than a pixel from its true location. From those point correspondences the fundamental matrix was estimated once again by the "Statistic" and "Euclidean" method. In addition, the fundamental matrix was computed using RANSAC implementation from OpenCV [8] and the HEIV method from [10]. In Fig. 11.6 the results for varying the total number of point correspondences are visualized, where the angle deviation compared to the ground truth for all four methods regarding the rotation components and the translation are shown.

It is shown that the "Statistic" method, which comprises the new criterion, performs very well as soon as sufficient ($n > 100$) point correspondences are available. The reason is that the computation of the covariance matrix requires the estimation of additional parameters, which is less accurate for a low number of point correspondences. In Fig. 11.7 the RMS-error as well as the mean epipolar distance of all methods are displayed.

The "Statistic" method performs well in this context. The direct comparison to the "Euclidean" method shows that, especially for a higher number of point correspondences, the new distance measure leads to a better outlier removal. An important observation is that the RMS-error and the mean epipolar distance are not always consistent with the ground truth comparison, which was assumed above and is an additional justification for a new measurement criterion. E.g., the HEIV method is

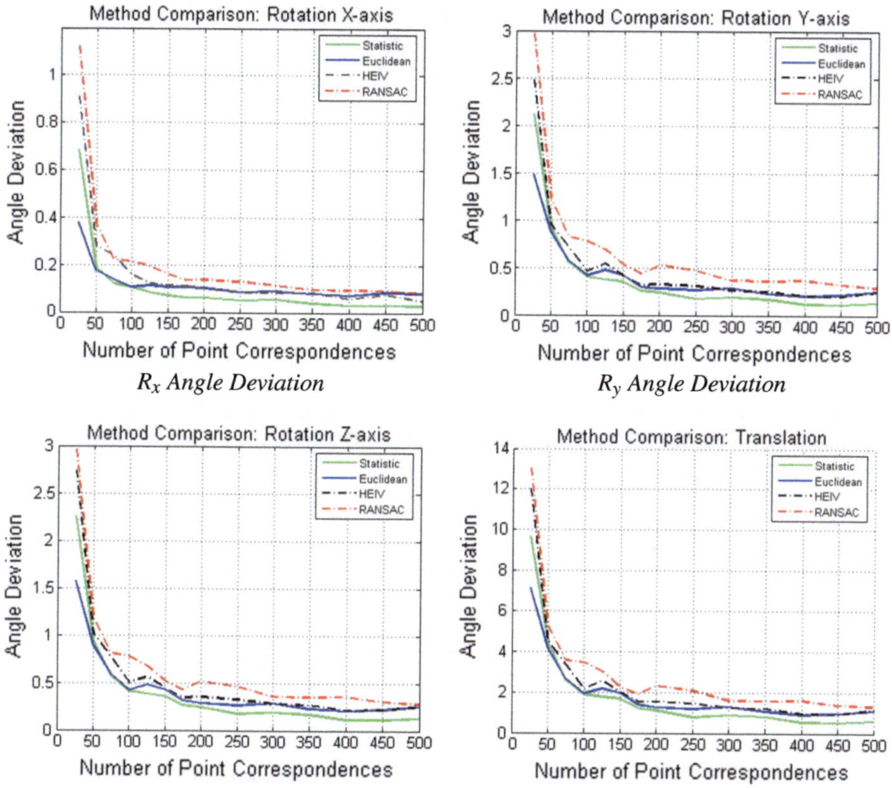

R_x Angle Deviation | *R_y Angle Deviation*

R_z Angle Deviation | *T Angle Deviation*

Fig. 11.6 Fundamental matrix decomposition for different numbers of point correspondences: angle deviation in degree from ground truth for the R_x, R_y, R_z rotation and translation components for 3-D points located evenly on two planes

rated better in most cases than the "Euclidean" method by RMS and the mean epipolar distance, but the comparison with the ground truth reveals that this evidence can not be supported. Also the relative distances between the other methods among each other behave differently.

In order to analyze the behavior of the four methods on structural changes, the ratio of the number of points in thc background and on the floor is varied.

In Fig. 11.8 the results for varying the ratio is shown, using $n = 100$ point correspondences. The "Statistic" method performs well up to a ratio of 70 %, from where the other methods reach a similar or even better performance. Due to its algorithmic structure the RANSAC approach runs into heavy problems, if one 3-D plane is dominant over the other. Since the correspondences of one plane can be explained by a homography, the RANSAC chooses putative good subsets, which comprise correspondences of only one plane. Hence, the resulting fundamental matrix is seriously flawed. While the floor plane does not cover as much image space as the background

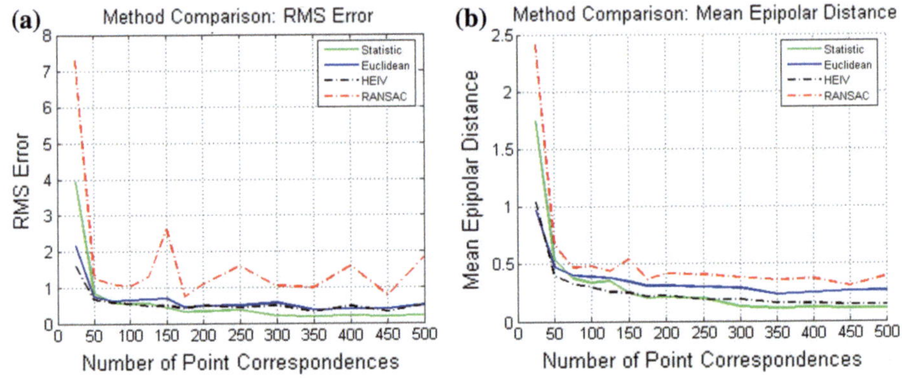

Fig. 11.7 Fundamental matrix evaluation varying the number of point correspondences: **a** RMS-error and **b** mean epipolar distance computed for 3-D points located evenly on two planes

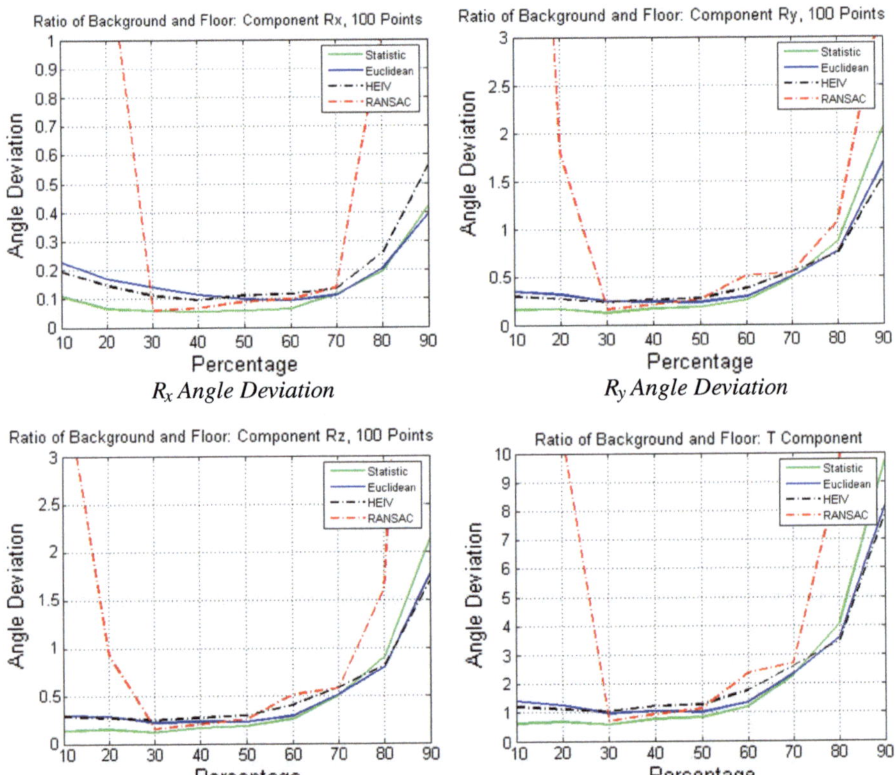

R_x Angle Deviation *R_y Angle Deviation*

R_z Angle Deviation *T Angle Deviation*

Fig. 11.8 Fundamental matrix decomposition for $N = 100$ point correspondences, varying the ratio of the two planes: angle deviation in degree from ground truth for the R_x, R_y, R_z rotation and translation components

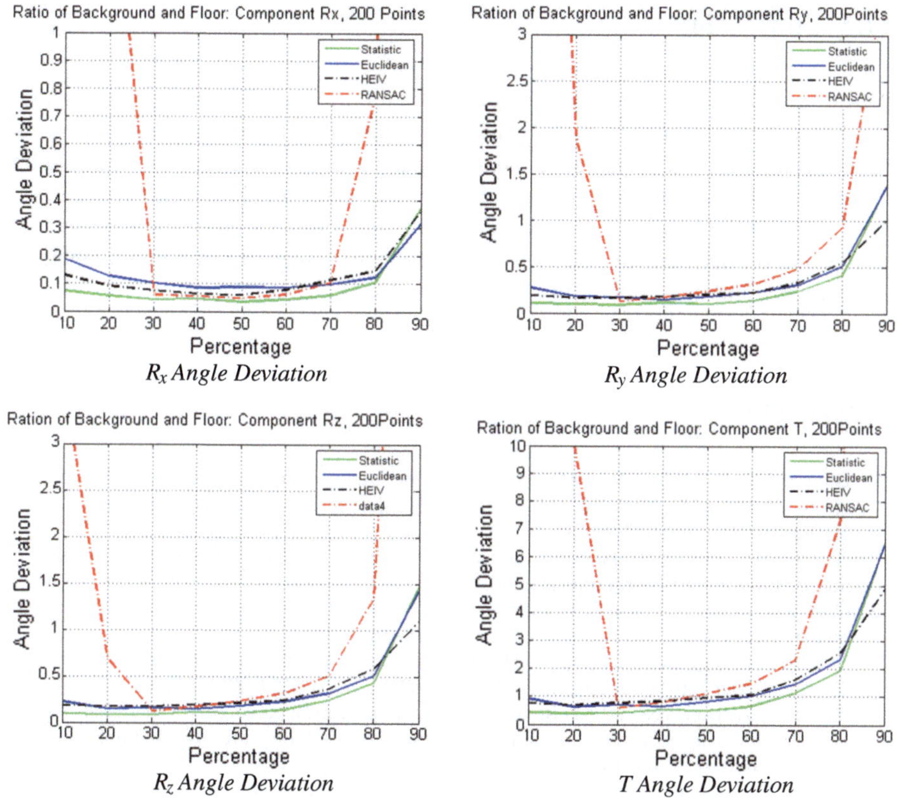

Fig. 11.9 Fundamental matrix decomposition for $N = 200$ point correspondences, varying the ratio of the two planes: angle deviation in degree from ground truth for the R_x, R_y, R_z rotation and translation components

plane, the other methods tend to neglect point correspondences from the floor plane as the ratio of background to floor exceeds 70 %. The most robust method to this effect seems to be the HEIV method. Since it does not really reject points as outlier, but tries to correct them according to the heteroscedasticity of the estimation problem, it achieves quite good results in this context. If more than $n = 100$ point correspondences are used, one may expect that the performance of the "Statistic" method increases compared to the other methods.

In Fig. 11.9 the same experiment was made with $n = 200$ point correspondences. As expected, the performance of the "Statistic" method increased.

To conclude, the assignment of the squared Mahalanobis distance k^2 instead of the epipolar distance shows a significant benefit for the task of outlier removal and indirectly for camera calibration. The direction of the translation as well as all angles of the camera rotation show a significant lower deviation compared to the ground truth.

11.5 Conclusion

We presented a novel criterion to evaluate point correspondences in stereo images. Our main contribution is to show that the uncertainty of the fundamental matrix can be integrated into a novel distance measure that can help evaluating point correspondences in stereo images by taking both geometry and probability into account. We demonstrated the benefits of our criterion for the computation of the fundamental matrix and outlier removal. Several other application scenarios are imaginable, but these are beyond the scope of this chapter.

References

1. Bradski GR, Pisarevsky V (2000) Intel's computer vision library: applications in calibration, stereo segmentation, tracking, gesture, face and object recognition. Proc IEEE Conf Comput Vis Pattern Recognit. 2:796–797
2. Brandt SS (2008) On the probabilistic epipolar geometry. Image Vis Comput 26(3):405–414. http://dx.doi.org/10.1016/j.imavis.2006.12.002
3. Csurka G, Zeller C, Zhang Z, Faugeras O (1997) Characterizing the uncertainty of the fundamental matrix. Comput Vis Image Underst 68:18–36
4. Erickson J, Har-Peled S, Mount DM (2006) On the least median square problem. Discret Comput Geom 36(4):593–607. http://dx.doi.org/10.1007/s00454-006-1267-6
5. Fischler MA, Bolles RC (1987) Random sample consensus: a paradigm for model fitting with applications to image analysis and automated, cartography. Morgan Kaufmann Publishers Inc., San Francisco, pp 726–740
6. Hartley R, Zisserman A (2003) Multiple view geometry in computer vision. Cambridge University Press, New York
7. Hartley RI (1997) In defense of the eight-point algorithm. IEEE Trans Pattern Anal Mach Intell 19(6):580–593. http://dx.doi.org/10.1109/34.601246
8. Intel: Open source computer vision library (http://www.intel.com/technology/computing/opencv)
9. Kay SM (1993) Fundamentals of statistical signal processing: estimation theory. Prentice-Hall, Inc., Upper Saddle River
10. Leedan Y, Meer P (2000) Heteroscedastic regression in computer vision: problems with bilinear constraint. Int J Comput Vis 37:127–150
11. Luong QT, Faugeras O (1995) The fundamental matrix: theory, algorithms, and stability analysis. Int J Comput Vis 17:43–75
12. Triggs B (2001) Joint feature distributions for image correspondence. In: Proceedings of the 8th international conference on computer vision, pp 201–208
13. Unger M, Stojanovic A (2010) A new evaluation criterion for point correspondences in stereo images. In: Proceedings of the international workshop on image analysis for multimedia interactive services. IEEE, Piscataway, Desenzano del Garda, Italy
14. Wang W, Tsui H (2000) An svd decomposition of essential matrix with eight solutions for the relative positions of two perspective cameras. In: ICPR '00: Proceedings of the international conference on pattern recognition, IEEE Computer Society, Washington, DC, USA, p 1362
15. Zhang Z (1998) Determining the epipolar geometry and its uncertainty: a review. Int J Comput Vis 27(2):161–195
16. Zhang Z, Deriche R, Faugeras O, Luong QT (1995) A robust technique for matching two uncalibrated images through the recovery of the unknown epipolar geometry. Artif. Intell. 78(1–2):87–119. http://dx.doi.org/10.1016/0004-3702(95)00022--4

Chapter 12
Local Homography Estimation Using Keypoint Descriptors

Alberto Del Bimbo, Fernando Franco and Federico Pernici

Abstract This chapter presents a novel learning-based approach to estimate local homography of points belong to a given surface and shows that it is more accurate than specific affine region detection methods. While others works attempt this by using iterative algorithms developed for template matching, our method introduces a direct estimation of the transformation. It performs the following steps. First, a training set of features captures geometry and appearance information about keypoints taken from multiple views of the surface. Then incoming keypoints are matched against the training set in order to retrieve a cluster of features representing their identity. Finally the retrieved clusters are used to estimate the local homography of the regions around keypoints. Thanks to the high accuracy, outliers and bad estimates are filtered out by multiscale Summed Square Difference (SSD) test.

Keywords Homography estimation · SIFT keypoints · Nearest neighbor · Robust matching · Scale and affine invariant features

12.1 Introduction

The last years have seen the development of many affine region detectors [11] that derive an approximation of the local image transformation around points of interest. Matched using a region descriptor, they provide to be very useful for many types of applications because getting rid of most of the complexity due to the image formation. More recently, a novel class of learning based approaches has been proposed [5–7]. This class of methods appears to be faster and more reliable, but relies on iterative refinements that makes it unqualified for very large image database. In this

---

A. D. Bimbo · F. Franco · F. Pernici (✉)
Media Integration and Communication Center (MICC),
University of Florence, Florence, Italy
e-mail: pernici@dsi.unifi.it

N. Adami et al. (eds.), *Analysis, Retrieval and Delivery of Multimedia Content*,
Lecture Notes in Electrical Engineering 158, DOI: 10.1007/978-1-4614-3831-1_12,

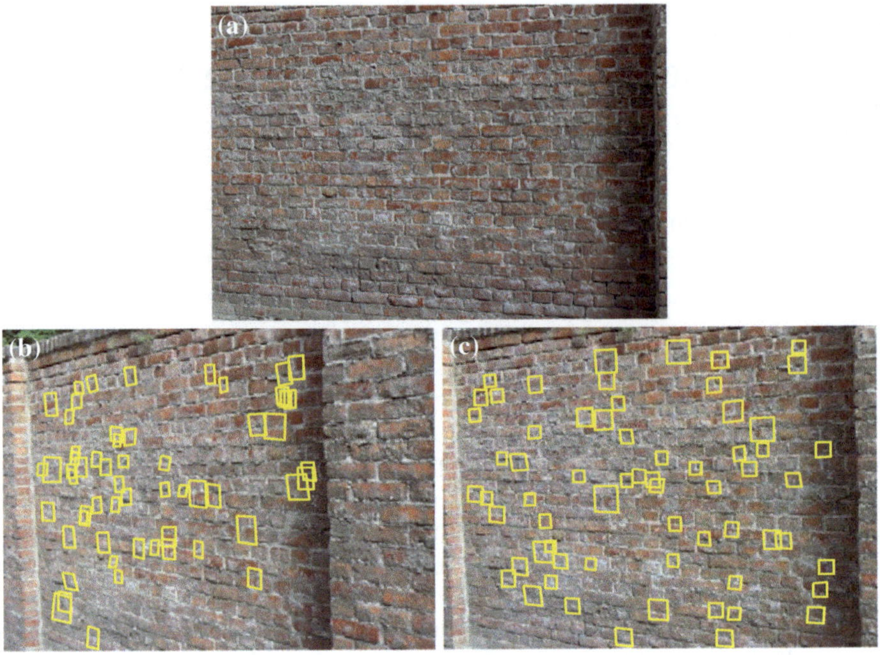

Fig. 12.1 Overview of the applications of the method and its performance on the Wall dataset. **a**: Reference sample image of the standard Wall dataset (Wall1). **b** and **c**: Homography estimation on the third image and fifth image of the Wall sequence

chapter, we propose a new learning-based approach that performs the other way around by direct estimation of the local homography around points of interest. Given a reference image of a smooth surface and an input image containing this surface, our method proceeds in three steps. We first generate a training set of features that compactly captures geometry and appearance information about multiple views of the same keypoints. Then input keypoints are associated with the correspondent sets of features by a matching process and a geometry consistency checking. At last, the informations related to keypoints in the sets are used to estimate the local perspective transformations. Outliers and bad estimates are filtered out using a multiscale SSD validation. As shown in Figs. 12.1 and 12.2, our approach avoids specific estimation of the transformation and gives us a more reliable estimate than affine region detectors for both planar and non-rigid surfaces. The rest of the chapter is organized as follows. Section 12.2 outlines the state of the art about local homography estimation. Section 12.3 contains an overview of the approach with details about the training set generation and the local homography estimation. In Sect. 12.4 experimental results are shown and discussed. Conclusions are drawn in Sect. 15.6.

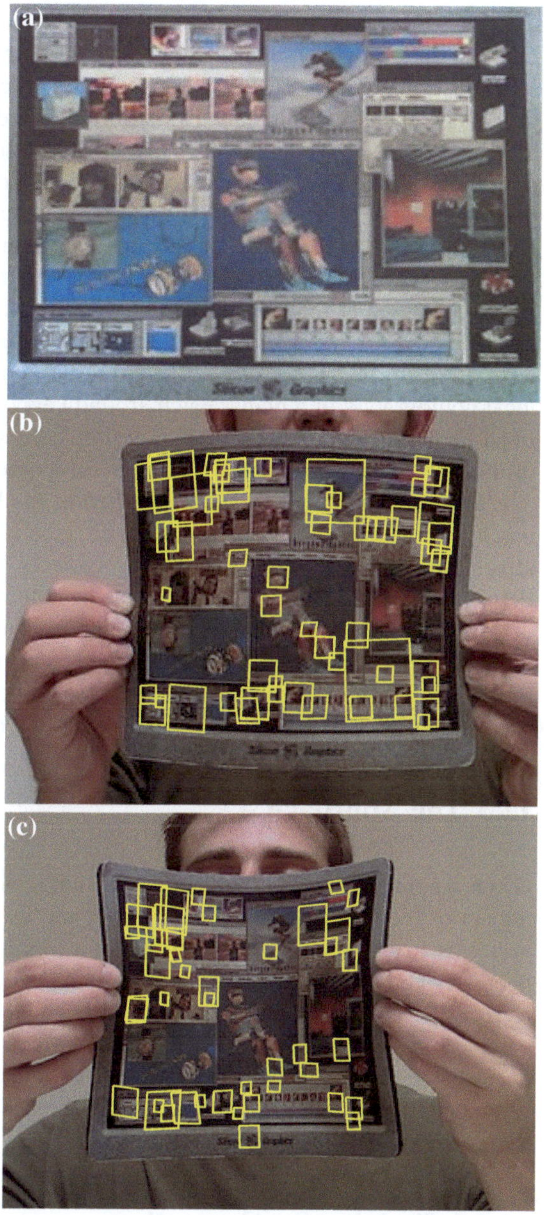

Fig. 12.2 Overview of the applications of the method and its performance on the mousepad dataset. **a**: Reference sample image of the mousepad dataset. **b** and **c**: Homography estimation on two different query images depicting the mousepad

12.2 Related Works

Many computer vision applications rely on the recovery of properties of interest points, or keypoints. For example, retrieving the poses of keypoints in addition to matching them is a fundamental task in vision-based robot localization [4], object recognition [3, 15] or image retrieval [14] to transform unconstrained problems into constrained ones. The standard approach proceeds by decoupling the matching process from the keypoint pose estimation. This is done by first using some particular affine region detectors and by then using SIFT descriptors computed on the rectified regions to match them. In the recent years many different detectors have been proposed to recognize keypoints under large perspective distortion. Among them, the Hessian-Affine detector of Mikolajczyk and Schmid [11] and the MSER detector by Matas et al. [10] have been shown to be the most reliable ones. However, they retrieve only an affine transformation without estimating the full perspective pose and often require handcrafting the descriptors to achieve invariance to specific distortion. Recently, a novel class of learning-based methods that attempts to compute local homography of a planar patch around keypoint has been developed [5–7, 13]. In particular the approaches of [5–7] mainly consist in two steps: the incoming point of interest is matched against a database of keypoints, each of which is associated to a coarse estimation of its pose (defined as the homography between a reference patch and the patch centered on the point); the coarse pose retrieved is hence iteratively refined by applying [8] and successively [2]. For the first step, [6] uses the Ferns classifier [12] while [5] and [7] relies on linear classifiers. In [13] keypoints are detected at 2D corners and matched to a pre-defined set of corners. Differently from the previous approaches, an estimation by regression is inserted here in the loop of the Ferns classifier for matching. In this way, the homographic transformation is directly checked during matching. Finer regression is only performed for close matches.

12.3 The Approach

Given a point of interest extracted at run-time, we want to match it against a training set of features and to accurately estimate its local homography. Our approach performs in three steps. The first builds a training set of features which captures geometry and appearance information about keypoints taken from multiple views of a given 3D object. The second step matches an incoming point of interest against the database in order to retrieve a cluster of features representing keypoint identity. In the last step the retrieved cluster is used to estimate the local patch homography. The estimated homography, outliers and bad estimates are then filtered out by multiscale Summed Square Difference (SSD) test.

12.3.1 Training Set Generation

Let us consider a set of m 3D points of interest $\boldsymbol{K} = \{\boldsymbol{k}_i\}_{i=1}^{m}$ lying on the surface of a given object. The aim is to build a large training set of features which captures geometry and appearance about different patches around these points extracted by multiple views of the object. According to this, an effective method to build the training set is to generate random synthetic views of the object using simple geometrical technique and extract SIFT keypoints from them. In this way, we can easily associate each keypoint with information about the patch around it and select keypoints that are more stable under noise and perspective distortion. We discuss below the construction of multiple views of the object given a reference image and then the process of extracting and selecting keypoints.

12.3.1.1 Multiple Views Sampling

Under the assumption of local smooth surface patches surrounding points of interest $\boldsymbol{k}_i$ can be considered as locally planar and their distortion under prospective projection can be represented by homographies. Therefore only one reference image $\boldsymbol{I}_r$ of the query object could be enough to generate the set of multiple views $\{\boldsymbol{I}_j\}$. Considering that for moderate foreshortening keypoints keep stable even under some viewpoint changes distorted image views are created from the reference image $\boldsymbol{I}_r$ taking a rectangular window of approximately one half the image area around each vertex of the reference image, selecting one point at random in each window, and assuming these points as the vertices of the newly generated image $\boldsymbol{I}_j$ where the original content is warped. Instead, since for strong foreshortening keypoints keep stable only for small variations of the viewing angle, in order to provide a finer sampling, the same procedure is applied to the vertices of already distorted images with windows of approximately one tenth of the image area. Figure 12.3 shows some views generated by this process.

12.3.1.2 Features Extraction

Once the multiple views $\{\boldsymbol{I}_j\}$ are sampled, we can extract SIFT keypoints from them in order to associate each feature with geometry and appearance information: $\boldsymbol{f}_i = \{\boldsymbol{d}_i, \mathtt{H}_{rj}\}$. Geometry information is captured by the homography $\mathtt{H}_{rj}$ between the reference image and the view $\boldsymbol{I}_j$ from which the keypoint is taken, while appearance information is represented by SIFT descriptor $\boldsymbol{d}_i$.

12.3.1.3 Features Selection

Because of noise and perspective distortion, the points lying on the object surface don't have the same probability $\boldsymbol{P}(\boldsymbol{k}_i)$ to be found in a query image $\boldsymbol{I}_t$ in which they are visible at runtime. In order to select the m keypoints with highest probability to be extracted, we proceed as follows. Let $\mathtt{H}_{rj}$ the homography which transforms the reference image $\boldsymbol{I}_r$ in the image $\boldsymbol{I}_j$ which contains the keypoint $\boldsymbol{k}_{ij}$. By applying $\mathtt{H}_{rj}^{-1}$ to $\boldsymbol{k}_{ij}$ the found 2D point is back-projected in the coordinate system of $\boldsymbol{I}_r$ and feed a 3D point accumulator which allows to estimate the probability $\boldsymbol{P}(\boldsymbol{k}_i)$ with which the corresponding 3D points can be detected in a new image. The 3D points

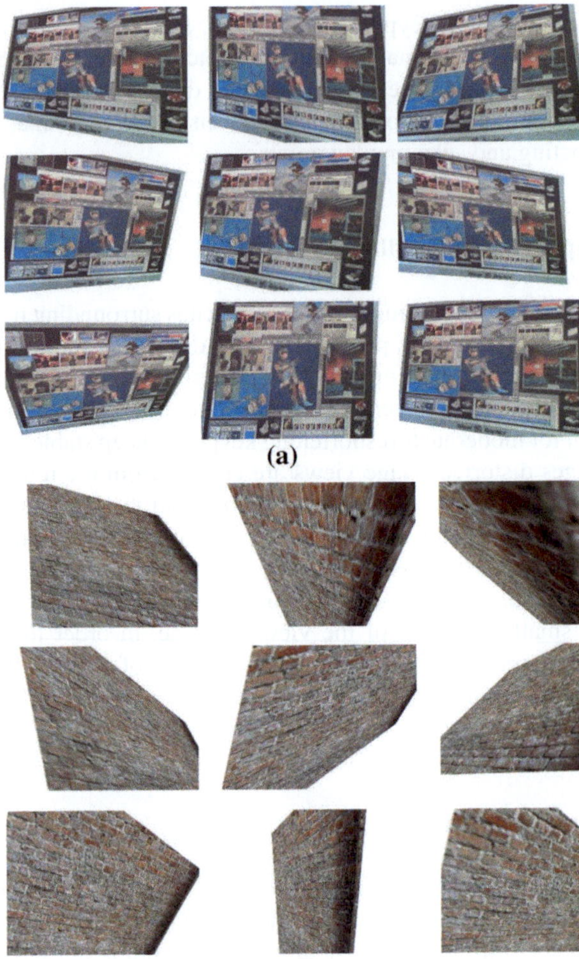

(a)

(b)

Fig. 12.3 Training set generation in the case of moderate **a** and in the case of strong foreshortening **b**. All views are respectively synthesized using the image depicted in Fig. 12.2 **a** and Fig. 12.1 **a** as reference image

accumulating most votes are retained as points of interest, having a large probability to be detected by SIFT in unknown query images.

12.3.2 Matching

Given a set $\mathcal{K}_{\mathcal{Q}}$ of SIFT extracted by a query image at runtime and the set of keypoints $\mathcal{K}_{\mathcal{G}}$ generated in the training phase, we want to retrieve the identities of keypoints lying on the surface of the query object and to obtain an estimation of their local homography. The problem of retrieving identity can be defined as a research for a function $\boldsymbol{B} : \mathcal{K}_{\mathcal{Q}} \rightarrow \mathcal{K}_{\mathcal{G}} \cup \boldsymbol{k}_0$ that assigns to every $\boldsymbol{k}_q \in \mathcal{K}_{\mathcal{Q}}$ either a cluster of features $\{\boldsymbol{C}_q\} \subset \mathcal{K}_{\mathcal{G}}$ or $\boldsymbol{k}_0$ representing no matching. According to this, since the training set contains multiple views of each 3D point of interest, each keypoint $\boldsymbol{k}_q \in \mathcal{K}_{\mathcal{Q}}$ is matched to its k nearest neighbors. This can be done in logarithmic time by using a kd-tree to find the approximate nearest neighbors [1]. We use $k = \frac{|\mathcal{K}_{\mathcal{G}}|}{m}$, where $|\mathcal{K}_{\mathcal{G}}|$ is the total number of keypoints extracted and m is the number of 3D points of interest views of which are contained in the training set. In particular the cluster $\{\boldsymbol{C}_q\}$ is associated to the corresponding keypoint $\boldsymbol{k}_q$ only if the descriptor of the second-closest neighbor is far enough ϵ to the descriptor of the first closest neighbor [1]:

$$\frac{\min_{d_i \in C_q} \|d_q - d_i\|_2}{\min_{d_i \in C_q \setminus B(d_q)} \|d_q - d_i\|_2} < \varepsilon\,, \tag{12.1}$$

where

$$B(d_q) = \arg \min_{d_i \in C_q} \|d_q - d_i\|_2 \tag{12.2}$$

is the Euclidean nearest neighbor of d_q (in our experiments we used a distance ratio greater than 0.75 as rejection criterion). Since each one of the retrieved keypoints in $\{C_q\}$ has its homography associated, matching also permits to obtain a number of coarse estimation of the keypoint local homography. However checking the geometry consistency of keypoints of cluster $\{C_q\}$ is necessary to filter out wrong matches.

12.3.2.1 Geometry Checking

Because of several self similar keypoints, each cluster $\boldsymbol{C}_q$ is processed in order to reject false matching with the corresponding $\boldsymbol{k}_q$. Since the closest neighbor $\boldsymbol{f}_1$ (i.e. 1-NN) corresponds with high probability to a different view of $\boldsymbol{k}_q$, we proceed as described in the following pseudo-code:

1. Back-project $\boldsymbol{f}_1$ in the coordinate system of $\boldsymbol{I}_r$ using $\mathtt{H}_1^{-1}$.
2. Define a circle c of radius 3 pixels centered on the back-projection of $\boldsymbol{f}_1$.
3. For each feature $\boldsymbol{f}_i$ $i = 2..k$ apply steps 4 and 5.
4. Back-project $\boldsymbol{f}_i$ in the coordinate system of $\boldsymbol{I}_r$ using $\mathtt{H}_i^{-1}$.
5. Discard $\boldsymbol{f}_i$ if its back-projection is outside the circle c.

Figure 12.4 shows an example of that process. This solution does not make any particular assumption about object rigidity but only exploits information of the keypoint local region. It allows therefore to apply the method also for matching keypoints of non-rigid objects. The approach differs from [15] where under the assumption of rigid objects the geometry checking is applied to all the pairs of keypoints (i.e. the existence of epipolar constraint and/or planarity relationship between 3D surface patches). Since only local information of the keypoint is used with no assumption on object rigidity, the method can be also applied to the case of non-rigid objects and computational requirements are drastically reduced.

12.3.3 Local Homography Estimation

For each sample keypoint $\boldsymbol{k}_q$ in $\boldsymbol{I}_q$ several valid keypoint correspondences with slightly different descriptors are found in the $\{\boldsymbol{C}_q\}$ cluster, with homographies that account for the differences in the viewing angle from which the object was observed. In particular we estimate the homography $\mathtt{H}_q$ from the patch around $\boldsymbol{k}_q$ to the reference image by proceeding as follows:

- The local image regions around valid corresponding keypoints $\boldsymbol{k}_i$ are aligned between them;
- The set of aligned homographies $\mathtt{H}_i$ and descriptors associated to these keypoints are exploited for the estimation;
- The estimation is validated by checking the right scale.

12.3.3.1 Region Alignment

Let's consider the set of valid corresponding keypoints that have been left in $\boldsymbol{C}_q$ after the removal of the wrong correspondences according to the geometry checking procedure reported in the previous section. For each keypoint $\boldsymbol{k}_i$ we calculate the shifts of scale (σ), orientation (θ) and position (u, v) with respect to the mean scale, orientation and position of the keypoints in $\boldsymbol{C}_q$ and define the similarity transformation $\mathtt{S}_i$:

$$\mathtt{S}_i = \begin{bmatrix} \sigma_i \cos\theta_i & -\sin\theta_i & u_i \\ \sin\theta_i & \sigma_i \cos\theta_i & v_i \\ 0 & 0 & 1 \end{bmatrix}. \tag{12.3}$$

The local regions around keypoints $\boldsymbol{k}_i$ can be hence aligned between them through the homographic transformation:

$$\mathtt{H'}_i = \mathtt{H}_i \; \mathtt{S}_i \, . \tag{12.4}$$

Figures 12.5 and 12.6 give evidence of the importance of this alignment step for the estimation of the homography. Both figures show a sample keypoint $\boldsymbol{k}_q$ and the valid corresponding keypoints left after projection and superimposed to the query image. The comparison of the two figures show that while in Fig. 12.6 the use of

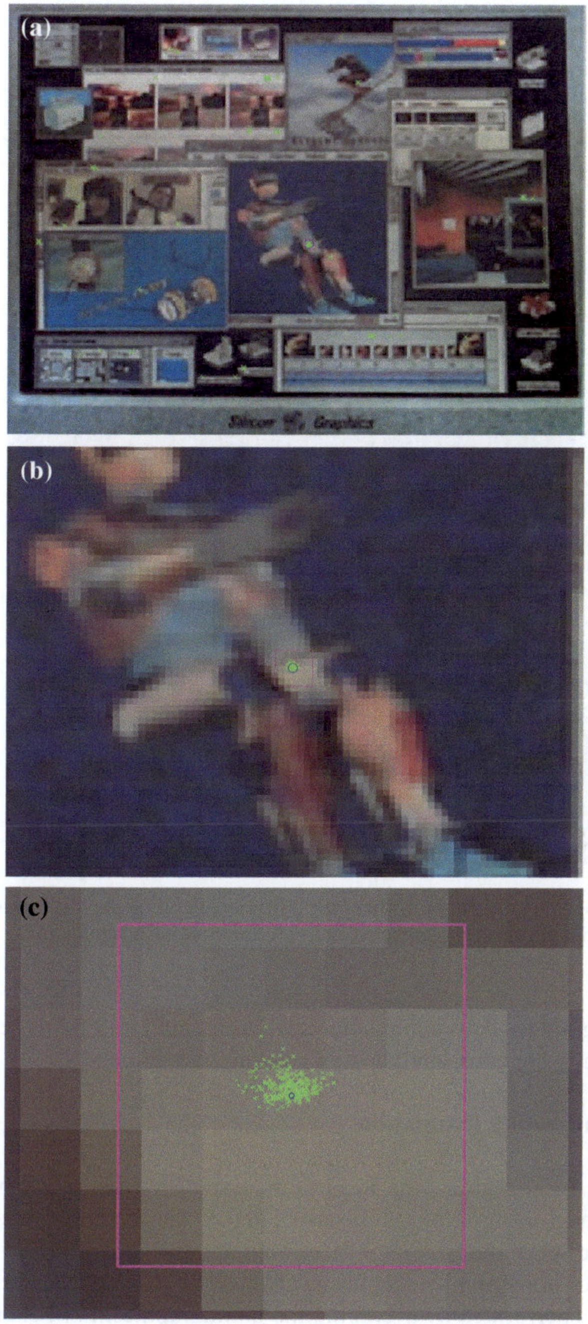

Fig. 12.4 Geometry checking process. **a**: The features belonging to cluster C_q before the application of the geometry checking. **b**: The features remaining in the cluster C_q before the application of the geometry checking **c**

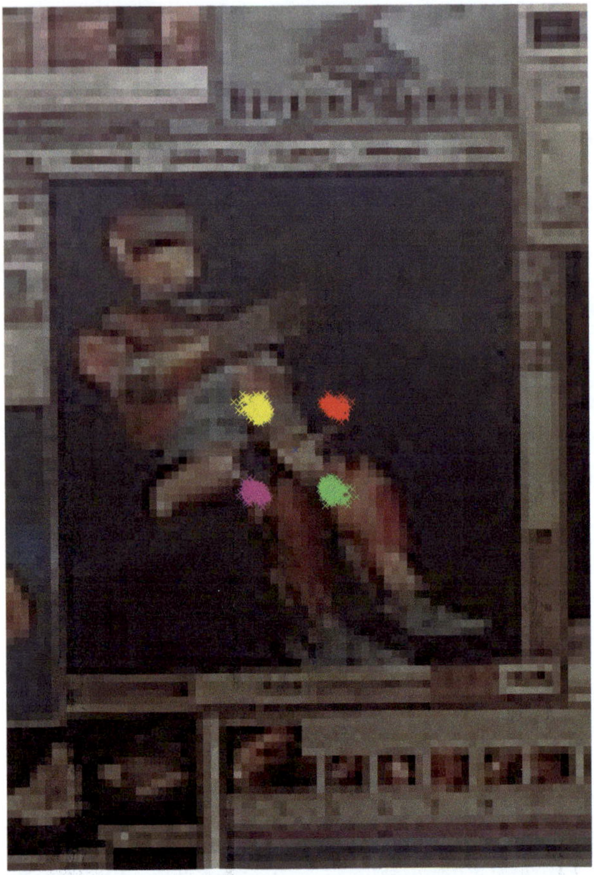

Fig. 12.5 Homography estimation without the alignment between the local regions around valid keypoints. Keypoints corners are projected and superimposed to the query image

aligned homographies determines some scattering, in Fig. 12.5 this scattering is lost and consequently are completely lost any perspective effects.

12.3.3.2 Homography Estimation

After aligning local regions, the local homography H_q of the patch around $\boldsymbol{k}_q$ is directly estimated using information associated to the features in the $\{\boldsymbol{C}_q\}$ cluster. Let $\{\boldsymbol{d}_i\}_{i=1}^n$ the set of descriptors representing appearance information and $\{\mathrm{H}_i\}_{i=1}^n$ the set of aligned homographies capturing geometry information about these features. The estimate is performed by averaging the homographies. For better accuracy, the contribution of each homography is weighed according to the distance between the relative descriptor and the descriptor $\boldsymbol{d}_q$:

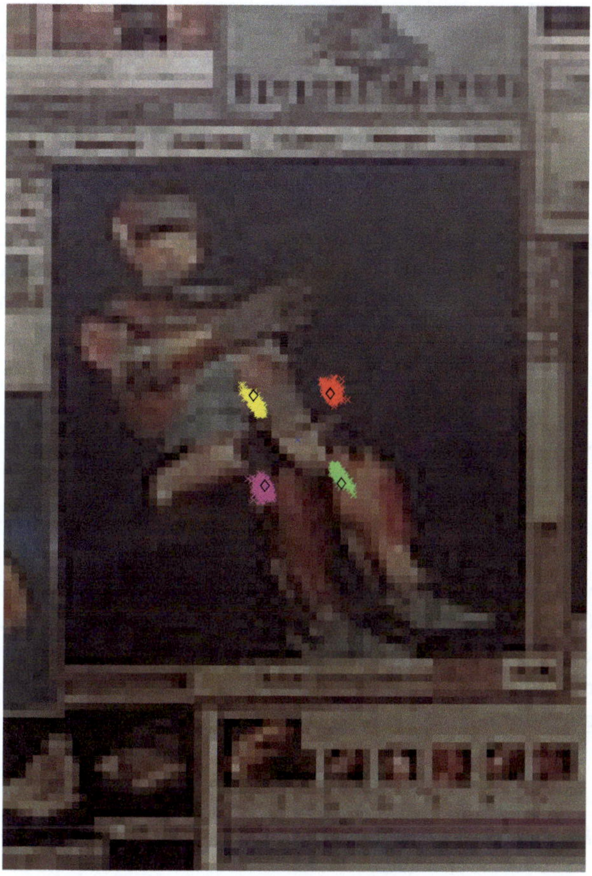

Fig. 12.6 Homography estimation with the alignment between the local regions around valid keypoints. Keypoints corners are projected and superimposed to the query image

$$\mathrm{H}_q = \frac{1}{n} \sum_{\boldsymbol{h}_i \in \boldsymbol{C}_q} \boldsymbol{w}_i \mathrm{H}_i \ , \tag{12.5}$$

where $\boldsymbol{w}_i = \|\boldsymbol{d}_q - \boldsymbol{d}_i\|_2$. Figures 12.1b, 12.1c, 12.2b and 12.2c show some applications of this estimation proccss.

12.3.3.3 Multiscale Validation

A final validation is needed to remove bad estimated keypoints. Thanks to the accuracy of the retrieved transformations, we are able to reject keypoints using the Summed Square Difference between the estimated patch and the warped patch in the reference image. We adopt a method similar to the one described in [9] in order to decide at which scale the reference patch should be warped to. In particular we apply warping using a matrix $\mathtt{A}$ computed from the Jacobian of the estimated homography evaluated at the keypoint:

$$\mathtt{A} = \begin{pmatrix} \frac{\partial \hat{\mathtt{H}}}{\partial x} & \frac{\partial \hat{\mathtt{H}}}{\partial y} \\ \frac{\partial \hat{\mathtt{H}}}{\partial x} & \frac{\partial \hat{\mathtt{H}}}{\partial y} \end{pmatrix}_{(x_0, y_0)} . \tag{12.6}$$

The determinant of matrix $\mathtt{A}$ corresponds to the area (in square pixel), that a single source pixel would occupy in the full-resolution image of the reference view. and the scale is chosen so that $det(\mathtt{A})/4^l$ is closest to unity. In our experiments we perform this SSD-based validation using a threshold of 0.9

12.4 Experimental Results

In the following we performed several experiments that asses the effectiveness of the method to recover the local homography of points belong to a given object. For all the experiments we build a large set of about 400,000 keypoints which correspond to a set of 800 3D points of interest detected in the reference image. In particular 220,000 keypoints are extracted by views at moderate foreshortening synthesized using the reference image, while the other keypoints are extracted by views at strong foreshortening synthesized using a set of already distorted images.

12.4.1 Robustness to Viewpoint Change

A set of experiments was run in order to assess the effectiveness of the method and compare it against specific affine region detectors. To this end, we generate synthetic views $\boldsymbol{I}_j$ with a factor of foreshortening ranging from 0 to 0.7. In this context, factor of foreshortening is a function of the homography $\mathtt{H}_j$ which transforms the vertices of the reference image in the new vertices: $K_{rj} = (\lambda_1^j \lambda_2^j - 1)$, where $\lambda_{1,2}^j$ are the two singular values of $\mathtt{H}_{rj}$ and represent the scaling values in two orthogonal directions. In particular K_{rj} is much greater then 0 as $\boldsymbol{I}_j$ is a more slanted version than the reference image under perspective transformation. For each view $\boldsymbol{I}_j$, we apply our method to identify approximately 50 keypoints and retrieve their homographies. We repeat this test 2,000 times for each cost and report the accuracy results in Fig. 12.7, in which our method is denoted by ‘SIFTHomography’. To create these graphs we proceed as follow. In the case of affine region detectors, we fit a square tangent to the normalized regions and warp this square back with the inverse transformation to get a quadrangle. In the case of our method, the quadrangle is simply taken to be the patch borders after warping the square on the reference image by the retrieved homography. In Fig. 12.7a we compare the average overlap between the quadrangles obtained using the ground truth homography and those obtained with our method and

Table 12.1 Average processing times for detecting 200 SIFT keypoints and estimating their homographies

Step	*Average time*
SIFT point extraction	0.025 s
Geometry consistency check	0.021 s
Kerypoint homography estimation	0.028 s

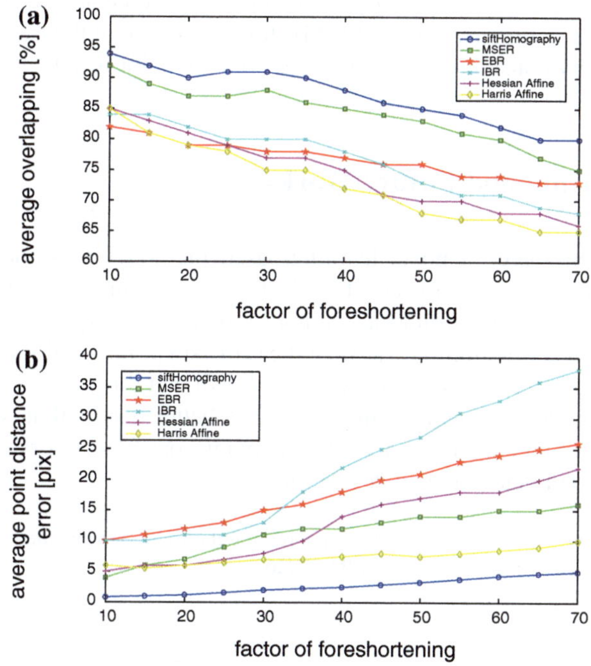

Fig. 12.7 Comparing our method against affine region detectors. **a** Average overlapping area of all correctly matched regions. **b** Average sum of the distances from the ground truth for the corner points. The factor of foreshortening in the abscissa is expressed in percentage values

with affine region detectors. This overlap is very close to 90 % for our method, about 5 % better than MSER and about 15 % than other affine region detectors. Figure 12.7b shows the comparison of the mean reprojection error for the quadrangle corners. The error of the patch corner is less than four pixel in average and outperforms other methods.

12.4.2 Computational Times and Memory Requirements

The present implementation runs at approx 15 frame per second with a keypoint set of 400.000 obtained from about 200 SIFT keypoints in the reference images, on a standard notebook with an Intel Centrino Core Duo with 2.4 GHz and 3 Gb RAM. The average processing time for the steps of the method are reported in Table 12.1. This compares favorably with both affine region detectors and other recently proposed state-of-the-art methods. Concerning the memory requirements, we observe that the method implementation uses about 300 MB for the storage of the training set, while the mean memory consumption required for each keypoint is only 121 Kb.

12.5 Conclusion and Future Works

This chapter introduced a novel learning-based method for estimating the local homography of a given 3D object. The effectiveness of our approach relies on two key ideas. First, the generation of a training set that captures geometry and appearance information about multiple views of the same keypoints, and second, the usage of this information for the estimate. We have shown that this process avoids specific estimation of the local transformation and gives better results than standard affine region detectors. Since we used only SIFT keypoints, our future work will investigate the use of different detectors and descriptors.

Acknowledgments This work is partially supported by the EU IST VidiVideo Project (Contract FP6-045547) and IM3I Project (Contract FP7-222267).

References

1. Beis JS, Lowe DG (1999) Indexing without invariants in 3d object recognition. IEEE Trans Pattern Anal Mach Intell 21(10):1000–1015
2. Benhimane S, Malis E (2007) Homography-based 2d visual tracking and servoing. Int J Rob Res 26(7):661–676
3. Ferrari V, Tuytelaars T, Gool L (2006) Simultaneous object recognition and segmentation from single or multiple model views. Int J Comput Vis 67(2):159–188
4. Goedeme T, Tuytelaars T, Gool LJV (2004) Fast wide baseline matching for visual navigation. In: CVPR (1). Washington, USA, pp 24–29
5. Hinterstoisser S, Benhimane S, Lepetit V, Navab N (2008) Simultaneous recognition and homography extraction of local patches with a simple linear classifier. In: British Machine Vision Conference
6. Hinterstoisser S, Benhimane S, Navab N, Fua P, Lepetit V (2008) Online learning of patch perspective rectification for efficient object detection. In: Conference on Computer Vision and Pattern Recognition
7. Hinterstoisser S, Kutter O, Navab N, Fua P, Lepetit V (2009) Real-time learning of accurate patch rectification. In: Conference on Computer Vision and Pattern Recognition

8. Jurie F, Dhome M (2002) Hyperplane approximation for template matching. IEEE Trans Pattern Anal Mach Intell 24(7):996–1000
9. Klein G, Murray D (2007) Parallel tracking and mapping for small AR workspaces. In: Proceedings of Sixth IEEE and ACM International Symposium on Mixed and Augmented Reality (ISMAR'07). Nara, Japan
10. Matas J, Chum O, Urban M, Pajdla T (2004) Robust wide baseline stereo from maximally stable extremal regions. In: Image and Vision Computing
11. Mikolajczyk K, Tuytelaars T, Schmid C, Zisserman A, Matas J, Schaffalitzky F, Kadir T, Gool LV (2005) A comparison of affine region detectors. Int J Comput Vis 65(1/2):43–72
12. Özuysal M, Fua P, Lepetit V (2007) Fast keypoint recognition in ten lines of code. In: Conference on Computer Vision and Pattern Recognition
13. Pagani A, Stricker D (2009) Learning local patch orientation with a cascade of sparse regressors. In: Proceedings of British Machine Vision Conference (BMVC 2009). London, UK
14. Philbin J, Chum O, Isard M, Sivic J, Zisserman A (2007) Object retrieval with large vocabularies and fast spatial matching. In: Conference on Computer Vision and Pattern Recognition
15. Rothganger F, Lazebnik S, Schmid C, Ponce J (2006) 3d object modeling and recognition using local affine-invariant image descriptors and multi-view spatial constraints. Int J Comput Vis 66(3):231–259

Chapter 13
A Cognitive Source Coding Scheme for Multiple Description 3DTV Transmission

Simone Milani and Giancarlo Calvagno

Abstract Multiple Description Coding has recently proved to be an effective solution for the robust transmission of 3D video sequences over unreliable channels. However, adapting the characteristics of the source coding strategy (Cognitive Source Coding) permits improving the quality of 3D visualization experienced by the end-user. This strategy has been successfully employed for standard video signals, but it can be applied to Multiple Description video coding for an effective transmission of 3D signals. The chapter presents a novel Cognitive Source Coding scheme that improves the performance of traditional Multiple Description Coding approaches by adaptively combining traditional predictive and Wyner-Ziv coders according to the characteristics of the video sequence and to the channel conditions. The approach is employed for video+depth 3D transmissions improving the average PSNR value up to 2.5 dB with respect to traditional MDC schemes.

Keywords Multiple description · 3DTV transmission · Distributed video coding · Cognitive source coding · DIBR video · Robust video coding

13.1 Introduction

According to the latest research trends and marketed products, the future of 3D video technology will not be limited to entertainment and gaming applications, as more and more telecommunication companies are looking to use the upcoming advancements in video technology to change the way people communicate. As a

S. Milani (✉) · G. Calvagno
Department of Information Engineering,
University of Padova, via Gradenigo 6/B, 35131 Padova, Italy
e-mail: simone.milani@dei.unipd.it

G. Calvagno
e-mail: calvagno@dei.unipd.it

N. Adami et al. (eds.), *Analysis, Retrieval and Delivery of Multimedia Content*, Lecture Notes in Electrical Engineering 158, DOI: 10.1007/978-1-4614-3831-1_13,

matter of fact, 3D video signals will be employed in most video communication systems, from IPTV broadcasts to immersive video conference and remote surveillance applications. However, the widespreading of these new 3D applications requires an adaptation of traditional coding and transmission strategies according to the characteristics of the signals [1]. In fact, the massive amount of 3D data, together with the strict Quality-of-Service (QoS) requirements that are proper of multimedia communications, make difficult to provide three-dimensional contents at a satisfying quality over heterogeneous networks. Protocols and coding architectures have to deal with independent sets of users operating with different devices and transmission capabilities. As a matter of fact, the format of the transmitted signal must permit an easy adaptation of the transmitted data to the characteristics of the 3D display and to the available transmission rate. Moreover, video data streams must prove to be robust in presence of data losses (due to channel noise, congestions, delays, etc...) enabling a fast recovery of the lost information and limiting the amount of channel distortion introduced in the displayed signal.

In order to effectively deal with these open problems, several solutions proposed in literature rely on a robust and flexible characterization of the data that permits both adapting the information stream to the network and mitigating the effects of losses. Some of the proposed approaches are based on scalable video coding architectures [2], which define several hierarchical streams that permit reconstructing the signal with a progressively-increasing quality. As a matter of fact, scalable solutions seem to fit the need of differentiating the transmitted video contents according the displaying device and the transmission capability. Dealing with the problem of robustness to packet losses, a significant research effort has been made on investigating effective Cross-Layer (CL) solutions that maximize the quality of the received video signal by allowing a synergic interaction between different protocol layers and varying the protection level of the coded data according to their significance in the decoding process [3]. In addition, new robust source coding strategies have been introduced like Distributed Video Coding (DVC), named also Wyner-Ziv video coding, which code the video signal such that a distortion-free reconstruction is still possible after data losses or corruption.

Each technique presents some advantages and disadvantages which make it more proper for certain transmission settings and applications. Among these, Multiple Description Coding (MDC) seems to conjugate both the need for an easily-adapting data stream and the need for a robust characterization of the transmitted signal. MDC techniques characterize the input signals via multiple independently-coded correlated streams. Each stream is transmitted to the end-users via separate channels. Whenever one stream gets lost, it is possible to estimate the missing information from the available data (i.e. the other streams correctly received). As a matter of fact, the more streams a user gets, the higher the quality of the reconstructed sequence is.

The characterization of the input signals via several streams permits a better differentiation of the transmitted stream and an increased robustness at high loss percentages (see [4]) with respect to Single Description Coding (SDC) solutions. This fact has contributed to its adoption in wireless 3DTV transmission schemes, like the one proposed in [5]. The performance of MDC schemes has been recently improved with the adoption of Wyner-Ziv coding paradigms in the traditional MDC structure [6]. These architectures have been called Multiple Description Distributed Video Coding (MDDVC) and increase the error-resiliency of the produced packet streams.

However, recent experimental results have shown that it is possible to improve the performance of traditional source coding paradigms by adaptively switching from one coding solution to another via a CL optimization. In this chapter we will refer to these solutions with the term Cognitive Source Coding (CSC) schemes in analogy to Cognitive Radio (CR) schemes [7] adopted for radio transmissions. CSC approaches can be considered a subset of CL solutions despite most of the CL solutions jointly adapt the transmission parameters at different layers but do not change the source coding strategy. A more careful analysis shows that CSC schemes present many features in common with CR solutions. As defined by Haykin in [8], "Cognitive radio is an intelligent wireless communication system that is aware of its surrounding environment (i.e., outside world), and uses the methodology of understanding-by-building to learn from the environment and adapt its internal states to statistical variations in the incoming RF stimuli by making corresponding changes in certain operating parameters (e.g., transmit-power, carrier-frequency, and modulation strategy) in real-time." CSC architectures implement many source coding strategies and adaptively switch from one to another depending on the channel state. In a similar way, CR systems implement many modulation schemes and can adaptively switch from one to another depending on which portion of the radio spectrum they want to use. Moreover, CSC schemes, as well as CR solutions, must sense the transmission environment in order to understand how many transmission channels are available and what their states are.

In this chapter we present a reconfigurable Multiple Description transmission scheme for 3DTV signals that adaptively switches from traditional predictive coding to Wyner-Ziv video coding according to the states of the transmission channels and the characteristics of the coded video signal. The proposed Cognitive Source Coding scheme is applied to a 3D video signal consisting in a video sequence and its related depth information, which permit a Depth Image Based Rendering (DIBR) of the 3D scene [9]. The proposed solution improves the quality of the reconstructed video sequence and depth maps allowing a much better 3D Quality-of-Experience (QoE). In the following, Sect. 13.2 overviews some of the solutions that have been proposed in literature for a robust 3D video transmission. Section 13.3 presents the structure of the coder, while Sect. 13.4 describes the cognitive optimization strategy. Experimental results are reported in Sect. 13.5, and conclusions are drawn in Sect. 13.6.

13.2 Related Works

During the last years, different works have been focusing on the reliable transmission of multimedia and 3D video data over unreliable networks. Several scalable coding solutions have been proposed in literature for stereoscopic signals, like in [10, 11]. Many approaches relies on a CL configuration of the transmission environment.

Among these it is possible to mention the solution proposed by Alregib et al. [12], which adopts a scalable compression for 3D models and applies an Unequal Error Protection (UEP) on the different layers in order to decrease the loss probability as the significance of the data in the decoding process increases. Another approach has been proposed by Balter et al. in [13], where the compression of the different signals is organized in order to maximize the quality of the final 3D scene rendered by the end user.

Other solution relies on characterizing the signal to be transmitted via multiple descriptions. One of the first examples is provided by the work [14], where a multiple description scheme is employed for stereoscopic video compression. Other solutions were proposed during the last years [11] employing scalable video coding solutions as well (see [5]).

A third set of strategies consider characterizing the 3D video signal using Wyner-Ziv coding solutions. One of this approaches can be found in [15] where the PRISM DVC coder [16] was adapted to the transmission of multiview video streams. Other works follow this strategy [17] since a distributed source coder that exploits the correlation existing between different camera views permits a successful decoding of the transmitted information from any side view.

At the same time, Distributed Video Coding paradigms have been applied to traditional MDC schemes. In this case, the correlation between different descriptions permits an error-free reconstruction of the coded data using any of the available description as reference. From the initial schemes in [18, 19], Multiple Description Distributed Video Coding (referenced in literature with the acronym MDDVC or MDVC) have so far evolved into a wide range of different approaches. Some of the proposed solutions separately generate a set of descriptions and replace the traditional predictive coding for their compression with a Wyner-Ziv coder (see [20]). Other solutions include the distributed source coding approach within the native MDC architecture, like the approaches in [21, 22]. Many MDDVC solutions rely on Wyner-Ziv coding strategies employing a feedback-channel [23] like most of the previous DVC video coders [24, 25]. Other solution rely on a PRISM-like characterization of the residual signal where no feedback information is needed [26, 27].

During the last years, MDDVC approaches have been applied to the transmission of 3D video signals. The approach in [28] presents an MDDVC strategy for the transmission of stereoscopic video sequences. Experimental results show that the employment of a Wyner-Ziv coding strategy permits mitigating the distortion propagation at high loss rates. These results have also shown that this strategy proves to be effective for geometry signals as well [29]. In fact, MDDVC significantly improves the transmission quality of DIBR video whenever the video signal presents strong correlation and the state of the network proves to be critical. From these preliminary results, it is possible to infer that performances can greatly benefit from an adaptation of the characteristics of the source coder to the features of the 3D data and of the network. As a matter of fact, within the existing cross-layer solutions, a subset of the proposed approaches (referenced here with the acronym CSC) adapt the chosen source coder to the characteristics of the transmitted video sequence and

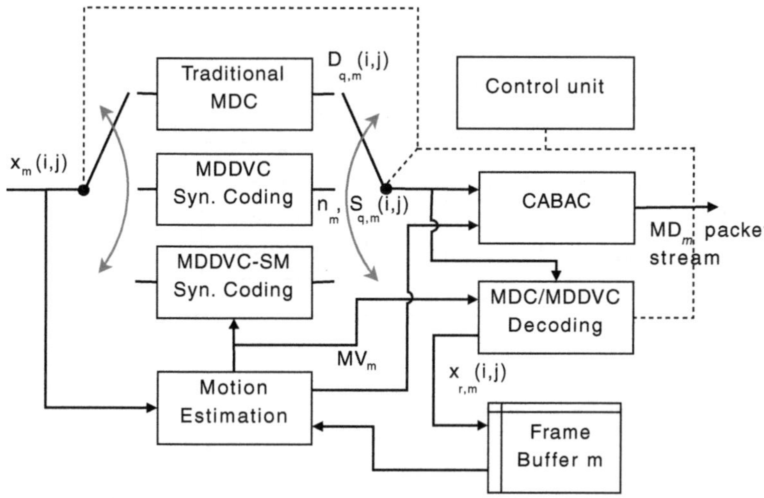

Fig. 13.1 Block diagram for the proposed encoder

to the network state [30]. In the following a Multiple Description CSC scheme will be presented showing how it can improve the quality of the 3D experience.

13.3 Structure of the Coding Scheme

The proposed multiple description strategy has been derived from a quite simple and general approach based on a polyphase subsampling of pixel rows in the input frames. The scheme was chosen since it is well-known in literature and the results can be extended to other MDC schemes as well. Each frame from both the input video and the depth signals is partitioned into odd and even lines of pixels creating two subsequences with halved vertical resolution. Each subsequence is coded independently by a CSC coder[1] as shown in Fig. 13.1, and the produced packets are multiplexed into two streams (i.e., two descriptions) associating the odd rows of the video sequence with the even rows of the depth signal and viceversa. Whenever two descriptions are received, the sequence can be reconstructed by the decoder shown in Fig. 13.2 assuming that there is no additional channel distortion. In case only one description is correctly received, it is possible to estimate the missing rows interpolating the available ones. Despite the error concealment strategy is the same, it is possible to obtain different performances in terms of source and channel distortions according to the chosen coding strategy for the residual signal after temporal prediction. As shown in Fig. 13.1, the coder of the CSC approach relies on processing the residual

[1] The basic coding engine was derived from a standard H.264/AVC codec.

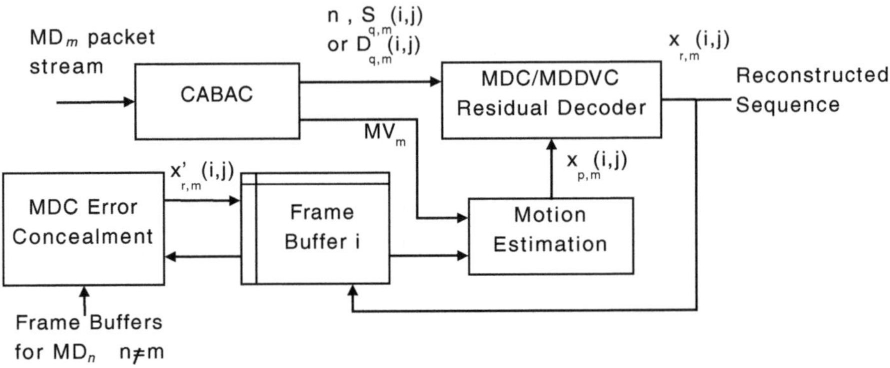

Fig. 13.2 Block diagram for the proposed decoder

signal after temporal prediction in different ways. In the following we will describe the possible alternatives in detail.

In the proposed scheme, each input field generated by the MDC subsampling (for both video and depth signals) is partitioned into blocks of 4×4 pixels. For each 4×4 block x_m of the description MDm ($m = 1, 2$), the Motion Estimation unit estimates a predictor block $x_{p,m}$ from the previously-coded fields.

From this point, different options can be chosen according to the state of the channel and the characteristics of the input signal.

13.3.1 Traditional Predictive Coding (MDC Mode)

In the traditional H.264/AVC coding, the prediction residual block $d_m = x_m - x_{p,m}$ is transformed by an approximated 4×4 DCT (see [31]), and the generated coefficients are quantized into a block of integer values that are coded in the output binary bit stream. In the reconstruction process, the decoded coefficients are dequantized and inversely transformed into the reconstructed prediction residual $d_{r,m}$, which differs from the original prediction error d_m for the additional distortion introduced by the quantization of transform coefficients, i.e. $d_{r,m} = d_m + e_{r,m}$. The coded field can be reconstructed adding $d_{r,m}$ to the predictor block $x_{p,m}$ and obtaining $x_{r,m} = d_{r,m} + x_{p,m}$. In case some packets get lost, an additional channel distortion has to be taken into account such that the reconstructed block is $x'_{r,m} = d_{r,m} + x_{p,m} + e_{c,m}$ where $e_{c,m}$ is the additional channel distortion produced by interpolation. The channel distortion then propagates until a complete refresh of the reference buffer state (i.e., the coding of an Intra frame) is performed.

13.3.2 Residual Coding using Syndromes (MDDVC Mode)

In order to mitigate the effects of the distortion propagation, it is possible to obtain a better error resiliency by adopting a more robust coding strategy for the prediction error after motion estimation. In the MDDVC coding setting, a distributed source coding strategy based on a Nested Scalar Quantization was adopted [32].

In fact, each pixel $x_m(i, j)$ of block x_m can be split into two components: a component correlated with its predictor $x_{p,m}(i, j)$ in $x_{p,m}$ (given by the most significant bits of the pixel value), and a component uncorrelated with $x_{p,m}(i, j)$ (given by the $n_m(i, j)$ least significant bits).

Like for the PRISM coder [33], the parameter $n_m(i, j)$ is computed via the equation

$$n_m(i, j) = \begin{cases} 0 & \text{if } \; d_m(i, j) < \delta, \\ 2 + \lfloor \log_2 (|d_m(i, j)|) \rfloor & \text{otherwise.} \end{cases} \tag{13.1}$$

The threshold δ is chosen in order to avoid the coding of syndromes whenever the $x_{p,m}(i, j)$ is close to $x_m(i, j)$ and prevent decoding errors (this latter motivation will be explained later). After analytical modelling of the decoding error and extensive coding trials on a wide set of sequences, the value δ has been set to $1/3$ of the quantization step Δ for the current block. The $n_m(i, j)$ least significant bits of $x_m(i, j)$ are referenced as

$$s_m(i, j) = x_m(i, j) \& 2^{n_m(i,j)-1}, \tag{13.2}$$

where & denotes a bitwise AND operator. The syndrome $s_m(i, j)$ permits identifying a quantizer characteristic which is centered on $s_m(i, j)$ with quantization step $2^{n_m(i,j)}$, and outputs the value $x_m(i, j)$ with respect to the side information $x_{p,m}(i, j)$. Figure 13.3 shows an example of the syndrome generation and decoding process for $x_m(i, j)$. In this case, the distance value $d_m(i, j) = 7$ leads the encoder to create a syndrome value $s_m(i, j) = 0100$ of 4 bits via Eqs. (13.1) and (13.2). Therefore, it is possible to recover the original $x_m(i, j)$ value quantizing $x_{p,m}(i, j)$ by means of the quantizer characteristic $Q_{s_m(i,j)} = Q_4$, which is centered on the value 0100 and has a quantization step value equal to $2^{n_m(i,j)}$. In the general case, the output values of the quantizer $Q_{s_m(i,j)}$ can be written as $s_m(i, j) + k \cdot 2^{n_m(i,j)}$, $k \in \mathbb{Z}$. Note that a correct decoding is possible using a different $x'_{p,m}(i, j) \neq x_{p,m}(i, j)$ such that the distance $d'_m(i, j) = x_m(i, j) - x'_{p,m}(i, j)$ via Eq. (13.1) leads to a number of syndrome bits $n'_m(i, j) \leq n_m(i, j)$.

The generation process for $s(i, j)$ corresponds to the syndrome generation strategy in [16] made exception for the facts that here it is performed in the pixel domain and the quantization step is 1 (i.e., no distortion has been introduced so far). After computing $n_m(i, j)$ for all the pixels in the block, the encoder chooses the maximum value

$$n_{\max,m} = \max_{i,j=0,\ldots,3} \{n_m(i, j)\} \tag{13.3}$$

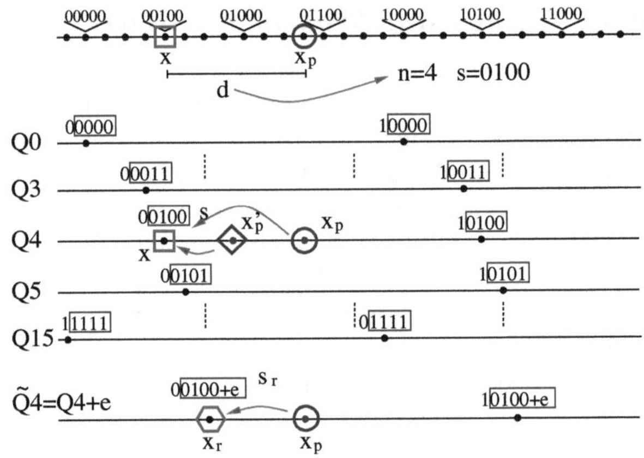

Fig. 13.3 Example of syndrome generation and decoding process (position indexes (i, j) and description index m are omitted for the sake of clarity). Assuming that $d = 7$ and $n = 4$, the syndrome $s = 0100$ identifies the quantizer $Q4$. After lossy encoding s into $s_r = 0100 + e$, the associated quantizer becomes $\tilde{Q}4$, and the reconstructed pixel is $s_r = s + e$ (under the assumption of additive quantization noise)

since it is possible to obtain a correct reconstruction of the pixel $x_m(i, j)$ with $n'_m(i, j)$ syndrome bits in case $n'_m(i, j) > n_m(i, j)$. In this way, the characterization of correlation level for the current block requires specifying only one parameter $n_{\max,m}$ and increases the robustness of the coded bit stream. The video encoder computes $s_m(i, j)$ via Eq. (13.2) replacing $n_m(i, j)$ with $n_{\max,m}$, and then the block of syndromes $\boldsymbol{s}_m$ is transformed into the block $\boldsymbol{S}_m$ via the 4×4 DCT defined within the standard H.264/AVC. The resulting coefficients $S_m(i, j)$ are then quantized into the values $S_{q,m}(i, j)$, which are coded in the bit stream together with the parameter $n_{\max,m}$ and the other coding parameters. The proposed architecture inherits the set of quantization steps defined within the standard H.264/AVC and identified by the quantization parameter QP.

After the block $\boldsymbol{S}_{q,m}$ has been created, the coded block needs to be reconstructed in order to be stored in the frame buffer. The coefficients $S_{q,m}(i, j)$ are dequantized and inversely-transformed into lossy syndromes

$$s_{r,m}(i, j) = s_m(i, j) + e_m(i, j), \tag{13.4}$$

where $e_m(i, j)$ is the distortion introduced by quantization. The quantizer characteristic related to $s_{r,m}(i, j)$ results slightly shifted with respect to $s_m(i, j)$, and as a consequence, the reconstruction $x_{r,m}(i, j)$ obtained from $x_{p,m}(i, j)$ using $s_{r,m}(i, j)$ in place of $s_m(i, j)$ differs from $x_m(i, j)$ in its least significant part, i.e. $x_{r,m}(i, j) = x_m(i, j) + e_m(i, j)$. As for the example in Fig. 13.3, the syndrome $s_{r,m}(i, j)$ identifies the quantizer characteristic $\tilde{Q}4$, which corresponds to the characteristic $Q4$ shifted by $e_m(i, j)$.

Note that the amount of shifting must be limited in order to avoid decoding errors. Setting the threshold δ to $\Delta/3$ permits satisfying the error-free decoding condition $\Delta < 5.6 \cdot 2^{n_{\max,m}}$ (which has been derived in Eq. (13.14) of the Appendix A). During the syndrome generation process (see Eq. (13.1)), the number $n_m(i, j)$ of syndrome bits is set to 0 if $|d_m(i, j)| < \delta = \Delta/3$, i.e. $\Delta > 3|d_m(i, j)| \simeq 1.12 \cdot 2^{n_m(i,j)}$, providing a sufficiently-robust margin for the non-zero syndromes since the error-free decoding condition $\Delta < 5.6 \cdot 2^{n_{\max,m}}$ (which has been derived in Eq. (13.14) of the Appendix) is satisfied. In fact, whenever $\Delta > 5.6 \cdot 2^{n_{\max,m}}$, using quantized syndromes in decoding does not bring a significant quality improvement with respect to the distortion related to the temporal correlation. As a matter of fact, it is possible to reconstruct the signal as if no additional residual information is provided after prediction.

In case $|d_m(i, j)| < \delta \; \forall (i, j)$ in the block, the correlation parameter $n_{\max,m}$ is reset to 0 and syndromes are not used at all for the decoding of the block; in the reconstruction process, the pixel $x_{r,m}(i, j)$ is set equal to $x_{p,m}(i, j)$.

Using a higher threshold δ permits reducing the amount of coded bits at the expense of a higher distortion level in the reconstructed residual signal. In the following paragraph, we will discuss how this choice can be useful for certain channel conditions.

13.3.3 Residual Coding Using Syndromes and Skip Mode (MDDVC-SM Mode)

The previous section has shown how decoding errors and bit rates can be controlled via the threshold δ. In fact, increasing the value δ makes possible to reduce the amount of coded syndromes $s_m(i, j)$ with $n_m(i, j) \neq 0$. As a result, the size of the coded bit stream is decreased since the coder avoids transmitting the residual signal for a greater number of 4×4 blocks.

At the same time, the probability of a wrong decoding is mitigated since a high packet loss probability induces a stronger channel distortion in the reconstructed video signal. As a matter of fact, stronger syndromes are needed since the correlation between the current pixel $x_m(i, j)$ and its predictor $x'_{p,m}$ (which is different from $x_{p,m}(i, j)$ because of corruption) is lower. A higher threshold $\delta' \gg \delta$ leads to zero syndromes whenever $\Delta > h \cdot 2^{n_{\max,m}}$ with $h \ll 5.6$ as Eq. (13.14) requires. As a consequence, only residual blocks with high $n_{\max,m}$ are coded minimizing the probability of decoding errors.

However, zero syndromes also reduce the quality of the reconstructed signal as, in case $n_{\max,m} = 0$, the predictor block $x_{p,m}$ simply replaces x_m without any motion compensation. As a matter of fact, the adoption of a higher threshold value must be considered according to the characteristics of the coded sequence and to the packet loss rates.

In the following, we will refer to this coding as MDDVC `Skip mode` or MDDVC-SM. In this residual coding unit, all the operations correspond to those described for the MDDVC mode made exception for an additional skipping strategy that avoids coding the current block $\boldsymbol{s}_{r,m}$ of syndromes (by signalling $n_m = 0$) whenever the average number of syndrome bits is lower than a threshold δ_n, i.e.

$$\overline{n}_m = \frac{1}{16} \sum_{i,j=0,\dots,3} n_m(i,j) < \delta_n. \tag{13.5}$$

13.4 Cognitive Source Coding of DIBR Video Sequences

Experimental results in Sect. 13.5 show that the effectiveness of MDC and MDDVC approaches varies according to the different network conditions and to the 3D video signal characteristics. As for the views captured by the camera, the performance of MDC with respect to the MDDVC depends on the temporal and spatial correlations of the coded signal. At low loss rates, signals with a strong temporal correlation can be effectively coded using MDDVC while video sequences that present fast-moving elements need to be coded effectively using a traditional MDC scheme. Whenever the loss rate increases (approximately for values of packet loss percentage higher than 10 %), MDDVC provides the best performance for all the coded video sequences. As for depth maps, their regular structures permit coding effectively the depth information with MDDVC most of the times. As a matter of fact, it is necessary an adaptive technique that chooses which coding solution is the best in each situation.

The first step of the presented CSC solution is the computation of a set of features that describe the current GOP of frames in terms of error resilience. In this computation, the capability of error concealment for MDC schemes has to be taken into consideration. As a matter of fact, spatial correlation must be parameterized numerically together with temporal correlation. This modelization permits estimating the quality of the reconstructed frames whenever even the MDC error concealment fails or performs poorly because of a low vertical correlation. Starting from these premises, the CSC algorithm computes for each frame of the current Group Of Picture (GOP) within the current video sequence the triplet of average gradients

$$g_v = [g_x, g_y, g_t], \tag{13.6}$$

where g_x and g_y are the average value of the vertical and horizontal Sobel operators for the current frame, and g_t is the average gradient computed between the pixels of the current frame and the corresponding pixels of the previous reference frame. The triplet g_v is then averaged for all the frames in the GOP computing

$$G_v = E[g_v]. \tag{13.7}$$

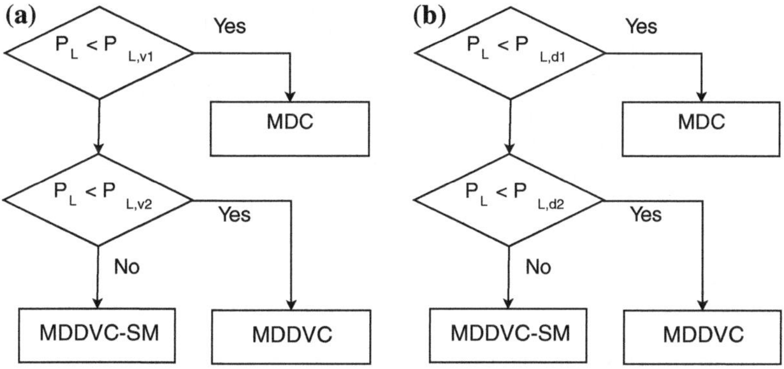

Fig. 13.4 Classification procedure to select the best coding mode. **a** Classifier for video signal. **b** Classifier for depth signal

The average descriptor G_v is then normalized using the weights w_v (which takes into consideration the different range of values and the different frame resolutions of the sequences) into the array $G'_v = w_v \circ G_v$ (where $\circ$ denotes an element-by-element multiplication) and classified via a K-means algorithm. Assuming that G'_v lies within the class c, the classification algorithm associates to the class c the loss probabilities $P_{L,v1}$ and $P_{L,v2}$. The packet loss probability $P_{L,v1}$ represents the probability threshold related to the current video sequence that separates the P_L values for which MDC is the best choice ($P_L < P_{L,v1}$) from those values for which MDDVC and MDDVC-SM are better. As for the threshold $P_{L,v2}$, it separates P_L values for which MDDVC performs better than MDDVC-SM ($P_L < P_{L,v2}$).

A similar descriptor G'_d is computed for the depth signal and classified using a different K-means classifier that outputs a different couples of loss probability thresholds ($P_{L,d1}$, $P_{L,d2}$).

As a matter of fact, the CSC algorithm computes the loss probability from RTCP control packets and identifies the best coding mode for the video signal and for the depth signal via the classification strategy represented by the binary trees in Fig. 13.4.

The whole algorithm can be summarized by the pseudo-code in the following page.

13.5 Experimental Results

The proposed CSC strategy has been tested on a wide set of video+depth 3D video sequences at different resolutions. Packet losses have been simulated using an independent Gilbert model with burst length $L_B = 4$ and varying loss probability P_{Li}

Algorithm 1: Pseudo-code for the proposed CSC algorithm

```
for each description MDi, i = 1, 2 do
  compute the packet loss rate P_Li for description MDi from RTCP packets;
  for the current GOP in the subsequence i do
    compute array G_v for the current GOP;
    classify G_v and finds out the thresholds P_L,v1 and P_L,v2;
    compute array G_d for the current GOP;
    classify G_d and finds out the thresholds P_L,d1 and P_L,d2;
    if P_Li < P_L,v1 then
      use MDC coder for the view signal;
    else
      if P_Li < P_L,v2 then
        use MDDVC coder for the view signal;
      else
        use MDDVC-SM coder for the view signal;
      end if
    end if.
    if P_Li < P_L,d1 then
      use MDC coder for the depth signal;
    else
      if P_Li < P_L,d2 then
        use MDDVC coder for the depth signal;
      else
        use MDDVC-SM coder for the depth signal;
      end if
    end if.
  end for
end for
```

for each independent channel CHi associated with description MDi. In our tests, we coded different sequences at different bit rates R_b with GOP structure IPPP, one slice per each row of macroblocks, and CABAC entropy coding. The adopted rate-distortion optimization strategy and the rate control algorithms for both the MDC and the MDDVC configurations are those defined within the JVT for the H.264/AVC coder. The set of training videos for the K-means algorithm includes the sequences `breakdancers`, `ballet` (format 1024×768) from Microsoft Research [34], the sequences `interview`, `orbi` (format 720×576), and `book arrival` (512×384 pixels) from FhG-HHI web-site [36]. As for the test sequences, we adopted the sequence `horse` and `car` (format 480×270) from the Mobile 3DTV project [35]. Since sequences present different formats, the values of G_v and G_d are equalized via the weights w_v and w_d according to the size of the frames. The plots in Figs. 13.5 and 13.6 show that the Cognitive approach permits improving the average PSNR values of the reconstructed views and reducing the average MSE of the received depth information. In our tests, the quality of the reconstructed video sequence has been evaluated using the PSNR and the SSIM quality metric, while the accuracy of the reconstructed depth map has been measured using MSE (since depth information is related to the geometry of the objects in the scene). In order to test the joint effects

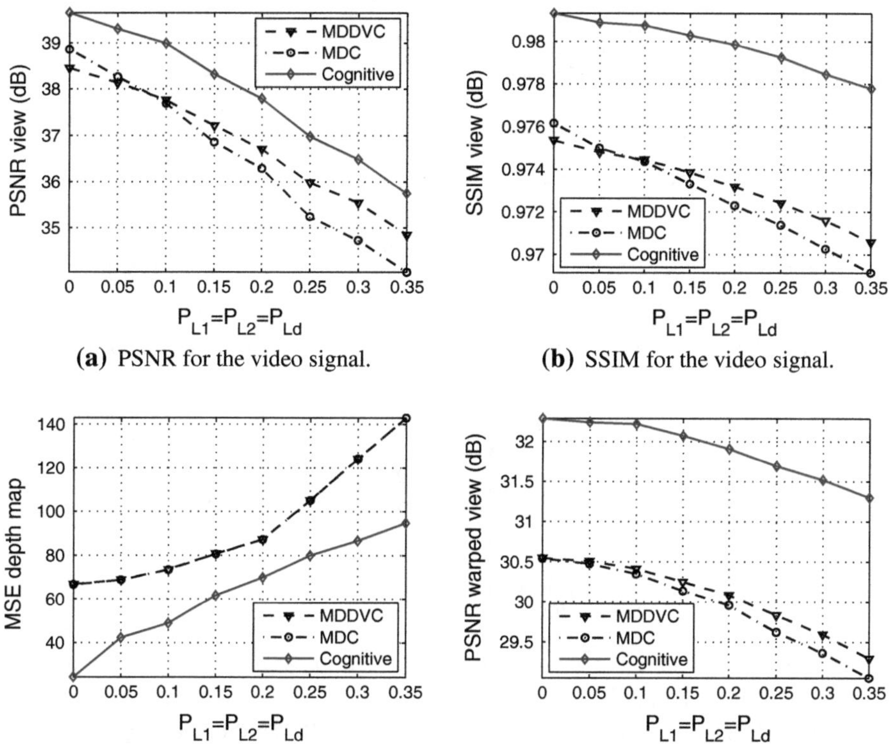

(a) PSNR for the video signal.

(b) SSIM for the video signal.

(c) MSE for the depth signal.

(d) PSNR of the warped view.

Fig. 13.5 Average PSNR, MSE, and SSIM metrics for the sequence `ballet` versus $P_{Lm} = P_{L1} = P_{L2}$

of the scheme on the final QoE level, we have also computed the PSNR between the warped view generated from the original video and depth signals and the warped view generated from their reconstructed versions. In this test, we generated a lateral right view with a principal axis shifted by 6 cm (required for a stereo visualization). The results show that the Cognitive approach is able to improve the PSNR value of the reconstructed view up to 2.5 dB (see the results in Fig. 13.6a for $P_{Li} = 0.2$). This improvement can also be verified considering the other metrics. As for the complexity increment, the only additional operations concern thc classification since the complexities of the MDC and the MDDVC schemes are approximately the same. However, the computation of gradients and their clustering into a set of classes does not require a great amount of calculation with respect to the computational load required by the coding operations.

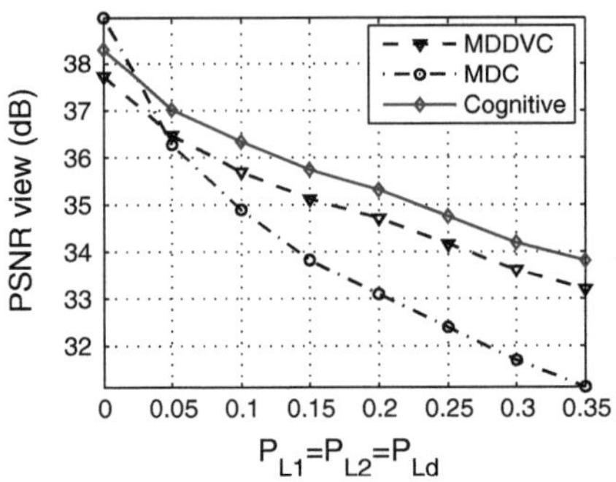

(a) PSNR for the video signal.

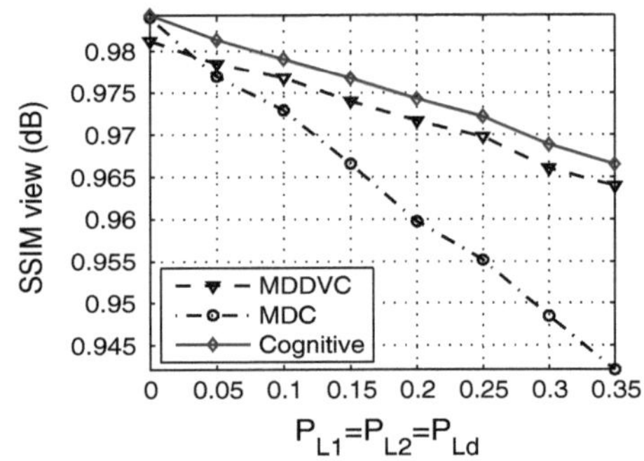

(b) SSIM for the video signal.

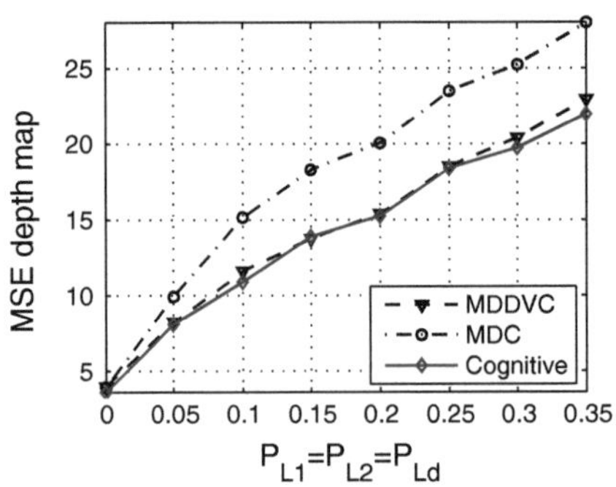

(c) MSE for the depth signal.

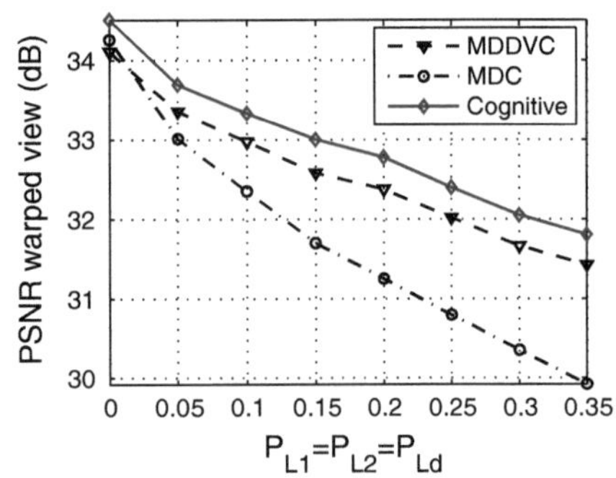

(d) PSNR of the warped view.

Fig. 13.6 Average PSNR, MSE, and SSIM metrics for the sequence `car` versus $P_{Lm} = P_{L1} = P_{L2}$

13.6 Conclusions

The chapter presented a Cognitive Source Coding scheme that adaptively chooses the most appropriate source coding strategy for the current video and depth signals according to the network conditions. The proposed solution increases the quality of the reconstructed sequence and of the received depth information improving the overall 3D Quality-of-Experience for a wide set of test sequences. Future work will be focused on including other robust video coding schemes (like single description video coding protected with additional FEC packets) in order to design a more flexible architecture.

Acknowledgments This work was partially supported by the PRIN 2008 project prot. 2008C59JNA founded by the Italian Ministry of University and Research (MIUR).

Appendix

Given the original pixel $x_m(i, j) = s_m(i, j) + k\ 2^{n_{\max,m}}$ ($k \in \mathbb{Z}$) and its reconstructed value $x_{r,m}(i, j) = s_m(i, j) + e_m(i, j) + k\ 2^{n_{\max,m}}$ after the quantization of the transformed syndromes, a wrongly-decoded pixel $x'_{r,m}(i, j)$ can be written as

$$\begin{aligned} x'_{r,m}(i, j) &= s_m(i, j) + e_m(i, j) + k' 2^{n_{\max,m}} \\ &= x_m(i, j) + e_m(i, j) + d_k\ 2^{n_{\max,m}}, \end{aligned} \tag{13.8}$$

with $k' \in \mathbb{Z}$, $k' \neq k$, and $d_k = k' - k$. In the following, we will omit pixel position indexes (i, j) and description index m for the sake of conciseness.

The probability of a wrong decoding is

$$P_W = P\left[x'_r \neq x_r\right] = P\left[|x_p - x'_r| < |x_p - x_r|\right], \tag{13.9}$$

which can be written as

$$P_W = P\left[| - d - e - d_k 2^{n_{\max}}| < | - d - e|\right], \tag{13.10}$$

where d is the difference between the current pixel and its predictor. Given d and d_k, the probability P_W becomes

$$P_W = \begin{cases} P[e \leq -d - d_k\ 2^{n_{\max}-1}] & \text{if} \quad d_k > 0, \\ P[e > -d - d_k\ 2^{n_{\max}-1}] & \text{if} \quad d_k \leq 0. \end{cases} \tag{13.11}$$

The error e is modelled with a normal distribution $\mathcal{N}(0, \sigma_{e,q})$ with mean 0 and variance $\sigma^2_{e,q} \simeq A^2 \Delta^2/12$ (where A is a scaling factor related to the adopted inverse transform since quantization is perform on the coefficients $\boldsymbol{S}$). The choice of a normal distribution is motivated by the fact that e is a linear combination of independent quantization errors generated in the transform domain and inversely-transformed. From Eq. (13.1) it is possible to infer that $2^{n_{\max}-2} \leq |d| < 2^{n_{\max}-1}$, and therefore, the probability of a wrong decoding becomes

$$P_W = Q\left(\frac{|d + d_k\ 2^{n_{\max}-1}|}{\sigma_{e,q}}\right). \tag{13.12}$$

From Eq. (13.12) it is possible to write the inequalities

$$\begin{aligned} P_W &\leq Q\left(\frac{2^{n_{\max}-2} + |d_k|\ 2^{n_{\max}-1}}{\sigma_{e,q}}\right) \\ &\leq Q\left(\frac{1.5\ 2^{n_{\max}-1}}{\sigma_{e,q}}\right) = Q\left(\frac{2.6\ 2^{n_{\max}}}{A\ \Delta}\right). \end{aligned} \tag{13.13}$$

As a matter of fact,

$$Q\left(\frac{2.6\ 2^{n_{\max}}}{A\ \Delta}\right) \leq 0.017 \quad \Rightarrow \quad \Delta < \frac{1.24 \cdot 2^{n_{\max}}}{A} = 5.6 \cdot 2^{n_{\max}} \qquad (13.14)$$

where A is assumed to be approximately equal to $1/4$ for the inverse 4×4 transform defined in the standard H.264/AVC (considering both transform amplification and rescalings).

References

1. Shi S, Jeon W, Nahrsted K, Campbell R (2009) M-TEEVE: Real-time 3D video interaction and broadcasting framework for mobile devices. In: Proceedings of the 2nd international conference on immersive telecommunications (IMMERSCOM '09), Berkeley, 2009
2. Schwarz H, Marpe D, Wiegand T (2007) Overview of the scalable video coding extension of the h.264/avc standard. IEEE Trans Circuits Syst Video Technol 17:1103–1120
3. Katsaggelos AK, Eisenberg Y, Zhai F, Berry R, Pappas TN (2005) Advances in efficient resource allocation for packet-based real-time video transmission. Proc IEEE 93:135–147
4. Milani S, Calvagno G, Bernardini R, Zontone, P (2008) Cross-layer joint optimization of FEC channel codes and multiple description coding for video delivery over IEEE 802.11e Links. In: Proceedings of the IEEE FMN (2008) Cardiff, Wales, September 2008. pp 472–478
5. Karim HA, Hewage CTER, Worral S, Kondoz AM (2008) Scalable multiple description video coding for stereoscopic 3D. IEEE Trans Consumer Electron 54:745–752
6. Crave O, Guillemot C, Pesquet-Popescu B, Tillier C (2007) Robust video transmission based on distributed multiple description coding. In: Proceedings of the EUSIPCO, Poznan, 2007. pp 1432–1436
7. Mitola J, Maguire GQ Jr (1999) Cognitive radio: making software radios more personal. IEEE Personal Commun Mag 6:13–18
8. Haykin S (2005) Cognitive radio: brain-empowered wireless communications. IEEE J Sel Areas Commun 23:201–220 (Invited)
9. Fehn C (2004) 3D-TV using depth-image-based rendering (DIBR). In: Proceedings of the PCS, San Francisco, December 2004
10. Aksay A, Bilen C, Kurutepe E, Ozcelebi T, Akar GB, Civanlar R, Tekalp M (2006) Temporal and spatial scaling for stereoscopic video compression. In: Proceedings of the 14th european signal processing conference (EUSIPCO 2006), Florence, September 2006
11. Karim HA, Hewage CTER, Yu AC, Worral S, Dogan S, Kondoz AM (2007) Scalable multiple description 3D video coding based on even and odd frame. In: Proceedings of the picture coding symposium, Lisbon, November 2007
12. Alregib G, Altunbasak Y, Rossignac J (2005) Error-resilient transmission of 3D models. ACM Trans Graph 24:182–208
13. Balter R, Gioia P, Morin L (2006) Scalable and efficient coding using 3D modeling. IEEE Trans Multimedia 8:1147–1155
14. Norkin A, Aksay A, Bilen C, Akar GB, Gotchev A, Astola J (2006) Schemes for multiple escription coding of stereoscopic 3D. Lecture notes in computer science, vol 4105. Springer, Heildelberg, pp 730–737
15. Yeo C, Ramchandran K (2007) Robust distributed multiview video compression for wireless camera networks. In: Proceedings of VCIP, San Jose, 2007. vol 6508, pp 65080P-1–65080P-9
16. Puri R, Ramchandran K (2002) PRISM: A new robust video coding architecture based on distributed compression principles. In: Proceedings of the 40th Allerton conference on communication, control and computing, Allerton, October 2002. pp 402–408
17. Adikari ABB, Fernando WAC, Weerakkody WARJ, Kondoz A, Martínez JL, Cuenca P (2008) DVC based stereoscopic video transmission in a mobile communication system. In:

Proceedings of the (2008) IEEE international conference on future multimedia networks (FMN 2008) (co-located with NGMAST2008), Cardiff, Wales, 2008. pp 439–443
18. Jagmohan A, Ahuja N (2003) Wyner-Ziv encoded predictive multiple descriptions. In: Proceedings of the data compression conference (DCC 2003) Snowbird, 2003. pp 213–222
19. Wu M, Vetro A, Chen CW (2004) Multiple description image coding with distributed source coding and side information. In: Proceedings of SPIE multimedia systems and applications VII, Philadelphia, October 2004. vol 5600, pp 120–127
20. Wang J, Wu X, Yu S, Sun, J (2006) Multiple descriptions in the Wyner-Ziv setting. In: Proceedings of the IEEE international symposium on information theory (ISIT 2006), Seattle, July 2006. pp 1584–1588
21. Fan Y, Wang J, Sun J, Wang P, Yu S (2003) A novel multiple description video codec based on Slepian-Wolf coding. In: Proceedings of the data compression conference (DCC 2008), Snowbird, 2003. p 515
22. Wang A, Zhao Y, Bai H (2009) Robust multiple description distributed video coding using optimized zero-padding. Sci China Ser F Inf Sci 52:206–214
23. Crave O, Guillemot C, Pesquet-Popescu B, Tillier C (2008) Multiple description source coding with side information. In: Proceedings of the 16th european signal processing conference (EUSIPCO 2008), Lausanne, 2008.
24. Aaron A, Zhang R, Girod, B.: Wyner-Ziv coding for motion video. In: Proceedings of asilomar conference on signals, systems and computers, Pacific Grove, 2002. vol 1, pp 240–244
25. Artigas X, Ascenso J, Dalai M, Klomp S, Kubasov D, Ouaret M (2007) The DISCOVER codec: architecture, techniques and evaluation. In: Proceedings of the 26th picture coding symposium (PCS 2007), Lisbon, 2007
26. Milani S, Calvagno G (2009) A distributed video coding approach for multiple description video transmission over lossy channels. In: 17th european signal processing conference 2009, pp 1824–1828. Glasgow, Scotland (2009)
27. Milani S, Calvagno G (2010) Multiple description distributed video coding using redundant Slices and Lossy syndromes. IEEE Signal Process Lett 17:51–54
28. Milani S, Calvagno G (2009) A distributed video coding approach for multiple escription video coding of stereo sequences. In: Proceedings of the 2009 GTTI annual meeting, Parma, 2009
29. Milani S, Calvagno G (2010) A cognitive source coding scheme for multiple description 3DTV transmission. In: Proceedings of the 11th international workshop on image analysis for multimedia interactive services (WIAMIS 2010), pp 581–590. Desenzano del Garda, Brescia (2010)
30. Milani S, Calvagno G (2010) A cognitive approach for effective coding and transmission of 3D video. In: Proceedings ACM multimedia 2010, Florence, 2010
31. Wiegand T (2004) Version 3 of H.264/AVC. In: Joint Video Team (JVT) of ISO/IEC MPEG & ITU-T VCEG (ISO/IEC JTC1/SC29/WG11 and ITU-T SG16 Q.6), 12th Meeting, Redmond, 2004
32. Sheng F, Li-Wei Z, Ling H (2007) An adaptive nested scalar quantization scheme for distributed video coding. In: Proceedings of the IEEE workshop on signal processing systems (SiPS 2007), Shanghai, 2007. pp 351–356
33. Puri R, Majumdar A, Ramchandran K (2007) PRISM: a video coding paradigm with motion estimation at the decoder. IEEE Trans Image Process 16:2436–2448
34. Kang SB (MSR 3D Video download) http://research.microsoft.com/en-us/um/people/sbkang/3dvideodownload
35. Mobite3DTV Project: (Repository of Mobile3DTV project: 3D Video database) http://sp.cs.tut.fi/mobile3dtv/stereo-video/
36. Smolic A (Repository FhG-HHI on 3DTV Network of Excellence Web Page) https://www.3dtv-research.org/3dav/3DAV_Demos/FHG_HHI/

Part V
Multimedia Delivery

Chapter 14
An Efficient Prefetching Strategy for Remote Browsing of JPEG 2000 Image Sequences

Juan Pablo García Ortiz, Vicente González Ruiz, Inmaculada García, Daniel Müller and George Dimitoglou

Abstract This chapter proposes an efficient prefetching strategy for interactive remote browsing of sequences of high resolution JPEG 2000 images. As a result of the inherent latency of client-server communication, the experiments of this study prove that a significant benefit, can be achieved, in terms of both quality and responsiveness, by anticipating certain data from the rest of the sequence while an image is being explored. In this work a model based on the quality progression of the image is proposed in order to estimate which percentage of the bandwidth will be dedicated to prefetching. This solution can be easily implemented on top of any existing remote browsing architecture.

Keywords JPEG 2000 · Remote browsing · Image sequences · JHelioviewer · Prefetching

14.1 Introduction

Some of the powerful features offered by the new JPEG 2000 standard [8] are very efficient lossless/lossy compression, random access to the compressed data streams, incremental decoding and high scalability. These characteristics make JPEG 2000 a state-of-the-art solution for remote browsing of high-resolution images. Using the JPIP protocol, defined in Part 9 [9] of the JPEG 2000 standard, clients can

J. P. G. Ortiz · V. G. Ruiz (✉) · I. García
Computer Architecture and Electronics Department,
University of Almería, 04120 Almería, Spain
e-mail: vicente.gonzalez.ruiz@gmail.com

D. Müller
European Space Agency ESTEC, Noordwijk, Netherlands

G. Dimitoglou
Dept. of Computer Science, Hood College, Frederick, MD 21701, USA

N. Adami et al. (eds.), *Analysis, Retrieval and Delivery of Multimedia Content*,
Lecture Notes in Electrical Engineering 158, DOI: 10.1007/978-1-4614-3831-1_14,

interactively explore remote image data by specifying a window of interest (WOI). This data exchange uses the available bandwidth efficiently and does not require any recoding or additional processes. The server extracts <only the required data from the images and transmits it to the clients.

JPEG 2000 has already been successfully used in many scientific areas; e.g., in tele-microscopy [23] or tele-medicine [11]. A promising application in space sciences is the JHelioviewer project [16], developed by the European Space Agency (ESA) in collaboration with the National Aeronautics and Space Administration (NASA). Its main goal is to provide an interactive data browsing, visualization and access platform to accommodate the staggering data volume of 1.4 TB of images per day that are returned by the Solar Dynamics Observatory [18]. Among other data products, SDO is providing full-disk images of the Sun taken every 12 s in ten different ultraviolet spectral bands with a resolution of 4096×4096 pixels. As of today, the combination of JPIP and JPEG 2000 seems to offer the best solution in order to efficiently browse image data sets of this magnitude.

The basic functionality of JHelioviewer allows users to explore the available data for a given point in time. A interesting extension of this functionality is to enable users to move smoothly through a sequence of time-coded solar images given a specific time range. This type of functionality could also prove to be very valuable in other domains such as tele-medicine and tele-microscopy.

However, viewing JPEG 2000 image sequences is both computationally and bandwidth-intensive and often compromises the quality of the viewing experience. This compromise manifests itself to the user as lack of responsiveness, i.e. choppy image rendering. To address these challenges, we propose a special prefetching strategy that enables users to view image sequences with smooth transitions and without experiencing any penalties in responsiveness or quality gaps.

This chapter is organized as follows: Sect. 14.2 gives a brief synopsis of related work, while Sect. 14.3 provides a detailed analysis of the problems related to remote browsing of JPEG 2000 image sequences. Section 14.4 is dedicated to explaining the proposed solution, which is then evaluated in Sect. 14.5. Section 14.6 concludes the chapter and discusses ideas for future work.

14.2 Related Work

The access and distribution efficiency of large image files over networks has been an active research topic for a long time, in particular because images account for a considerable fraction of the total network traffic. Caching has been historically recognized as one of the most promising techniques to reduce bandwidth usage, server load and to improve performance, and various caching and prefetching schemes and algorithms have been proposed to reduce network traffic and minimize access delays [1, 2, 7, 17]. More recently, image-specific caching techniques have been proposed to take advantage of the memory and processing capabilities of modern client systems and to expedite image retrieval [21, 22, 24]. The JPEG 2000 standard has introduced new capabilities that can also be leveraged to augment some of the existing caching techniques.

Extensive work has been done in the area of accessing JPEG 2000 images over HTTP to improve the user's interactive browsing experience [6]. One approach used a dynamic traffic regulating mechanism [14] and another one employed a virtual media protocol to prioritize the compressed bit stream of the region of interest (ROI) [13]. Other approaches have focused on prefetching techniques using prediction-based server scheduling and cache management algorithms [15] or quad-tree-based indexing techniques, which take advantage of the space-frequency localization property of wavelet transforms and support subregion access [19, 20].

This notion of subregion access for streaming a WOI seems promising in enabling efficient, demand-driven browsing, which allows clients to quickly access regions of interest from voluminous images. Still, given the inherent client/server exchange latency, the browsing experience of JPEG 2000 images can be further improved by early fetching of future WOI data. A number of different approaches have been proposed, such as introducing a new type of a prioritizing scheme that enables transmission of volume regions by their scene content [12]. Others have included the use of heuristic mechanisms that improve browsing responsiveness [15], using a formal rate-distortion (RD) framework [5], and taking advantage of a user navigation model to manage the client cache and to prefetch data [3].

When it comes to prefetching strategies for remote browsing of JPEG 2000 images, the work by Descampe et al. [4] provides a comprehensive view of the state of the art along with their own proposed solution. In their work, the authors propose and evaluate several solutions to anticipate future WOIs in order to achieve a better responsiveness. However, all the solutions only take into consideration user exploration of single large-scale images, discarding the dynamic nature of navigation along an image sequence.

For the case studied in this chapter, the resolution of the images is not as high as that used by Descampe et al. as it is more important to improve the responsiveness to user movements along an image sequence as opposed to browsing just one single image. Moreover, the solutions proposed by [4] require special scheduling of JPEG 2000 packets, which is rather difficult to implement given the existing standard.

To achieve an acceptable level of responsiveness and avoid disturbing quality gaps while navigating through an image sequence, a special prefetching procedure is required. In the following sections, we present an efficient prefetching strategy for remote browsing of JPEG 2000 image sequences that offers good performance and smooth transitions, as well as an easy implementation.

14.3 Problem Description

Client/server JPIP communication is based on the exchange of requests and responses. Within each request, clients specify, among other parameters, the remote file to explore and the WOI to be shown. Files may contain a sequence of N different images (in the case of JHelioviewer, they are related to a specific time range), so requests must also include the desired range of images $[a, b]$, with $0 \le a \le b \le N-1$. Without any additional user interaction, the same WOI is retrieved from all the images within the time range.

There are two kind of possible JPIP requests: stateless and session-oriented ones. Stateless requests are independent of each other, and no state is recorded during the exchange of messages between clients and servers. In the case of the session-oriented requests, all the requests related to the same remote image file are associated with the same session. This allows the server to remember which image parts have been sent to a client, thus avoiding redundant transmissions, e.g. for overlapping WOIs. Most of the remote browsing applications use this communication type.

JPIP servers assume by default that clients always retain all of the data received within a session. If a client needs to remove some of the data, for example to free up memory, the server should be notified. This study does not deal with client resource restrictions, and we therefore assume that there are none.

Session-oriented communications allow clients to control the data flow. For a certain WOI, a client can specify the maximum length L desired for the server response in his request and then retrieve the data of the WOI in increments of L by simply repeating the same request several times. In the case of image sequences, the response data should be uniformly distributed by the server over the requested time range.

For each request, the value L may be adapted depending on specific requirements. The most common scheme balances the usage of the available bandwidth and the response time to WOI changes by the user. Notice that for large values of L, the data of a new WOI takes a long time to be received, especially at low bandwidths, after receiving that of the previous WOI. This generates long response times for the user interaction. On the contrary, a faster response time is achieved when a smaller value of L is used, even though additional overhead is generated due to the increase in the number of transmitted HTTP headers.

Some client applications, like kdu_show [10], adapt the L value according to the relation between the round-trip time (RTT)[1] and the time T taken to extract the message data from the communication link, for each server response. Then, L is modified with the aim of equaling RTT$/T$ to a certain target ratio (L is decreased if the ratio is higher and increased if it is lower), just after retrieving a server response and before performing the next request. This method is implemented by the clients in the solution proposed in this chapter.

This work focuses on JPIP applications, such as JHelioviewer, designed to explore remote image sequences. Figure 14.1 shows an example of five sequential times during a remote browsing session using JHelioviewer. Once the user has selected a time range, the server builds a virtual JPEG 2000 file which only contains links to those images whose time stamp belongs to that time range. The client starts a JPIP session for that file and requests the first image, displayed at time t_0. The user can

[1] The round-trip time is the elapsed time between the instant when the client generates the request and the instant when the first bit of the reply arrives to the client.

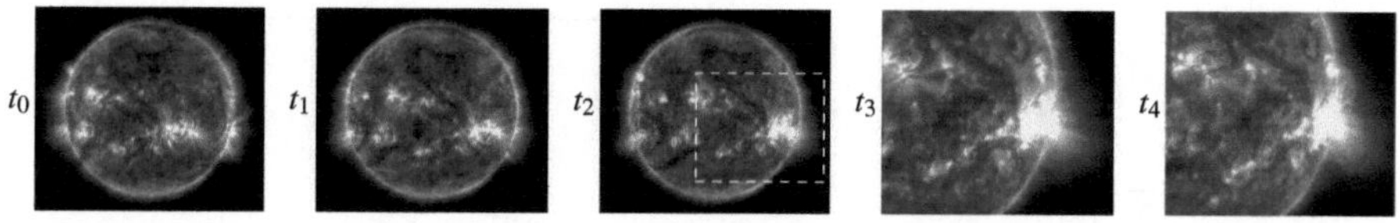

Fig. 14.1 An example of different points in time of a remote browsing session. In the beginning, the user specifies a data source and a time range, and the first image is presented at time t_0. At time t_1, where $t_i > t_{i-1}$, the image produced by the Sun has changed, due mainly to its rotation. At time t_2 the user specifies a window of interest (WOI). t_3 and t_4 are times during the remainder of remote visualization session where only the selected WOI is displayed and transmitted

then watch the sequence of images belonging to the time range one by one. In this example, at time t_2 the user zooms in on a certain region, thus changing the current WOI. From this new WOI, the user continues moving forward through the sequence, to times t_3 and t_4.

This type of interactive browsing requires a new communication scheme capable of offering smooth transitions while maintaining good responsiveness, and designed to be implemented over session-oriented JPIP communications. An added value is that this scheme is easy to implement on the client-side by simply combining the parameters L and $[a, b]$ of every request, and does not require any server or protocol modifications.

14.4 The Proposed Prefetching Strategy

The method presented here assumes that the images have been encoded using a suitable collection of encoding parameters. These parameters should allow spatial and quality scalability, and the definition of WOIs without any transcoding on the server side. It is also assumed that, for every WOI request, the JPIP server delivers the associated data minimizing the distortion of the displayed imagery.

As described in the previous section, a typical JHelioviewer user will spend some time displaying a concrete WOI (that could be the entire image) of a given single image and some time reproducing the entire image sequence, and frequently repeat these steps several times for the same image sequence (see Fig. 14.1), pausing the movie mode at any image.

In the simplest transmission strategy (without prefetching), the data requested by the client belongs only to the WOI of currently displayed image. Therefore, the quality of a given WOI of each of the images of the sequence will be proportional to the amount of time that each of the images have been displayed and the available band-width of the transmission link during this time.

A rate-distortion curve of a typical image of the Sun (see Fig. 14.2) indicates that, for a constant transmission rate, the visual quality of the images increases much faster in the beginning of the transmission. Therefore, in order to maximize the quality of

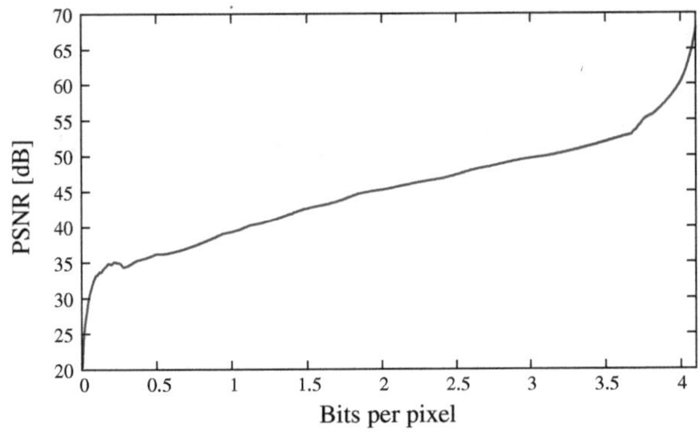

Fig. 14.2 Rate-distortion curve for the SDO/AIA image *aia_test.lev1.304A_2010-06-12T09_29_26.13Z.image_lev1*. The reversible path of JPEG 2000 has been used

the entire image sequence, at any time of the visualization, a given part of the bit budget should be dedicated to the currently displayed WOI w and the rest of the budget should be used for prefetching the same w of subset of the remaining the images of the time range.

To build our prefetching procedure, some definitions are required. Let $E_w(b)$ be the distortion between w and $w^{(b)}$, where $w^{(b)}$ is the reconstructed WOI after receiving b bits of w, calculated by means of the Mean Square Error

$$E_w(b) = \text{MSE}(w, w^{(b)}).$$

$E_w(b)$ will exhibit an approximately exponential behavior (in this case with a negative exponent). Thus, when b bits have been received for a given WOI w, $E_w(b)$ can be used to decide for the next request, which percentage of L is assigned to refine i_c (incrementing its quality) and which percentage of L is used for prefetching of other images from the time range.

A central issue related to the calculation of $E_w(b)$ is its dependence on the content of w, i.e. image data that is only known at the end of the transmission. However, taking into account the exponential trend of $E_w(b)$, this behavior is fairly similar to the behavior of a function that only considers the differential quality increments of $w^{(b)}$, i.e.

$$dE_w^K(b) = \text{MSE}(w^{(b)}, w^{(b-K)}),$$

for a certain constant increment of received bits, K.

Note that the values of $dE_w^K(b)$ will decrease along the transmission of an image, but faster in the beginning of the transmission than in the end. Therefore, $dE_w^K(b)$ can be used to decide if the data requested next should be part of the currently displayed image i_c or should be dedicated to prefetching data from the other images within the time range. The idea is that, if the $dE_w^K(b)$ value is close to zero, then most of the requested data should be dedicated to the time range prefetching.

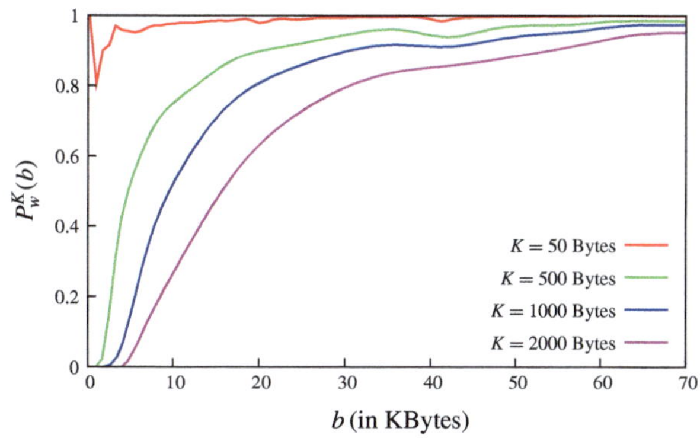

Fig. 14.3 Impact of K on the prefetching model function $P_w^K(b)$

With this idea in mind, the normalized percentage of each request to be dedicated to the prefetching can be modeled as

$$P_w^K(b) = \sigma e^{-d_w^K E(b)},$$

where the value $0 \leq \sigma \leq 1$ is used to limit this percentage. Figure 14.3 shows the effect of K on the calculation of this percentage along the transmission of an image (with $\sigma = 1$ for all the cases and also for the rest of the figures of this document). It can be seen that the larger the value K is, the smaller the percentage $P_w^K(b)$ and, therefore, the less aggressive the prefetching will be. It should be also noted that too small values of K (less that 50 bytes) could lead to the current and the next WOIs being identical, which will erroneously increase the prefetching of data, and that too big values of K will produce a slow prefetching scheme that, if the available bandwidth is also small, could disable the prefetching of data altogether. Our experiments show that values close to 1,000 bytes are suitable for a common remote browsing scenario.

According to $P_w^K(b)$, when the L value has been adjusted after receiving a server response, the next request to the server would be divided into

$$L_c = \left(1 - P_w^K(b)\right)L$$

bytes for the WOI within the current image i_c, and

$$L_p = P_w^K(b)L$$

bytes for prefetching of the rest of images of the time range. In our proposal, because the time range could potentially include too many images to be prefetched, the L_p budget is dedicated only to those images of the time range that are, at most,

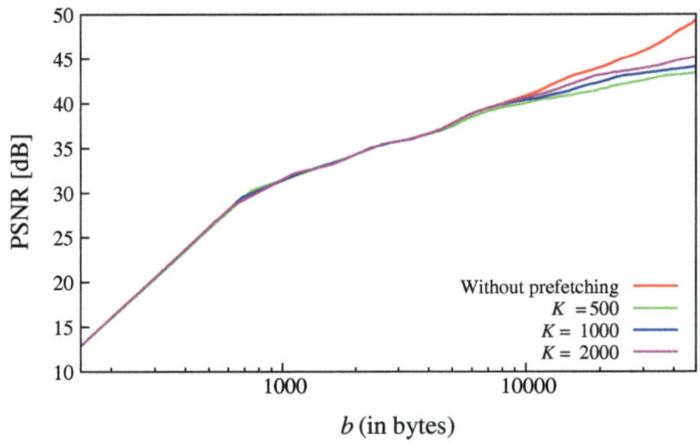

Fig. 14.4 Quality of a window of interest of i_c (the currently displayed image) retrieved with and without prefetching

located λ images away from i_c. λ therefore defines the size of a window of images $\{i_{c-\lambda}, \cdots, i_{c-1}, i_{c+1}, \cdots, i_{c+\lambda}\}$ that is centered on i_c but does not include it.

In order to measure the loss of quality of i_c due to the prefetching, Fig. 14.4 shows a rate/distortion comparison, in terms of Peak Signal-to-Noise Ratio (PSNR) in decibels, between a WOI of i_c retrieved with and without prefetching and for different values of K. The figure shows that the loss of quality produced by the time range prefetching is negligible in the beginning of the transmission and visually insignificant in the end, when a user could hardly differentiate between both WOIs.

Finally, from an implementation perspective, several points should be clarified. JPIP does not allow the specification of different values of L within the same request. Due to this limitation, in order to carry out the prefetching of the time range, each request must be divided into two requests: one with L_c and another one for the prefetching with L_p. These requests should be sent continusly, profiting from the transmission pipelining, in order to avoid any communication delays.

Since an independent request must be used for i_c as well as for prefetching, it is necessary to control the overhead generated by the protocol. Experience leads to the assumption that, on average, JPIP servers include around $H = 300$ additional bytes within each response due to headers. Once the values L_c and L_p have been obtained, they must be modified depending on the ratio H/L_p. If this ratio is above a certain threshold, L_c is set to L, thus discarding the second prefetching request. Experimental results show that a good value for this threshold is 0.5 or less. Changes in the value of H might also be taken into account during the communication for a better overall performance.

Once the current WOI has been completely received, the client should continue prefetching data using $L_p = L$. This makes it possible to also exploit the time spent by the user to analyze the content of the image.

14.5 Efficiency of the Prefetching in JHelioviewer

This proposed solution has been implemented in the JHelioviewer client. A random user browsing session composed of 200 consecutive movements over a remote file that contains a sequence of eighty-eight 4096×4096 solar images has been generated. The same session was simulated twice, once with prefetching and another time without it, for each condition evaluated. The Kakadu JPIP server [10] has been used for these experiments. The client/server bandwidth has been fixed to 1 Mbit/s, with a RTT of 1 s.

A slightly modified version of the user model proposed by Descampe et al. has been used for generating the random user browsing session of the 200 movements. The possible user movements have been reduced to five: panning, zooming in, zooming out, moving forward and moving backward. The first movement consists of changing the position of the current WOI to a new random position within the same resolution level, with a distance of 128 pixels. After a panning movement, the WOI is fully overlapped by the precinct partition (128×128 for every resolution level) without gaps. The zooming movements change the resolution level of the WOI one by one. The zooming-in movement is limited so that the minimum allowed resolution level is 512×512. The last two movement types allow the moving through the sequence one image at a time. None of these movements modify the WOI dimension, which is always 512×512 pixels.

It is assumed that the user behavior is defined by a first order Markov process, so given a movement of one type, the probability of choosing the same movement as the next one is δ, while the probability of choosing a different one is $(1 - \delta)/4$. The experiments in this study have used a value of $\delta = 0.4$. From the point of view of these experiments, the value of δ is not critical for evaluating our proposal since the distribution of movements is homogeneous. In Descampe's work this value had to be evaluated because its was associated to the predictability of the user behavior, a factor that affected the prefetching scheduler, but this is not the case for our solution. We have therefore chosen an intermediate value for δ, corresponding to a user behavior between almost deterministic ($\delta = 0.9$) and fully random ($\delta = 0.2$).

During the simulated browsing session, as soon as the quality of the current reconstructed WOI achieves a quality better than a certain PSNR threshold Θ, the next movement is triggered after a reaction delay B_τ, expressed in bytes. As in the Descampe's work, we have evaluated the values 30, 35 and 40 for Θ. For all values of Θ we have assumed $B_\tau = 0$, with the exception of $\Theta = 40$, for which we have also used the values 10 and 20 kB.

A total of 40 different conditions have been evaluated, varying σ from 0.1 to 1, in steps of 0.1, with values of 1, 2, 5 and 10 for λ. The average difference of the PSNR obtained in the remote browsing session, with and without prefetching, has been calculated for each condition. This difference is expressed as a percentage relative to the first session. The value used for K was 1,000 bytes.

Figures 14.5, 14.6 and 14.7 show that the best results are obtained for $\Theta = 40$. For $\Theta = 30$ the improvement is hardly noticeable because the WOI is moved before

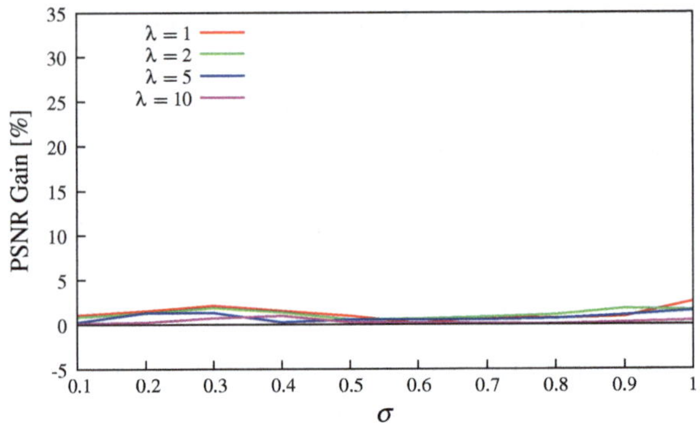

Fig. 14.5 Experimental results for $\Theta = 30\,\text{dB}$ and $B_\tau = 0\,\text{bytes}$

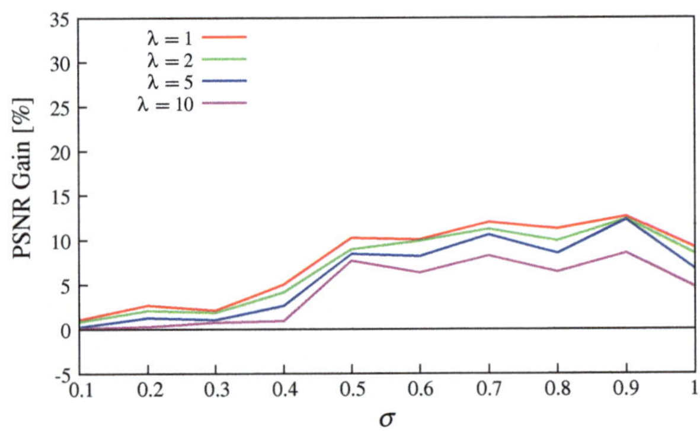

Fig. 14.6 Experimental results for $\Theta = 35\,\text{dB}$ and $B_\tau = 0\,\text{bytes}$

a significant stabilization of the differential quality has happened, and prefetching can therefore not be applied before the next user movement.

Figure 14.7 shows that with our solution there is always an improvement in the average PSNR independently of the values used for σ and λ. Nevertheless, the maximum improvement is achieved around $\sigma = 0.9$.

In all cases, the higher the value of λ, the less improvement is achieved. Taking into account the chosen user model, with one-by-one movements through the image sequence, this result is to be expected. It is perceivable that with another user model, with different degrees of freedom of movement through the sequence (e.g. allowing the user to skip images), the impact of λ might be different.

Taking into account that the best results have been obtained with $\Theta = 40$, we have also evaluated this threshold value with two different reaction delays $B_\tau = 10$ and $B_\tau = 20$, in Kbytes, as shown in Figs. 14.8 and 14.9, respectively. It is observed that the performance worsens with increasing value of B_τ, and it is even possible to obtain worse PSNR values with high σ values. These negative values are achieved due to the PSNR metric. After 40 dB, the differences in the visual quality of the sun images

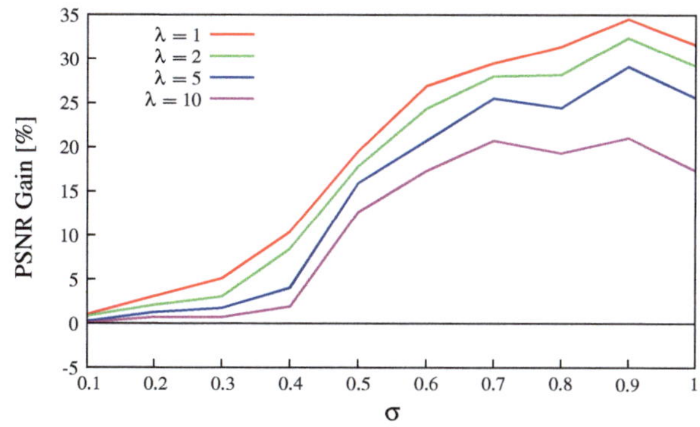

Fig. 14.7 Experimental results for $\Theta = 40\,\mathrm{dB}$ and $B_\tau = 0\,\mathrm{kB}$

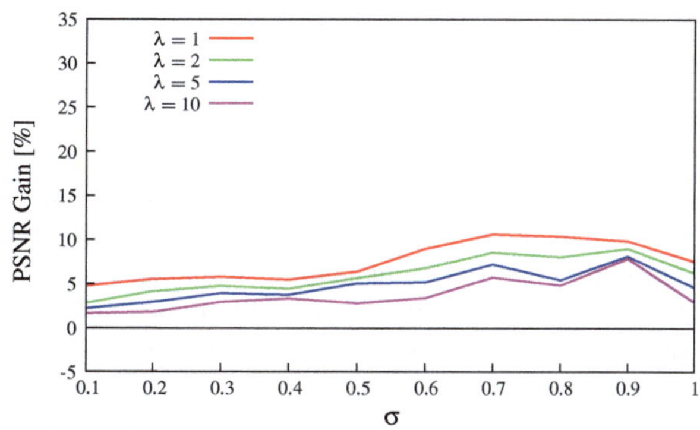

Fig. 14.8 Experimental results for $\Theta = 40\,\mathrm{dB}$ and $B_\tau = 10\,\mathrm{kB}$

is hardly noticeable. At these high σ values, the client bandwidth is therefore almost completely dedicated to prefetching. The PSNR is thus incremented very slowly, while the non-prefetching solution continues to increase the PSNR. Although this difference cannot be noticed by the user, the exact results are affected.

The results presented in this chapter show that the proposed solution improves the general user experience, defined in terms of PSNR, when browsing remote sequences of JPEG 2000 images. The best value for σ is associated to the speed of the user movements: for fast movements, the best results are obtained with high σ values; for slow movements, it is better to use low σ values that guarantee prefetching without significantly affecting the download of the current WOI.

The client application might even adjust the value of σ dynamically depending on the user behavior. When the user is moving through a sequence looking for a specific image, the movements are usually quite fast. However, once the user has located an interesting image, the movements become slow, with high reaction delays.

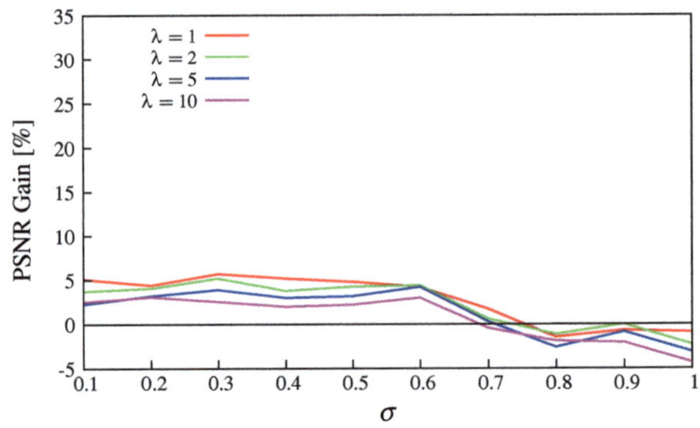

Fig. 14.9 Experimental results for $\Theta = 40\,\text{dB}$ and $B_\tau = 20\,\text{kB}$

14.6 Conclusions

In this work, a new efficient prefetching technique for interactive remote browsing of JPEG 2000 image sequences is proposed. Its most relevant characteristics are: (i) it offers an easy implementation that can be added to any existing JPIP client/server architecture; (ii) from the client/server bandwidth available, a certain fraction is allocated to prefetching, which is estimated using a differential quality model function; and (iii) an average improvement of the reconstructed WOI is always achieved, independently of how much fine-tuning is carried out. As a continuation of this study, future work will analyze an extension of this technique that utilizes sequences of images for video streaming.

References

1. Chae Y, Guo K, Buddhikot M, Suri S, Zegura E (2002) Silo, rainbow, and caching token: schemes for scalable, fault tolerant stream caching. IEEE J Sel Areas Commun 20(7): 1328–1344. doi:10.1109/JSAC.2002.802062
2. Chankhunthod A, Danzig PB, Neerdaels C, Schwartz MF, Worrell KJ (1995) A hierarchical internet object cache. In: Proceedings of the 1996 USENIX technical conference, San Diego, January 1996. pp 153–163
3. Descampe A, Ou J, Chevalier P, Macq B (2005) Data prefetching for smooth navigation of large scale JPEG 2000 images. IEEE international conference on multimedia and expo, Amsterdam, July 2005. p 4 doi:10.1109/ICME.2005.1521571
4. Descampe A, Vleeschouwer CD, Iregui M, Macq B, Marques F (2007) Prefetching and caching strategies for remote and interactive browsing of JPEG 2000 images. IEEE Trans Image Process 16(5):1339–1354
5. Descampe A, Vleeschouwer CD, Iregui M, Macq B, Marqus F, Stvin B, Levant PD (2005) Pre-fetching and caching strategies for remote interactive browsing of JPEG 2000 images. In: Technical Report TELE - UCL

6. Deshpande S, Zeng W (2001) HTTP streaming of JPEG 2000 images. International conference on information technology: coding and computing, Las Vegas, April 2001, p 15. doi:10.1109/ITCC.2001.918758
7. Fan L, Cao P, Almeida J, Broder AZ (1998) Summary cache: A scalable wide-area web cache sharing protocol. In. IEEE/ACM transactions on networking, 1998. pp 254–265
8. International Organization for Standardization: Information Technology-JPEG 2000 Image Coding System-Core Coding System. ISO/IEC 15444–1:2004 (2004)
9. International Organization for Standardization: Information Technology-JPEG 2000 Image Coding System-Interactivity Tools, APIs and Protocols. ISO/IEC 15444–9:2005 (2005)
10. Kakadu JPEG 2000 SDK. http://www.kakadusoftware.com
11. Krishnan K, Marcellin M, Bilgin A, Nadar M (2006) Efficient transmission of compressed data for remote volume visualization. IEEE Trans Med Imaging 25:1189–1199
12. Krishnan K, Marcellin MW, Bilgin A, Nadar MS (2006) Efficient transmission of compressed data for remote volume visualization. IEEE Trans Med Imaging 25(9):1189–1199
13. Li J, Sun HH (2003) On interactive browsing of large images. IEEE Trans Multimedia 5: 581–590
14. Liang ST, Chang TS (2005) A bandwidth effective streaming of JPEG2 000 images using hypertext transfer protocol. In: IEEE international conference on multimedia and expo, Lausanne, August 2002. pp 525–528
15. Lin C, Zheng YF (1999) Fast browsing of large scale images using server prefetching and client cache techniques. In: Applications of digital image processing XXII SPIE, Denver, July 1999. pp 376–387
16. Müller D, Fleck B, Dimitoglou G, Caplins BW, Amadigwe DE, Ortiz JPG, Wamsler B, Alexanderian A, Hughitt VK, Ireland J (2009) JHelioviewer: visualizing large sets of solar images using JPEG 2000. Comput. Sci. Eng. 11(5):38–47
17. Ortiz J, Ruiz V, López M, García I (2008) Interactive transmission of JPEG2000 images using Web proxy caching. IEEE Trans Multimedia 10(4):629–636. doi:10.1109/TMM.2008.921738
18. Pesnell W (2008) The solar dynamics observatory: your eye on the Sun. In: 37th COSPAR scientific assembly, COSPAR, Plenary Meeting, Montreal, July 2008. vol. 37, pp 2412-+
19. Poulakidas A, Srinivasan A, Egecioglu O, Ibarra O, Yang T (1996) Experimental studies on a compact storage scheme for wavelet-based multiresolution subregion retrieval. In: Proceedings of NASA 1996 Combined industry, space and Earth science data compression workshop, Utah, April 1996. pp 61–70
20. Strobel N, Mitra SK, Manjunath B, An approach to efficient storage, retrieval, and browsing of large scale image databases. In: Photonics East, proceedings of the SPIE–the international society for, optical engineering. pp 324–335
21. Su Z, Washizawa T, Katto J, Yasuda Y (2001) Performance improvement of graceful image caching by using request frequency based prefetching algorithms. In: Proceedings of IEEE region 10 international conference on electrical and electronic technology. vol 1, pp 370–376. doi:10.1109/TENCON.2001.949616
22. Su Z, Washizawa T, Katto J, Yasuda Y (2002) Hierarchical image caching in content distribution networks. In: Proceedings of IEEE region 10 conference on computers, communications, control and power engineering. vol 2, pp 786–790. doi:10.1109/TENCON.2002.1180239
23. Tuominen V, Isola J (2007) The application of JPEG 2000 in virtual microscopy. J Digit Imaging 22(3):250–258
24. Weidmann C, Vetterli M, Ortega A, Carignano F (1997) Soft caching: image caching in a rate-distortion framework. In: Proceedings of the International Conference on Image Processing, 1997. vol 2, pp 696–699. doi:10.1109/ICIP.1997.638591

Chapter 15
Comparing Spatial Masking Modelling in Just Noticeable Distortion Controlled H.264/AVC Video Coding

Matteo Naccari and Fernando Pereira

Abstract This chapter studies the integration of a just noticeable distortion model in the H.264/AVC standard video codec to improve the final rate-distortion performance. Three masking aspects related to lossy transform coding and natural video contents are considered: frequency band decomposition, luminance component variations and pattern masking. For the latter aspect, three alternative models are considered, namely the Foley–Boynton, Foley–Boynton adaptive and Wei–Ngan models. Their performance, measured for high definition video contents, and reported in terms of bitrate improvement and objective quality loss, reveals that the Foley–Boynton and its adaptive version provide the best performance with up to 35.6 % bitrate reduction at the cost of at most 1.4 % objective quality loss.

Keywords Human visual system · Just noticeable distortion modeling · Pattern masking · Perceptual video coding

15.1 Introduction

Nowadays, video technologies allow capturing and displaying visual contents at high resolutions with prices affordable for consumer usage. Therefore, it is expected that the amount of high definition video stored and/or transmitted will rapidly increase, demanding for more storing capacity and/or bandwidth. In this scenario, it is important to design new video compression algorithms able to further improve the compression efficiency beyond the state-of-the-art H.264/AVC (Advanced Video Coding) standard [17]. The Call for Proposals issued by both ITU and MPEG in

M. Naccari (✉) · F. Pereira
Instituto Superior Técnico - Instituto de Telecomunicações, 1049-001 Lispon, Portugal
e-mail: matteo.naccari@lx.it.pt

F. Pereira
e-mail: fernando.pereira@lx.it.pt

N. Adami et al. (eds.), *Analysis, Retrieval and Delivery of Multimedia Content*,
Lecture Notes in Electrical Engineering 158, DOI: 10.1007/978-1-4614-3831-1_15,

January 2010 on Video Compression Technology and the following running standardization activities on the High Efficiency Video Coding project express well these emerging needs [5].

One approach to reduce the video bitrate for a certain target quality may rely on the exploitation of the characteristics of the Human Visual System (HVS) which is always the last *quality judge*. In fact, the distortion induced by the usual quantization tools is not uniformly perceived by human beings but it rather varies, notably due to some video content masking capabilities/effects [18]. From this initial observation, several studies have been conducted to model the Just Noticeable Distortion (JND) corresponding to the minimum visibility threshold below which no change can be perceived by the HVS [18]. JND in video contents usually depends on three main spatial perceptual aspects: (1) the type of frequency representation (i.e. the transform) used; (2) the luminance variations; and (3) the presence of some patterns such as textured regions which boost masking effects. In the context of video coding, the main purpose of a JND model is to drive the quantization process by performing a coarser quantization in the image regions where the JND is higher since this should allow saving rate resources without subjective quality reduction. As an example, if the quantization is performed in the Discrete Cosine Transform (DCT) domain, then a different quantization step might be used for the various DCT bands in different image regions. Naturally, to correctly decode the coded video, these quantization steps must be transmitted and/or stored as side information which would increase the rate. While some authors claim that the bitrate associated to this side information may be rather high [4], it is also possible to estimate the JND model, and thus the associated quantization steps, at decoding time, thus eliminating the rate increase.

As an essential intermediary step to a complete perceptual video coding solution, the main objective of this chapter is to identify a good solution for the pattern masking component of the JND model by comparing solutions available in the literature. With this purpose, this chapter integrates a complete spatial JND model into a H.264/AVC High profile video codec to evaluate the bitrate reductions obtained for a certain subjective quality. The adopted JND model accounts for all the three aforementioned perceptual masking aspects and works in the DCT domain. Three pattern masking models proposed in the literature are considered: the Foley–Boynton [3] and Wei–Ngan [16] models, and an adaptive version of the Foley–Boynton model. The result of this comparison will drive the selection of the most suitable spatial JND model, notably the one bringing the best trade-off between bitrate reduction and side information bitrate increase when a perceptual coding approach is adopted.

The remainder of this chapter is organized as follows: Sect. 15.2 provides a brief overview on some of the most relevant JND models proposed in the literature. Section 15.3 describes the adopted and integrated spatial JND model and the three pattern masking models under evaluation. Section 15.4 presents the codec architecture modified with the JND model integration. Moreover, it is also discussed how the perceptual thresholds are considered in the quantization process. The experimental setup, as well as the measured performance, are provided and discussed in Sect. 15.5, while Sect. 15.6 concludes the chapter.

15.2 Background Overview

One of the first attempts to derive a JND model in the DCT domain, taking into account also the luminance variations in different image areas, dates back to the early nineties with the work of Ahumada and Peterson [1]. The results of this seminal study have been successfully applied to JPEG image coding in the DCTune algorithm [15] which provides graceful quality degradation by allowing the optimization of the quantization matrices for a given target rate. By taking into account also pattern masking effects, Höntsch and Karam in [4] adapted the Foley–Boynton model [3] to the DCT domain and proposed an estimation of the JND model at the decoder side. The authors showed that the performance obtained with the estimated JND model is comparable with the performance that would be achieved if the "*original*" JND thresholds would be available at decoding time. Finally, the authors also show that their model can reduce up to 23.9 % the bitrate required by the DCTune algorithm for the same perceptual quality. The main reason for the better performance of the model in [4] is related to the explicit modeling of the pattern masking mechanism which is not taken into account in the DCTune algorithm [15]. The original Foley–Boynton model in [3] can be further improved by considering inter-band masking [18]. In analogy with the audio case, inter-band effects refer to the masking provided by one DCT band with respect to the others. Inter-band masking has been modeled in the literature by an empirical weighting of the JND thresholds computed according to the Foley–Boynton model [3]. The weighting might be driven by either the absolute energy of the DCT coefficients as in the work by Zhang et al. [20] or some image features as proposed by Wei and Ngan [16]. Thanks to inter-band masking, the model in [20] outperforms the DCTune algorithm for JPEG image coding based on subjective assessment. For the same settings, the model in [16] shows a better performance than both the DCTune and the model in [20]. In particular, for the same subjective quality, the model in [16] can tolerate more distortion (i.e. lower Peak Signal-to-Noise-Ratio values).

Naturally, also the wavelet domain has been targeted by the research on JND modeling. As an example, Liu et al. proposed a perceptual image encoder which is fully compliant with the JPEG 2000 standard and achieves a bitrate reduction up to 21.2 % when compared with conventional JPEG 2000 coding [9]. In [8], Leung and Taubman propose to adaptively modulate the exponent of the model in [3] to better preserve image edges. The method is integrated into a wavelet based scalable video codec and its performance is reported both in terms of objective and subjective measurements. For the same bitrate, the sequence reconstructed with the proposed method achieves lower Perceptual Distortion Metric (PDM) values [8]. Similar conclusions are also achieved with subjective testing.

Finally, also the JND modeling in the pixel domain has been addressed in the literature notably for subband and MPEG-2 Video coding. In [2], Chou and Li propose a pixel domain JND model for subband image coding which nonlinearly blends the variance and the average of the luminance background. The model is used to suppress the DCT coefficients whose magnitude is less than the corresponding JND

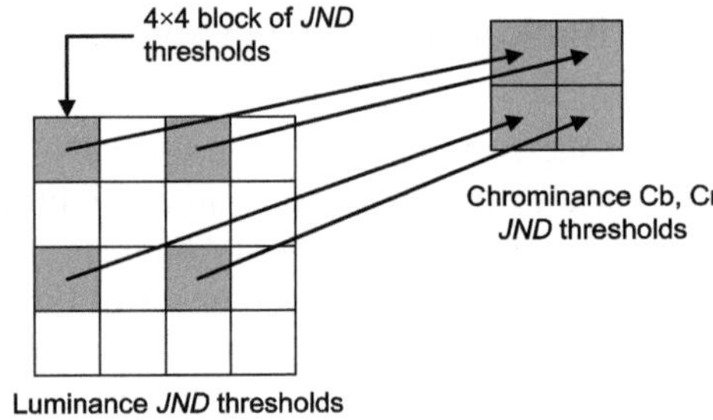

Fig. 15.1 JND thresholds selection for the chrominance components at the macroblock level

threshold. The designed image codec has been compared with the JPEG standard for grey level image coding using the Peak-Signal-to-Perceptible-Noise-Ratio (PSPNR) as objective metric [2]. Experimental results show that the proposed perceptual codec outperforms the JPEG standard by up to 10 dB PSPNR at the same rate. The model in [2] has been later extended to the chrominance components by Yang et al. for MPEG-2 Video coding [19]. The masking capabilities of the proposed JND model have been assessed both for images and video. For images, the authors conducted subjective tests showing that their model achieves the same perceptual quality of [2] but with less 2 dB PSNR. Conversely, for the video case, the designed JND model has been integrated in a MPEG-2 Video encoder to suppress the prediction residuals with absolute values lower than their corresponding JND thresholds. The authors showed that, at the same rate, the modified MPEG-2 Video encoder increases the PSPNR up to 0.25 dB against conventional MPEG-2 Video coding.

15.3 Spatial and Model Description

This section presents the spatial JND model to be integrated in the H.264/AVC encoder. First, the general form of the model is introduced and, after, all the terms modeling the three (spatial) aspects mentioned in Sect. 15.1 are presented. The overall spatial JND model for the (i, j) DCT band in the k-th $B \times B$ DCT block is given by:

$$JND(i, j, k) = JND_{band}(i, j) \cdot JND_{lum}(k) \cdot JND_{pat}(i, j, k), \qquad (15.1)$$

where $JND_{band}(i, j)$ denotes the model for the DCT band decomposition masking, $JND_{lum}(k)$ denotes the model for the luminance variations masking in the DCT block and, finally, $JND_{pat}(i, j, k)$ denotes the model for the pattern masking at block level. The JND threshold in Eq. 15.1 is computed over the luminance component of each *original* video frame. The JND thresholds for the chrominance components are obtained by selecting, at 4×4 level, and with the same chrominance subsampling ratio, the even luminance JND thresholds as shown in Fig. 15.1. Finally, this chapter considers a 4×4 DCT block size as this is the basic H.264/AVC transform size; the extension towards the 8×8 size is currently ongoing work.

15.3.1 Frequency Band Decomposition Masking Model

This model accounts for the different sensitivity of the human eye to the noise introduced at different spatial frequencies (i.e. different DCT bands). In particular, the human eye shows a band-pass behavior which can be approximated by a parabolic function with downward concavity [18]. Closed form solutions for this approximation have been proposed in [1, 16] but their parameterization, carried out by means of subjective tests, regards only the 8×8 floating point DCT. Therefore, an extension of these models for the 4×4 and 8×8 integer DCT used in H.264/AVC would require new designs and subjective testing. In this chapter, the DCT masking model is constituted by the default perceptual weighting matrices adopted in the H.264/AVC reference software [7] since they were properly designed to quantify the error visibility for each DCT band [13]. These perceptual weighting matrices differ for Intra and Inter coded macroblocks as reported in Eqs. 15.2 and 15.3.

$$JND_{band}^{intra}(i, j) = \begin{bmatrix} 6 & 13 & 20 & 28 \\ 13 & 20 & 28 & 32 \\ 20 & 28 & 32 & 37 \\ 28 & 32 & 37 & 42 \end{bmatrix}, \tag{15.2}$$

$$JND_{band}^{inter}(i, j) = \begin{bmatrix} 10 & 14 & 20 & 24 \\ 14 & 20 & 24 & 27 \\ 20 & 24 & 27 & 30 \\ 24 & 27 & 30 & 34 \end{bmatrix}. \tag{15.3}$$

The values in Eqs. 15.2 and 15.3 include a multiplication by 16 in order to preserve the 16-bit integer arithmetic adopted by the H.264/AVC integer DCT [10] and also to obtain perceptual weights lower than 1 which reduce the quantization step. For example, a value 6 in the JND_{band} matrix corresponds to a perceptual weight of $6/16 \simeq 0.37$, i.e. a weight which reduces the quantization step for a given DCT coefficient.

15.3.2 Luminance Variations Masking Model

This model accounts for the masking effect provoked by luminance variations in different image regions. The rationale behind this masking is that quantization errors are less visible in darker and brighter regions. The starting point is the Weber–Fechner law which states that the minimal brightness difference which may be perceived increases with the background brightness [18]. The work in [16] proposes a numerical approximation of this law. Considering the H.264/AVC standard, this approximation comes as follows:

$$JND_{lum}(k) = \begin{cases} \frac{-3\cdot\bar{L}(k)}{62} + 4 & \bar{L}(k) \leq 62 \\ 1 & 62 < \bar{L}(k) < 115 \\ \frac{3\cdot\bar{L}(k)-205}{140} & \bar{L}(k) \geq 115 \end{cases}, \tag{15.4}$$

where the term $\bar{L}(k)$ denotes the average luminance intensity in block k belonging to the *original* frame. Regarding [16], Eq. 15.4 has been modified by changing the range where $JND_{lum}(k) = 1$ and scaling its values. The former change is necessary since experimental tests revealed that the original range was too wide. The latter change is a consequence of the different frequency band masking term adopted in this chapter; in fact, the JND_{lum} values have been rescaled to guarantee the same increment as observed when the model in Eq. 15.4 is applied to the JND_{band} model used in [16]. The varying JND thresholds due to luminance masking are shown in Fig. 15.2 for the first frame of the *Crew* sequence with a luminance spatial resolution of 720×1280 and 4×4 block size. As it may be noted, brighter areas as the wall and the door frame have the higher JND_{lum} threshold values. The same trend is also observed in darker areas, such as the men's shoes and the man behind the one on the left.

15.3.3 Pattern Masking Model

This model accounts for the spatial masking effects corresponding to some specific patterns in the image. In fact, it is well known that quantization errors are less noticeable on patterns such as textures or corrugate regions [18]. These pattern masking effects have an impact on the JND thresholds. The work in [3] measured the JND threshold variations operated by masker signals such as Gabor and sinusoidal patterns. The measurements led to a nonlinear relation between the JND thresholds and the normalized masking energy $E(i, j, k)$ defined as the squared ratio between the masker (i.e. the image) and the masking signal (i.e. the quantization error):

$$E(i, j, k) = \left| \frac{C(i, j, k)}{JND_{band}(i, j) \cdot JND_{lum}(k)} \right|^2, \tag{15.5}$$

where $C(i, j, k)$ denotes the DCT coefficient for the (i, j) frequency band and the k-th image block belonging to the *original* frame. This nonlinear relationship constitutes the starting point for the three pattern masking solutions to be compared in this chapter as described in the following.

Foley–Boynton Model

The nonlinear relationship between the normalized contrast energy $E(i, j, k)$ and the JND thresholds, firstly introduced in [3], has been adapted to the DCT case by Höntsch and Karam in [4] as:

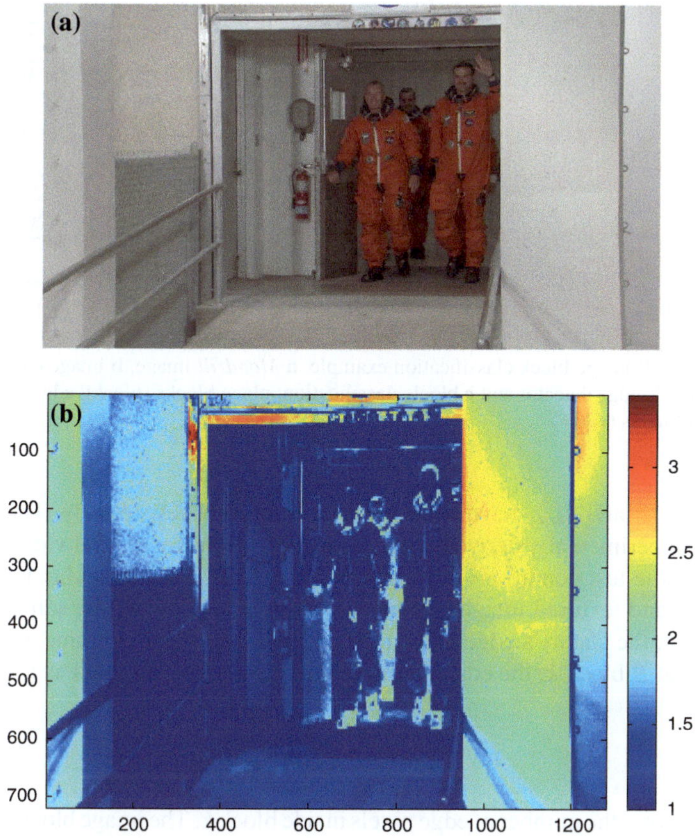

Fig. 15.2 First frame of the Crew sequence (**a**) and JND thresholds for the JND_{lum} term (**b**)

$$JND_{pat}(i, j, k) = \begin{cases} 1 & \text{if } i, j = 0 \\ \max(1, E(i, j, k)^{\varepsilon}) & \text{otherwise} \end{cases}, \tag{15.6}$$

where the exponent ε is equal to 0.6. Considering the H.264/AVC standard, the values of $JND_{pat}(i, j, k)$ have been further clipped to 1.2 to guarantee the same increment in the JND thresholds observed for the model in [4] with 4×4 floating point DCT.

Adaptive Foley–Boynton Model

The exponent ε in Eq. 15.6 can be made adaptive at the $B \times B$ block level. The rationale for this change comes from recent studies showing how spatio-temporal edges (rather than interior regions) generate the strongest HVS responses [8]. From this result, the exponent ε can be set to either ε_{min} for image blocks which contain edges and plane regions or to ε_{max} for textured image blocks. In this chapter, the

Fig. 15.3 $B \times B$ image block classification example. **a** *Mandrill* image, **b** image edges obtained with the Canny's edge detector and **c** block classification: plane blocks (*black*); edge blocks (*grey*) and textured blocks (*white*)

values for ε_{min} and ε_{max} have been set to 0.3 and 0.6, respectively. These values have been experimentally derived after visual inspections for several video contents, under the same test conditions described in Sect. 15.5. The block classification for edge, plane and textured image regions is performed following the same approach in [16]. First, the Canny's edge detector is applied over the n-th frame $F(n)$. Then, for each $B \times B$ block k, the edge pixels fraction, $\rho(k)$, is computed as follows:

$$\rho(k) = \frac{\Lambda}{B^2}, \tag{15.7}$$

where Λ denotes the number of edge pixels inside block k. The image block k is classified as *plane* if $\rho(k) = \alpha$, as *edge* if $\alpha < \rho(k) \leq \beta$ and as *textured*, otherwise. Differently from the work in [16], the thresholds (α, β) are adaptively computed using the normalized histogram $h_\rho(F(n))$ of the ρ values for $F(n)$. In particular, the threshold α is computed as the centroid of $h_\rho(F(n))$ while β is computed as the centroid of the histogram $\bar{h}_\rho(F(n))$ defined as: $\{\bar{h}_\rho(F(n)) = h_\rho(F(n)), \forall k' \text{ such that } \rho(k') > \alpha\}$. An example of the block classification for the image "*Mandrill*" is shown in Fig. 15.3.

Wei–Ngan Model

The work in [16] uses the model in 15.6 with ε equal to 0.36 and performs an adaptive weighting of $JND_{pat}(i, j, k)$. The adaptive weight ψ is selected depending on the aforementioned block classification and the band location (i, j). In particular, ψ is set to 1 for edge and plane blocks, while for texture blocks its value depends also on the frequency band location (i, j):

$$\psi = \begin{cases} 2.25 & \text{if } i^2 + j^2 \leq 4 \\ 1.25 & \text{otherwise} \end{cases}. \tag{15.8}$$

This weighting considers that for textured blocks the quantization can be coarser in lower frequencies and smoother for higher frequencies. Finally, the $JND_{pat}(i,j,k)$ term is given as follows:

$$JND_{pat}(i,j,k) = \begin{cases} \psi & i^2 + j^2 \leq 4 \text{ and plane or edge} \\ \psi \cdot \min(4, \max(1, E(i,j,k)^{0.36})) & \text{otherwise} \end{cases}, \quad (15.9)$$

In summary, the three presented pattern masking models can be characterized by:

1. They are all based on the general Foley–Boynton's model depicted in Eq. 15.6.
2. Both the last two models are adaptive and require to code some additional side information associated to the block classification.
3. While the adaptive Foley–Boynton adapts to the image content to preserve image edges, the Wei–Ngan model adaptive weighting preserves better the texture details.

15.4 H.264/AVC JND Model Integration

This section describes how the adopted spatial JND model has been integrated in the H.264/AVC encoder architecture. To this end, Fig. 15.4 illustrates the overall encoder architecture, highlighting the JND model related blocks. As it may be noticed, the JND model output controls the quantization of the transformed residuals, after spatial or temporal prediction. As stated in the Introduction, the purpose of a JND model is to perceptually modulate the quantization step for each DCT coefficient. More specifically, the non-perceptul quantization step $Q(i,j,k)$ is increased or decreased by a multiplying factor which corresponds to the JND threshold $JND(i,j,k)$. Therefore, the perceptual quantization step Q_{JND} becomes:

$$Q_{JND}(i,j,k) = Q(i,j,k) \cdot JND(i,j,k). \quad (15.10)$$

To implement Eq. 15.10, the Multiplication Factor (MF) [10] for the DCT coefficient at frequency band (i,j) is modified as follows:

$$MF_{JND}(i,j) = \frac{(MF(i,j,QP(k)\%6) \cdot 16)}{JND(i,j,k)}, \quad (15.11)$$

where $QP(k)$ denotes the Quantization Parameter (QP) for the k-th image block and % denotes the division remainder. The multiplication by 16 is made to compensate the multiplication by 16 of the JND_{band} values as described in Section 15.3.1. Equation 15.11 highlights the need for the threshold $JND(i,j,k)$ at the decoder side in order to correctly decode the compressed bitstream. Furthermore, as already stated at the end of Sect. 15.3.3, if either the adaptive Foley–Boynton or the Wei–Ngan

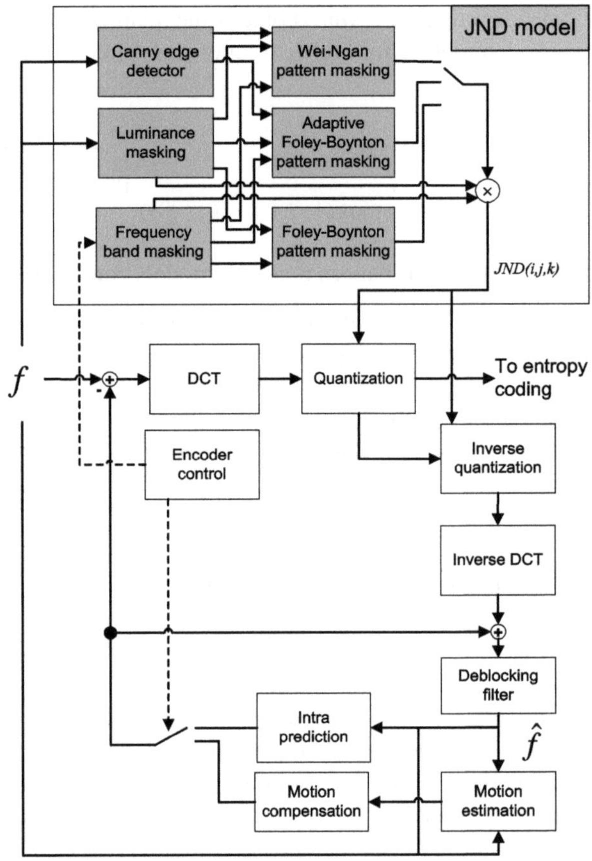

Fig. 15.4 H.264/AVC JND enabled encoder architecture

model are used as the JND_{pat} term in Eq. 15.1, the information related to block k classification must be also transmitted to the decoder side.

The overall processing performed by the architecture in Fig. 15.4 can be described by the sequence of steps listed in Algorithm 1.

15.5 Experimental Results and Discussion

This section presents the experiments conducted to assess the rate-distortion performance improvement obtained by integrating the adopted JND models. The selected test videos have spatial resolutions of 1280 × 720 and 1920 × 1080 with frame rates of 60 and 24, respectively, and 4:2:0 chrominance subsampling ratio. They have been downloaded from the repository indicated in [14] which contains 15 sequences for the former resolution and 6 for the latter. The results presented in this chapter refer

Algorithm 1: Frame level processing performed by the H.264/AVC JND enabled encoder architecture in Fig. 15.4.

for each frame $F(n)$ of the video sequence **do**
if JND_{pat} model is adaptive Foley–Boynton or Wei–Ngan **then**
Perform Canny's edge detection over $F(n)$
for each 4×4 block k **do**
Compute the edge pixel density $\rho(k)$ according to Eq. 15.7
Go to next block k
end for
Compute the histogram $h_{\rho}(F(n))$ and the thresholds α and β
for each 4×4 block k **do**
Classify block k according to (α, β) as described in Sect. 15.3.3
Go to next block k
end for
end if
for each macroblock MB belonging to $F(n)$ **do**
for each coding mode $m \in \{intra, inter\}$ **do**
for each 4×4 block k **do**
Set $JND_{band}(k)$ to Eq. 15.2 or 15.3 depending on m
Compute $JND_{lum}(k)$ according to Eq. 15.4
if JND_{pat} model is Adaptive Foley–Boynton **then**
Compute JND_{pat} according to Eq. 15.6 and block k classification (plane, edge or texture)
else if JND_{pat} model is Wei–Ngan **then**
Compute JND_{pat} according to Eq. 15.9
else
Compute JND_{pat} according to Eq. 15.6
end if
Compute $JND(i, j, k)$ according to Eq. 15.1
Compute $MF_{JND}(i, j)$ according to equation 15.11
Perform H.264/AVC integer DCT, forward quantization and entropy coding
Go to next block k
end for
Go to next coding mode m
end for
Go to next macroblock MB
end for
Go to next frame n
end for

only to three representatives sequences for each tested resolution.[1] To choose the representatives video sequences, the Spatial Index (SI) and Temporal Index (TI), expressing how a sequence would be difficult to encode, have been computed according to the ITU specifications given in [6]. The SI and TI indexes for each sequence are reported in Fig. 15.5, gathered into three clusters for each resolution. From the clusters depicted in Fig. 15.5, the representatives selected for the 1280×720 resolution are "*SpinCalendar*", "*Raven*" and "*Preakness*", while for the 1920×1080 resolution

[1] More results available at http://amalia.img.lx.it.pt/temp/wiamis_addendum.pdf

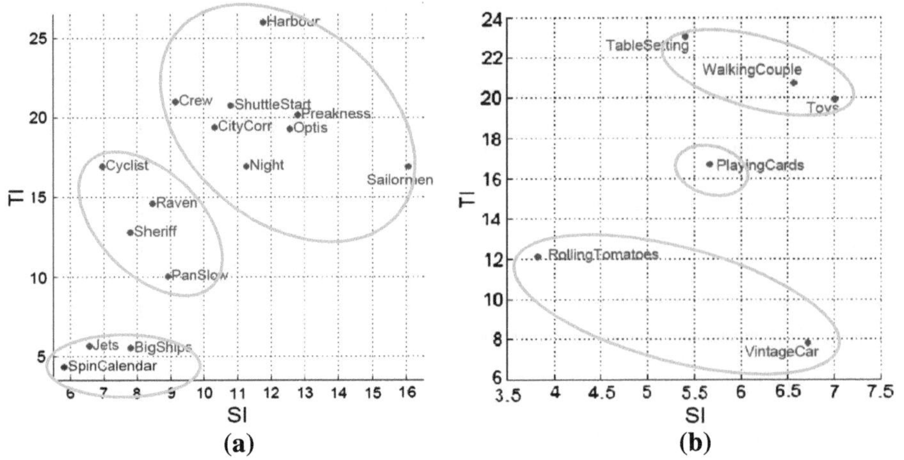

Fig. 15.5 SI and TI indexes for the 1280 × 720 resolution (**a**) and 1920 × 1080 resolution (**b**) sequences

are "*TableSetting*", "*PlayingCards*" and "*VintageCar*". These video sequences have been coded with the H.264/AVC High profile according to the parameters indicated in [14], except for the DCT size which is only set to 4 × 4. Beside the coding parameters, [14] indicates also four QP values to be used for the Intra (I), Predicted (P) and Bidirectional predicted (B) frames. The QP for I frames (QPI) can assume the following values: 22, 27, 32 and 37 while the QP for P and B frames (QPP and QPB) are computed as QPI + 1 and QPI + 2, respectively. Hereafter, each combination (QPI, QPP, QPB) is referred to as Group i (Gi, $i = 1, \cdots, 4$). For each test video, the bitrates required by the JND model enabled codec and the H.264/AVC High profile codec (hereafter denoted as HP) are compared. The HP solution only employs the perceptual matrices as in Eqs. 15.2 and 15.3. The bitrate related to the JND based solution does not account for the rate spent for each *JND* (i, j, k) threshold assuming the *JND*(i, j, k) thresholds are available at the decoder side without taking into account the relative rate. From this assumption, the assessment presented in the following should, thus, be understood as the asymptotic performance.

Together with the bitrate improvement, the objective quality of each reconstructed sequence is measured by means of the Multiple Scale-Structural SIMilarity (MS-SSIM) index [11] computed over the luminance component of each frame and averaged for the entire sequence. This objective quality metric has been chosen given the promising results of a recent study [11, 12] which reports a Pearson's correlation coefficient of up to 0.69 with subjective scores.

For the test sequences, both the bitrate and the MS-SSIM reductions in percentage of the JND based solution regarding the HP solution are reported in Table 15.1. As a first comment, one can notice that there are significant bitrate reductions (up to 35.63 %) against negligible losses in objective quality (with a maximum of 1.38 % for the Foley–Boynton model with the *VintageCar* sequence). Moreover, all the

Table 15.1 Bitrate and MS-SSIM comparison between the HP codec and the three alternative JND enabled codecs

Sequence	Group	Foley–Boynton		Adaptive Foley–Boynton		Wei–Ngan	
		Δ Rate(%)	ΔMS SSIM(%)	ΔRate(%)	ΔMS SSIM(%)	ΔRate (%)	ΔMS SSIM(%)
SpinCalendar	G1	−30.79	−0.09	−30.77	−0.09	−25.23	−0.08
	G2	−16.85	−0.15	−17.03	−0.15	−12.03	−0.13
	G3	−11.67	−0.29	−11.57	−0.29	−8.59	−0.25
	G4	−9.28	−0.67	−9.16	−0.68	−7.45	−0.56
Raven	G1	−24.27	−0.15	−24.31	−0.15	−22.03	−0.18
	G2	−20.44	−0.29	−20.35	−0.29	−20.17	−0.36
	G3	−17.41	−0.56	−17.32	−0.56	−17.48	−0.71
	G4	−13.80	−0.87	−13.52	−0.88	−14.39	−1.06
Preakness	G1	−29.44	−0.08	−29.32	−0.08	−17.55	−0.09
	G2	−27.14	−0.15	−27.10	−0.15	−24.25	−0.13
	G3	−19.65	−0.27	−19.55	−0.27	−16.01	−0.21
	G4	−15.52	−0.54	−15.55	−0.55	−11.77	−0.40
TableSetting	G1	−35.59	−0.30	−35.63	−0.30	−33.40	−0.31
	G2	−22.94	−0.27	−22.96	−0.27	−21.49	−0.31
	G3	−17.10	−0.39	−17.21	−0.39	−16.34	−0.45
	G4	−13.18	−0.39	−13.18	−0.38	−13.47	−0.42
PlayingCards	G1	−30.13	−0.11	−30.17	−0.11	−24.98	−0.13
	G2	−22.06	−0.17	−21.99	−0.17	−19.58	−0.22
	G3	−18.58	−0.32	−18.75	−0.32	−17.04	−0.38
	G4	−15.64	−0.55	−15.77	−0.59	−15.41	−0.67
VintageCar	G1	−32.68	−0.26	−32.71	−0.26	−28.38	−0.22
	G2	−21.82	−0.40	−21.85	−0.40	−16.24	−0.30
	G3	−16.18	−0.77	−16.18	−0.77	−12.24	−0.55
	G4	−16.85	−1.38	−17.13	−1.35	−13.01	−0.85
Average		−20.79	−0.39	−20.79	−0.39	−17.85	−0.37

reconstructed sequences have been visually inspected[2] by the authors leading to the general conclusion that the JND models additional artifacts are hardly noticeable. Regarding the pattern masking models, the Foley–Boynton and its adaptive version provide the highest bitrate reductions. The sequences reconstructed with these two models are almost undistinguishable with the exception of some details which are better preserved with the adaptive model. Conversely, the Wei–Ngan model provides the smallest bitrate reductions and sometimes the objective quality reduction is higher (for example, G1 for *PlayingCards*). Furthermore, this model over smoothes some image details (e.g. in *SpinCalendar*[3]). From these results, the selection of the best

[2] More results available at http://amalia.img.lx.it.pt/temp/wiamis_addendum.pdf

[3] More results available at http://amalia.img.lx.it.pt/temp/wiamis_addendum.pdf

pattern masking model for a perceptual video codec is to be made between the Foley–Boynton model and its adaptive version. As stated at the end of Sect. 15.3.3, the adaptive Foley–Boynton model requires additional bitrate for the block classification side information. Therefore, in order to trade-off high bitrate reductions and low side information related bitrates, the Foley–Boynton original model (see Eq. 15.6) seems to be a suitable pattern masking model for the adopted spatial JND model.

15.6 Conclusion and Future Work

This chapter has studied and compared the benefits of integrating different pattern masking models in the H.264/AVC video coding standard. The complete JND model takes into account frequency decomposition, luminance variations and pattern masking. For this last aspect, three models proposed in the literature have been compared. The performance comparison revealed that the Foley–Boynton model and its adaptive version provide the best performance. Between these two models, the former is more suitable for integration into a practical perceptual video coding scheme since it does not require additional rate for a similar quality. However, in order to obtain a performance closer to the one achieved with the Foley–Boynton model, the JND thresholds ($JND(i, j, k)$) need to be estimated at the decoder side. By looking at the formulations for the JND_{band}, JND_{lum} and JND_{pat} models, it is easy to recognize that the JND_{band} model does not need any decoder side estimation as the perceptual weighting matrices (Eqs. 15.2 and 15.3) can be transmitted only once for each video sequence. Conversely, for the JND_{lum} and JND_{pat} models, the $\bar{L}(k)$ and $C(i, j, k)$ terms need to be estimated at the decoder side. It should be noted that estimating these terms would turn into estimating the original video frame $F(n)$ (see Eqs. 15.4 and 15.6). However, in motion compensated predictive video coding scheme, a good approximation of $F(n)$ is constituted by the predictor $P(k)$ used for the predictive coding of block k. Therefore, the ongoing work targets the design of a decoder side estimation mechanism for $\bar{L}(k)$ and $C(i, j, k)$ based on the block k predictor $P(k)$ as well as the extension of the overall spatial JND model towards the 8×8 integer DCT used in the H.264/AVC High profile. Finally, also the human visual system temporal masking [18] due to the motion activity present in a video sequence will be taken into account.

References

1. Ahumada AJ, Peterson HA (1992) Luminance-model-based dct quantization for color image compression. In: SPIE: Human vision, visual processing and digital display III, San Jose, February 1992
2. Chou CH, Li YC (1995) A perceptually tuned subband image coder based on the measure of just-noticeable distortion profile. IEEE Trans Circuits Syst Video Technol 5(6):467–476
3. Foley JM, Boynton GM (1994) A new model of human luminance pattern vision mechanism: analysis of the effects of pattern orientation, spatial phase and temporal frequency. In: SPIE:

Computational vision based on neuro-biology, 1994
4. Höntsch I, Karam LJ (2002) Adaptive image coding with perceptual distortion control. IEEE Trans Image Process 11(3):312–222
5. ISO/IEC JTC1/SC29/WG11: Joint call for proposal on video compression technology. Technicla Report. N11113, MPEG (2010)
6. ITU-T: Recommendation ITU-R P 910 (1999). Subjective video quality assessment methods for multimedia applications
7. (JVT), J.V.T.: H.264/AVC reference software version JM16.0. http://iphome.hhi.de/suehring/tml/download/
8. Leung R, Taubman D (2009) Perceptual optimization for scalable video compression based on visual masking principles. IEEE Trans Circuits Syst Video Technol 19(3):337–346
9. Liu Z, Karam LJ, Watson AB (2006) JPEG2000 encoding with perceptual distortion control. IEEE Trans Image Process 15(7):1763–1778
10. Malvar HS, Hallapuro A, Karczewicz M, Kerofsky L (2003) Low-complexity transform and quantization in H.264/AVC. IEEE Trans Circuits Syst Video Technol 13(7):598–603
11. Seshadrinathan K, Soundararajan R, Bovik AC, Cormack LK (2010) Study of subjective and objective quality assessment of video. IEEE Trans Image Process 19(6):1427–1441
12. Seshadrinathan K, Soundararajan R, Bovik AC, Cormack LK (2010) A subjective study to evaluate video quality assessment algorithms. In: SPIE: Human vision and electronic imaging, San Jose, 2010
13. Suzuki T, Sato K, Yagasaki Y (2002) Weighting matrix for JVT codec. Technical Report. JVT-C053r1, JVT 2002
14. Tan T, Sullivan G, Wedi T (2008) Recommended simulation common conditions for coding efficiency experiments, revision 2. Technical Report. VCEG-AH10r3, JVT 2008
15. Watson AB (1993) Dctune: a technique for visual optimization of DCT quantization matrices for individual images. J Soc Inf Disp Dig Tech Papers 24:946–949
16. Wei Z, Ngan KN (2009) Spatio-temporal just noticeable distortion profile from grey scale image/video in DCT domain. IEEE Trans Circuits Syst for Video Technol 19(3):337–346
17. Wiegand T, Sullivan GJ, Bjontegaard G, Luthra A (2003) Overview of the H.264/AVC video coding standard. IEEE Trans Circuits Syst Video Technol 13(7):560–576
18. Wu HR, Rao KR (2005) Digital video image quality and perceptual coding. CRC press, Boca Raton
19. Yang X, Ling W, Lu Z, Ong E, Yao S (2005) Just noticeable distortion model and its applications in video coding. Sig Process Image Commun 20(7):662–680
20. Zhang X, Lin W, Xue P (2005) Improved estimation for just-noticeable visual distortion. Sig Process 85(4):795–808

Chapter 16
Coherent Video Reconstruction with Motion Estimation at the Decoder

Claudia Tonoli and Marco Dalai

Abstract In traditional motion compensated predictive video coding, both the motion vector and the prediction residue are encoded and stored or sent for every predicted block. The motion vector brings displacement information with respect to a reference frame while the residue represents what we really consider to be the innovation of the current block with respect to that reference frame. This encoding scheme has proved to be extremely effective in terms of rate distortion performance. Nevertheless, one may argue that full description of motion and residue could be avoided if the decoder could be made able to exploit a proper a priori model for the signal to be reconstructed. In particular, it was recently shown that a smart enough decoder could exploit such an a priori model to partially infer motion information for a single block given only neighboring blocks and the innovation of that block. This chapter presents an improvement over the single-block method. In particular, it is shown that higher performance can be achieved by simultaneously reconstructing a frame region composed of several blocks, rather than reconstructing those blocks separately. A trellis based algorithm is developed in order to make a global decision on many motion vectors at a time instead of many single separate decisions on different vectors.

Keywords Video coding · Decoder side motion estimation

C. Tonoli (✉) · M. Dalai
Department of Information Engineering,
University of Brescia via Branze 38,
25123 Brescia, BS, Italy
e-mail: claudia.tonoli@ing.unibs.it

M. Dalai
e-mail: marco.dalai@ing.unibs.it

N. Adami et al. (eds.), *Analysis, Retrieval and Delivery of Multimedia Content*, Lecture Notes in Electrical Engineering 158, DOI: 10.1007/978-1-4614-3831-1_16,

16.1 Introduction

In predictive video coding schemes very high compression efficiency is obtained thanks to motion compensated prediction. The basic idea of this approach is to exploit the temporal redundancy across frames to produce a prediction of the current frame to be encoded relying on reference frames which are already available to the decoder. This is done by estimating relative motion between current and reference frames. The motion compensated prediction is then subtracted from the current frame to obtain the so called residue. Thus, the encoding of the current frame is obtained by encoding both the estimated motion field and the residue. The decoder, obviously, reverts those operations. It reconstructs the frame by means of a motion compensated prediction from the reference frame using the received motion field and it simply adds the received prediction residue to the compensated frame.

In this scheme, the decoder does not take into account any additional information aside from the received encoded stream. In particular, it does not take into account any a priori information on the nature of the encoded signal, even if such information could carry important knowledge about the possible results that one might expect from the decoding process. For example, if a natural scene is being decoded, one would expect to find as a result of the decoding process a relatively piecewise smooth signal with relatively regular contours. In particular, given that the partitioning of the frame in blocks is fixed and independent from the signal content, one does not expect the reconstructed signal to have a high degree of irregularity across blocks boundaries. Consider now what happens if a wrong motion vector is used for one predicted block. In that case, the reconstructed frame would probably have high discontinuities along the borders of that block. This could be interpreted by a smart decoder as an indication of probable error in the motion vector. So, the motion vector of a block is partially predictable given only neighboring blocks in the current frame and the innovation of that block, and thus its description could be avoided in some cases.

The possibility of omitting the transmission of motion information has recently attracted an increasing interest. For example in [1] an algorithm for motion derivation at the decoder side for the H.264/AVC codec is presented. Such algorithm is based on the L-shaped causal past of each block, called "context". At the decoder, the context is searched for in the reference frame, obtaining its displacement. The same displacement is taken as a motion vector for the current block.

In [2] an algorithm dealing with H.264/AVC B-frames is presented. According to this algorithm, the decoder estimates the motion between two known key frames and interpolates the obtained motion fields to estimate the motion field of the intermediate B frame.

In [3] a block-wise algorithm for motion compensated prediction of the current frame at the decoder side, performed in absence of motion information, is presented. In that work, for every block to be predicted a set of possible candidates is analyzed and the most appropriate predictor is chosen based on a LSE principle. The choice is made block by block, independently, with a causal scanning of blocks. A problem observed in that case is that an erroneous choice for the motion vector of one block rapidly propagates to adjacent blocks.

In this chapter we propose an algorithm that relies on the contextual decoding of a region composed of several blocks. In particular, we propose a Viterbi-like algorithm for the simultaneous estimation of the motion vectors associated to a row of blocks. To this aim, in order to reduce as much as possible the computational complexity, we introduce a new regularity parameter, with respect to [3], based on the energy distribution in the DCT domain. This allows a faster construction and analysis of the trellis that would require a much higher complexity with the regularity criterion adopted in [3].

The chapter is structured as follows. In Sect. 16.2 motion estimation principles are recalled and decoder based motion estimation is introduced. In Sect. 16.3 a model based on side information and the spatial coherence principle is introduced and a parameter for the evaluation of spatial coherence is described. In Sect. 16.4 a blockwise motion estimation algorithm is briefly summarized, whereas in Sect. 16.5 the proposed region-based algorithm is presented. Simulation results are presented and discussed in Sect. 16.6. Finally, concluding remarks are given in Sect. 16.7.

16.2 Motion Estimation at the Decoder

Predictive video coding is based on motion estimation at the encoder and motion compensation at the decoder. For a complete description of the topics related to motion estimation and its applications in state-of-art video coding, we refer the reader to [4–6]. In this section, instead, the basic ideas of motion estimation coding are briefly recalled, while introducing the notation that will be used throughout the chapter. Then, the idea of motion estimation at the decoder side is formalized.

In the following, the notation is referred to the scheme in Fig. 16.1. Let $X_{m,n}$, with $0 \leq m < M$ and $0 \leq n < N$, be the $B \times B$ block having $X(mB, nB)$ as the top-left pixel. Let W be the search window in the reference frame X^{ref}, i.e., the previous frame. W is centered in (mB, nB). A predictor for $X_{m,n}$ is searched for among all the blocks contained in W. Each possible predictor $\hat{X}_i$ is identified through the displacement from (m, n), i.e., its motion vector $\boldsymbol{v_i} = (v_y, v_x)$. Precisely, $\hat{X}_{m,n,\boldsymbol{v_i}}$ is the $B \times B$ block having $X^{ref}(mB + v_y, nB + v_x)$ as the top-left pixel. Define now the prediction error

$$R_{m,n,\boldsymbol{v}} = X_{m,n} - \hat{X}_{m,n,\boldsymbol{v}},$$

where the difference is performed pixel by pixel. For the given block $X_{m,n}$ the selected optimal motion vector $\overline{\boldsymbol{v}}$ is chosen so as to minimize, according to a given metric, the residue norm, that is

$$\overline{\boldsymbol{v}} = \underset{\boldsymbol{v}}{\operatorname{argmin}} \left\| R_{m,n,\boldsymbol{v}} \right\| .$$

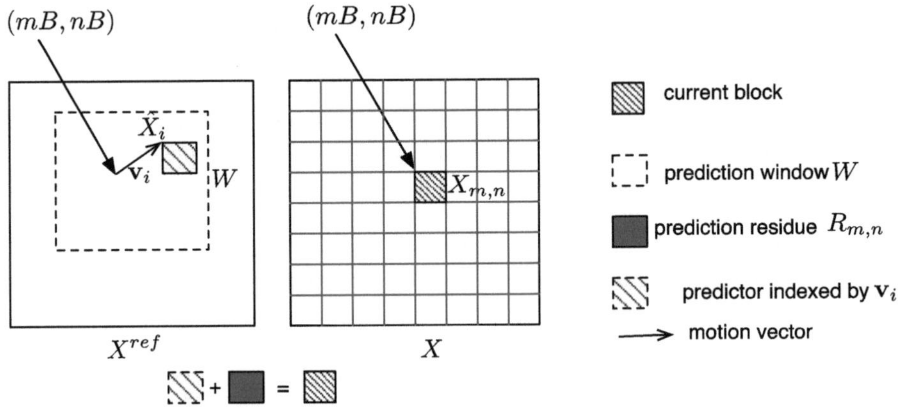

Fig. 16.1 Candidate set generation

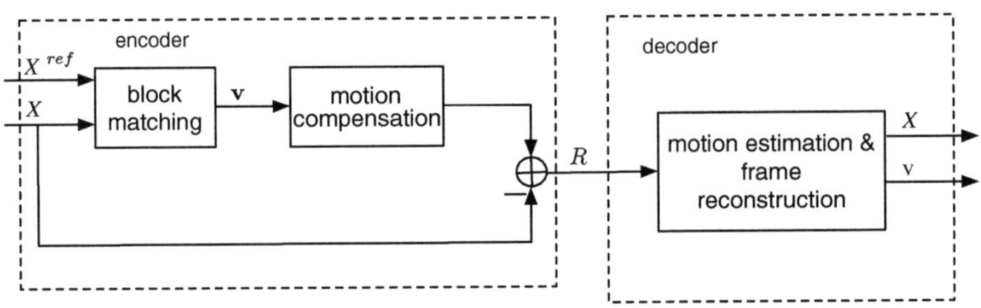

Fig. 16.2 Scheme of a coding system based on motion estimation at the decoder side

In traditional coding, for every block the motion vector $\overline{v}$ and the block residues $R_{m,n,\overline{v}}$ are suitably entropy encoded, and they are transmitted to the decoder. In this case, at the decoder each predictor can be identified very easily, by the use of the motion vector as an index. So each block is reconstructed as:

$$\overline{X}_{m,n} = \hat{X}_{m,n,\overline{v}} + R_{m,n,\overline{v}} \,. \tag{16.1}$$

The decoder reconstructs each frame operating in a strictly blockwise mode, since each block is reconstructed independently from its neighbors. It is worth remarking that the frame encoding order is such that the each frame is always decoded after its reference frame.

In the scheme considered in this chapter, depicted in Fig. 16.2, only $R_{m,n,\overline{v}}$ is known at the decoder, whereas $\overline{v}$ is not transmitted. Therefore, in this scenario the main challenge is to infer the vector $\overline{v}$, where the only available information consists of causally neighboring pixel values and the residue $R_{m,n,\overline{v}}$.

The idea is to simply consider all possible motion vectors v and to build a set of corresponding possible reconstructions for the block $X_{m,n}$. Since the decoder is assumed to know the search window W, it can generate the set of all candidate reconstructions for block $X_{m,n}$ as

$$\mathscr{C}_{m,n} = \left\{\hat{X}_{m,n,v_k} + R_{m,n}\right\} \tag{16.2a}$$

$$= \left\{X_{i,j}^{ref} + R_{m,n} \quad | \; X_{i,j}^{ref} \in W\right\} . \tag{16.2b}$$

In other words, each block in the search window W is added the known residue and is considered as a possible reconstruction for block $X_{m,n}$.

In the following, the generic element of the candidate set $\mathscr{C}_{m,n}$ is referred to as $C_i^{(m,n)}$, whereas K is the cardinality of $\mathscr{C}_{m,n}$. In order to determine the best candidate $\overline{C}_{m,n}$ and the corresponding $\overline{v}_{m,n}$, a model based on spatial coherence is now formulated, which involves the use of a spatial coherence parameter.

16.3 Side Information and Spatial Coherence

In video coding, the phrase *side information* refers, to a general extent, to pieces of information correlated to the signal to be trasmitted. For the purposes of the model presented in this chapter, we define as *side information* the information that the decoder knows and which is correlated to the part of the frame currently being decoded.

In particular, our scheme inherits the decoding order from predictive coding. The decoding order is such that, when decoding a given frame, its reference frame has been decoded previously, and hence it is completely known. Besides the reference frames, which can be referred to as *inter-frame* side information, our model relies on another type of side information, i.e., *intra-frame* side information. Intra-frame side information is composed of the already decoded blocks in the current frame. For example, let us suppose that, when $X_{m,n}$ is being decoded, the row above is known, which is the case considered in Sect. 16.5. With this hypothesis, the intra-frame side information for $X_{m,n}$ is

$$\mathscr{I}_{m,n} = \left\{X_{i,j} \quad | \quad 0 \leq i < m, \; 0 \leq j < N\right\} . \tag{16.3}$$

The key idea underlying this work is the practical exploitation of the intra-frame side information, through the introduction of an assumption on the frame structure. We assume that the frame content is characterized by edges that preserve their continuity across the block boundaries and we define this property *spatial coherence*. This means that, given the neighborhood of a block, it is possible to infer that the more suitable predictor in a candidate set will be the one that matches at best the neighborhood edges. In Fig. 16.3b, c an example of a good match and one of a bad match,

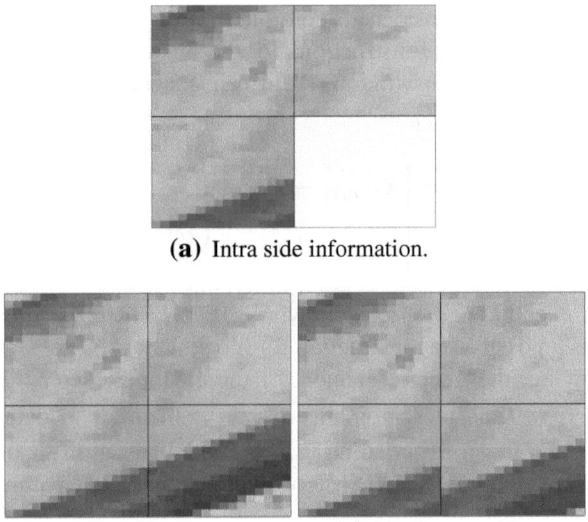

(**a**) Intra side information.

(**b**) Well matching predictor.

(**c**) Badly matching predictor.

Fig. 16.3 Spatial coherence at block edges

respectively, are shown. This example highlights the importance of the information carried by the neighborhood: the side information (Fig. 16.3a) allows to predict that the border contained in the unknown block is more likely to be structured as shown in Fig. 16.3b rather than as shown in Fig. 16.3a. The introduced hypothesis is loose, as the block boundaries are decided for coding convenience, regardless of the frame content. In the following, a coding scheme based on this model is presented.

16.3.1 Spatial Coherence Parameter

In order to apply the ideas described in Sect. 16.3, our hypotheses about the spatial coherence need to be formalized. To this extent, let us assume that, given any macroblock composed of 2×2 neighboring blocks, its "amount of spatial coherence" can be assessed through a one dimensional feature of the blocks. From now on such feature will be referred to as Spatial Coherence Parameter (SCP).

In more detail, let Y be a macroblock of size $2B \times 2B$ be composed of 4 $B \times B$ blocks. The spatial coherence parameter p is a positive valued one-dimensional feature of Y:

$$p : \mathbb{R}^{2B} \times \mathbb{R}^{2B} \rightarrow \mathbb{R}^{+}$$
$$Y \mapsto p(Y).$$

The function $p(Y)$ can be interpreted as a penalty and should ideally satisfy the empirical requirement that given two macroblocks Y_1 and Y_2, $p(Y_1) < p(Y_2)$ if the blocks forming Y_1 "match better" than the ones forming Y_2.

The Spatial Coherence Parameter used in this work is different from that used in [3] and is based on the properties of the Discrete Cosine Transform. This is principally motivated by two different reasons. First, frequency domain techniques which exploit the already decoded neighborhood to predict a missing block are used in error concealment methods (see for example [7]). Second, as will be clarified later, the introduced parameter will allow the construction of the trellis with a much smaller computational complexity than the parameter proposed in [3] would require.

Let Y be a $2B \times 2B$ macroblock composed of four neighboring blocks, and consider its DCT transform. In the following, some remarks about how its spatial coherence properties reflect in the frequency domain are presented. Then, a parameter "measuring" the level of coherence of the macroblock is introduced. The presence of a spatial discontinuity, i.e., a non matching edge across a block boundary, introduces high frequencies, that do not belong to the original image. Thus, when this happens, in the DCT domain the energy is distributed on a large number of coefficients, some of them located in the higher frequency range. On the contrary, when edges match properly, in the DCT domain the energy should be very concentrated on few coefficients. For example, in Fig. 16.4 the squared moduli of the DCT coefficients, for two macroblocks are reported. Whereas the side information is the same for both the macroblocks, the candidates are different: the macroblock whose DCT squared modulus is reported in Fig. 16.4a contains the correct candidate, whereas the DCT squred modulus in Fig. 16.4b corresponds to the macroblock with the wrong candidate. More specifically, the two macroblocks are the ones reported in Fig. 16.3b, c, respectively. The squared moduli are represented in logarithmic scale, in gray tones, where the lighter is the gray, the higher is the coefficient value. It can be seen that the wrong candidate produces higher coefficients spread on almost all the frequencies, whereas the energy in the DCT of the candidate containing the correct block is much more concentrated: indeed, the coefficient in the central area in Fig. 16.4a are darker, and thus smaller, than those in the corresponding area in Fig. 16.4b.

Given a macroblock Y, composed of 4 $B \times B$ blocks as described above, its DCT Spatial Coherence Parameter $p(Y)$ is computed according to the following steps:

- $Z_Y = DCT(Y)$ is computed and normalized in order to have unitary energy.
- DCT coefficients are sorted in descending squared modulus magnitude order; let $\tilde{Z}_Y(k)$ be the $k-th$ coefficient in such order.
- given a fixed threshold T, $p(Y)$ is defined as the minimum value of k that verifies the following condition: $\sum_{j=1}^{k} |\tilde{Z}_Y(j)|^2 \geq T$

In other words, the square moduli of the coefficients are added in descending order, until the sum reaches a fixed threshold T; the parameter p of the macroblock y is the number of coefficients needed to obtain a value greater than the threshold.

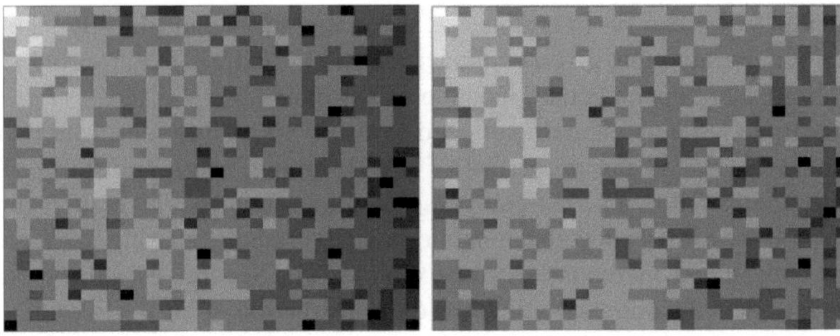

(a) DCT of the macroblock with correct candidate. **(b)** DCT of the macroblock with wrong candidate.

Fig. 16.4 Squared moduli (in log scale) of the DCTs of two macroblocks with the same side information but different candidates

16.4 Block-Wise Selection Algorithm

The Spatial Coherence Parameter can be used as a test to determine whether a block fits the given neighborhood. In [3] an algorithm for motion estimation at the decoder based on this principle, but using a different SCP, is described. This algorithm, which operates on each block separately, in the following is referred to as block-wise algorithm and it is used as a reference for the performance assessment of the row-wise algorithm presented in this chapter. According to this algorithm, first the candidate set for the current block $X_{m,n}$ is generated, as described in Sect. 16.2. Then, a ranking of the candidates is obtained, to determine which candidate fits best the known causal neighborhood, i.e., the upper-left, upper and left neighbors. For each $C_i^{(m,n)}$, the macroblock $Y_{C_i}^{(m,n)}$, composed of the current candidate $C_i^{(m,n)}$ and the three known causal neighbors, is constructed as:

$$Y_{C_i}^{(m,n)} = \begin{bmatrix} \overline{X}_{m-1,n-1} & \overline{X}_{m-1,n} \\ \overline{X}_{m,n-1} & C_i^{(m,n)} \end{bmatrix}. \tag{16.4}$$

Then, the SCP $p(Y_{C_i}^{(m,n)})$ is computed. The selected predictor $\overline{C}^{(m,n)}$ is the one that minimizes p:

$$\overline{C}^{(m,n)} = \operatorname*{argmin}_{C_i \in \mathscr{C}_{mn}} p(Y_{C_i}^{(m,n)}). \tag{16.5}$$

Hence, the estimated motion vector is the one associated with $\overline{C}^{(m,n)}$. This algorithms works well as long as the hypothesis that the true original block is always the most coherent one holds.

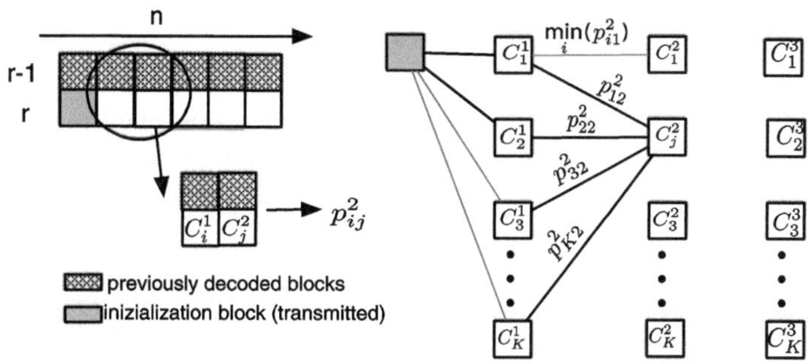

Fig. 16.5 Computation of the p_{ij}^n parameters and trellis construction

16.5 Region Based Model for Motion Estimation at the Decoder

In this section a novel algorithm for motion estimation at the decoder is introduced. This algorithm is based on the contextual decoding of several blocks belonging to a given region of the frame. All the possible combinations of blocks for the region (or at least the more likely ones, depending on whether any optimization is carried out, see Sect. 16.5.2) are generated, combining the candidates of each block in the region.

The algorithm presented in this chapter is designed to overcome one of the main drawbacks of the block-wise algorithm, i.e., the impossibility of detecting when the Spatial Coherence Parameter fails to identify the correct original block. In fact, errors occur when one or more candidates happen to have a SCP smaller than the original block SCP, i.e., when the condition (16.5) does not provide the correct candidate. A wrong block is likely to induce an error on the next block when it is used as a neighbor for the latter, and so on, allowing the error to propagate across the frame. A global algorithm, taking into account a region of adjacent blocks, can exploit the propagation of a wrong choice. Indeed, the main feature of this algorithm is an average of the SCP over the blocks of the entire region. Let us consider the case in which the correct block is not the minimum SCP block, so a wrong candidate is chosen. Since the neighbor of the next block contains such wrong candidate, all the computed parameters will be high, because no good match with a wrong neighbor can be found. So the averaging can be thought of as a balancing of the presence of a not minimal SCP correct block with the effect of the error propagation on other blocks. In the following, instead of an average a non-normalized sum will be performed; this does not affect the just discussed property.

For the sake of simplicity, the algorithm has been implemented taking as a region a single row of blocks. This special region shape simplifies the interdependency of unknown blocks, allowing for a Viterbi-based minimization of the SCP sum.

16.5.1 Row-Wise SCP Sum Minimization

In this section, the algorithm previously introduced is described in detail. In this work, we have chosen the simplest possible region shape, which is an entire row of blocks:

$$T_{\overline{m}} = \left\{X_{(\overline{m},n)},\ \ 0 \leq n < N\right\}. \tag{16.6}$$

The rows are decoded from the top one (T_1) to the bottom one (T_{N-1}). The first row, T_0, is assumed to be completely transmitted, because it is required for the algorithm inizialization. So, a whole row is decoded at once, under the hypothesis that the decoding of the row above has already been performed. Such hypothesis guarantees that, for each block, its upper and upper-left neighbors are known. On the contrary, the left neighbor is uncertain, since it still belongs to the current row and it is being decoded too. Hence, for each candidate $C^{\overline{m},n_k}$ of the n-th block, K coherence parameters p^n_{ij} need to be computed, each one using a different candidate $C^{\overline{m},n-1}_i$ as left neighbor.

Let the region predictor $\underline{L}_{\boldsymbol{q}} = (C^1_{q1}, C^2_{q2}, \cdots, C^N_{qN})$, indexed by the index vector q, be the row predictor composed of the candidate C^1_{q1} for the first block, the candidate C^2_{q2} for the second block, and so on. As depicted in Fig. 16.5, the parameters p^n_{ij} are the SCP of the macroblock Y^n_{ij}, defined as follows:

$$Y^n_{ij} = \begin{bmatrix} X_{\overline{m}-1,n-1} & X_{\overline{m}-1,n} \\ C^{(\overline{m},n-1)}_i & C^{(\overline{m},n)}_j \end{bmatrix}. \tag{16.7}$$

The sum $\sigma(L_{\boldsymbol{q}})$ of the SCPs of all the involved blocks is considered, for each combination $\underline{L}_{\boldsymbol{q}}$:

$$\sigma(\underline{L}_{\boldsymbol{q}}) = \sum_{n=1}^{N} p^n_{q(n-1)q_n}. \tag{16.8}$$

In order to track the combinations with lower SCP Sum σ, a Viterbi-like minimization (see for example [8]) is used: each path on the trellis represents a combination of candidates, i.e., a candidate row $\underline{L}_{\boldsymbol{q}}$. In Fig. 16.6 this model is depicted; the highlighted path represents the combination composed of candidate C^1_1 for the first block, C^2_2 for the first block, and so on. Being each candidate univocally indexed by a motion vector, each path represents a row of the motion field, as well. At each step, the selected candidate $\overline{C}^{n-1}_l$ is such that

$$p^n_{lj} = \min_i(p^n_{ij}), \tag{16.9}$$

where the hypotesis is $C^n = C^n_j$. Thanks to the application of the Viterbi algorithm, the combination with higher σ are automatically discarded.

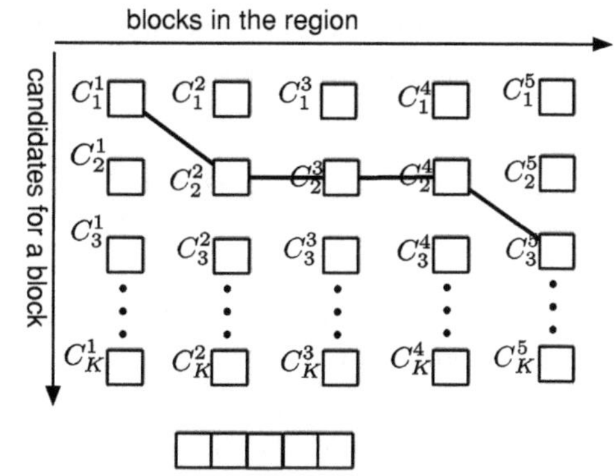

Fig. 16.6 Trellis contruction: the path represents the combination $(C_1^1, C_2^2, C_2^3, C_2^4, C_3^5)$

16.5.2 *Computational Complexity Reduction*

The main drawback of this scheme is that, despite the use of the Viterbi algorithm, the decoding complexity remains quite high. This is important to be noted, especially because the encoder is as complex as in predictive coding. In the following, the methods used to reduce the computational load of the decoder are described.

16.5.2.1 Motion Smoothness

Introducing some hypothesis on the structure of the motion field can reduce considerably the computational complexity while preventing unlikely combinations from being erroneously selected. In this implementation, we have supposed that the motion field is smooth, so as to discard the combinations leading to a motion field that varies abruptly from one block to the next. Since each transition in the trellis corresponds to a motion vector, reducing the set of admissible motion vectors also reduces the number of SCP to be computed. This principle has been implemented by weighting the SCPs of each candidate by means of a weight function depending on the difference between the motion vector associated to the candidate and the motion vector associated to the neighbor and thresholding the obtained distance.

16.5.2.2 Parameter Decomposition

Since the SCP parameter is a scalar index of how well the four blocks in the macroblock match with one another, the computation of an SCP can never be completely decomposed into the separate computation of a feature of each block. Nevertheless, if the parameter is easily deduced from a linear function of the pixel values, as is the DCT, the number of operations required for the computations of the SCP used in the trellis can be strongly reduced.

Let us consider the computation of the parameter p^n_{ij}, and let us focus the structure of the associated macroblock Y^{mn}_{ij}

$$Y^{mn}_{ij} = \begin{bmatrix} \overline{X}_{m-1,n-1} & \overline{X}_{m-1,n} \\ C_i^{(m,n-1)} & C_j^{(m,n)} \end{bmatrix}.$$

Note that $C_i^{(m,n-1)}$ and $C_j^{(m,n)}$ are generic candidates for blocks $X_{m,n-1}$ and $X_{m,n}$, and they both thus run over a set of K possible values. Since the evaluation of p^n_{ij} requires the computation of the DCT of Y^{mn}_{ij}, K^2 DCT evaluation would be required for the computation of all optimal trellis transitions from block $X_{m,n-1}$ to $X_{m,n}$. Note now that we can write $Y^n_{ij} = Y^n_{i0} + Y^n_{0j}$, where

$$Y^{mn}_{i0} = \begin{bmatrix} \overline{X}_{m-1,n-1} & \overline{X}_{m-1,n} \\ C_i^{(m,n-1)} & 0_{B\times B} \end{bmatrix}$$

and

$$Y^{mn}_{0j} = \begin{bmatrix} \overline{X}_{m-1,n-1} & \overline{X}_{m-1,n} \\ 0_{B\times B} & C_j^{(m,n)} \end{bmatrix},$$

$0_{B\times B}$ being the null $B \times B$ matrix. For the linearity of the DCT we clearly obtain that $DCT(Y^{mn}_{ij}) = DCT(Y^{mn}_{i0}) + DCT(Y^{mn}_{0j})$. This implies that the evaluations of all the K^2 DCTs can be reduced to only $2K$ DCTs evaluations and K^2 matrix sums with an obvious noticeable reduction of the complexity.

16.6 Experimental Results

In this section the Spatial Coherence Parameter behavior if first analyzed, then a performance comparison between the proposed Viterbi-based decoding algorithm and the blockwise one is presented.

16.6.1 SCP Behavior Assessment

As described in Sect. 16.3.1, the computation of the spatial coherence parameter requires that a threshold is set. Such threshold value has been decided experimentally, by evaluating the behavior of the blockwise algorithm at different thresholds, for different video sequences. In Fig. 16.7 an example is reported. The number of correct blocks in the first frame of the *Mobile* sequence is plotted vs. the considered threshold values. It can be seen that, as the threshold increases, the number of correctly selected

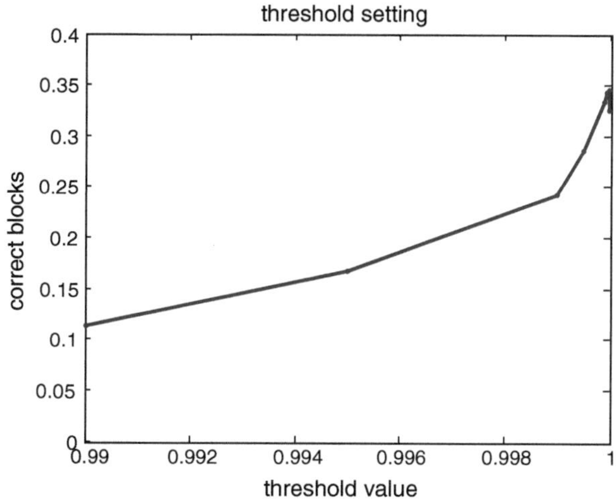

Fig. 16.7 Percentage of correct blocks at different threshold values

candidates keeps increasing until the threshold is very close to 1, then it decreases slightly. As a similar behavior has been observed for all the sequences, the threshold has been set to 0.9999.

In Fig. 16.8 the cumulative energy of macroblock DCT is shown for several candidate of the same block, namely block (3, 15) in the second frame of the *Mobile* sequence, for increasing number of coefficients considered in the sum. To generate the curve, the coefficients are added in square modulus magnitude order and the cumulative energy is normalized to the sum of all the coefficients, as required for SCP computation (see Sect. 16.3.1). In this example, for two wrong candidates the cumulative energy on the first 100 coefficients is higher than the one obtained for the correct candidate. Only when more than 100 coefficients are added and the cumulative energy is about 0.99 the correct candidate is the one associated to the most concentrated energy. Thus, when the threshold value, i.e., 0.9999 is reached, the candidate with lower SCP is the correct one. It has been verified experimentally that the behaviour is the same for almost all blocks, even though the number of coefficients that are needed to reach a given cumulative energy value may vary substantially from one block to another.

Finally, in Fig. 16.9 an example of behavior of the DCT based Spatial Coherence Parameter with the selected threshold is reported, in the case of blockwise selection of the candidate. For different block positions (horizontal axis), the interval of SCP values associated to all its candidates is reported. The circle corresponds to the SCP of the correct block. For the sake of plot clarity, only a subset of the blocks in the considered frame (frame n. 2 of the *Mobile* sequence) is reported. It can be seen that the SCP can assume very different values, depending on the characteristics of the blocks. The SCPs of the candidates in the set can be concentrated in a narrow range or spread in a wider range, depending on the block characteristics.

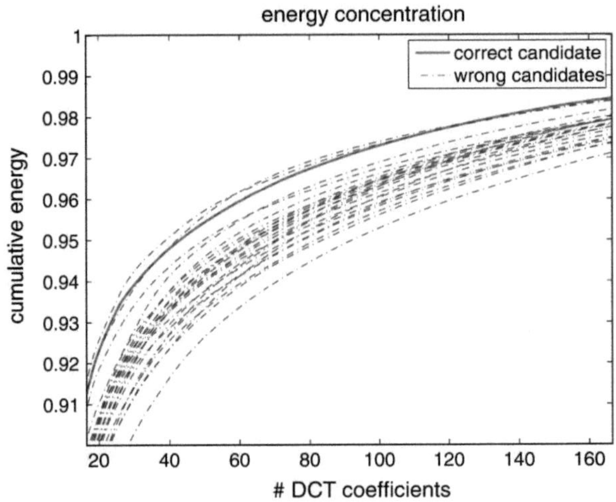

Fig. 16.8 Example of cumulative energy for the block in position (3, 15) in the first frame of the *Mobile* sequence

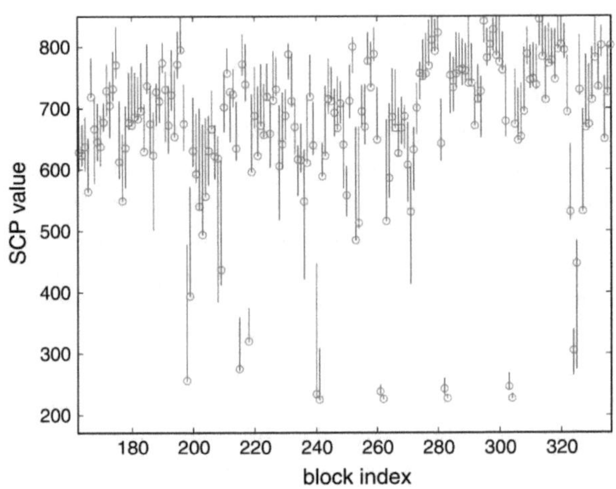

Fig. 16.9 SCP value for different candidates and different blocks

16.6.2 Performance Evaluation

The performance of the proposed algorithm has been evaluated by assessing the improvement with respect to the block-wise algorithm, using the DCT based SCP for both algorithms. The percentage of correctly reconstructed blocks, i.e., of correctly estimated motion vectors, versus the PSNR of the encoded sequence are plotted in Fig. 16.10 for the first 50 frames of the *Foreman*, *Mobile*, *Highway*, *Bus*, *Coastguard* and *Soccer* sequences. From the obtained results, it emerges that the region-based

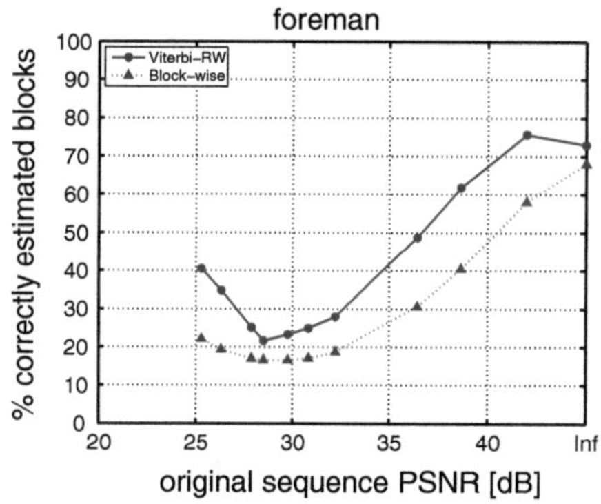

(a) *Foreman* sequence.

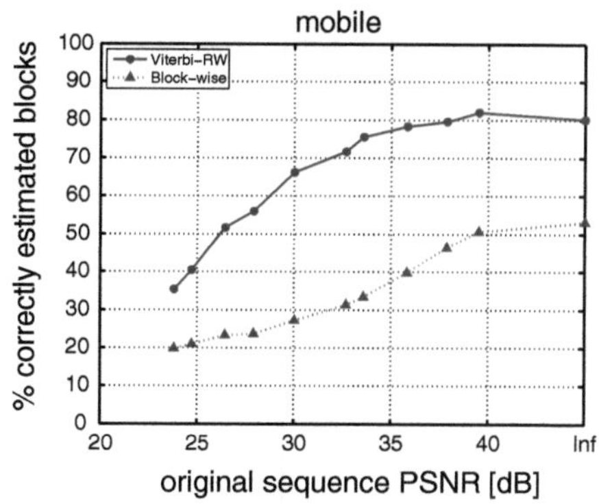

(b) *Mobile* sequence.

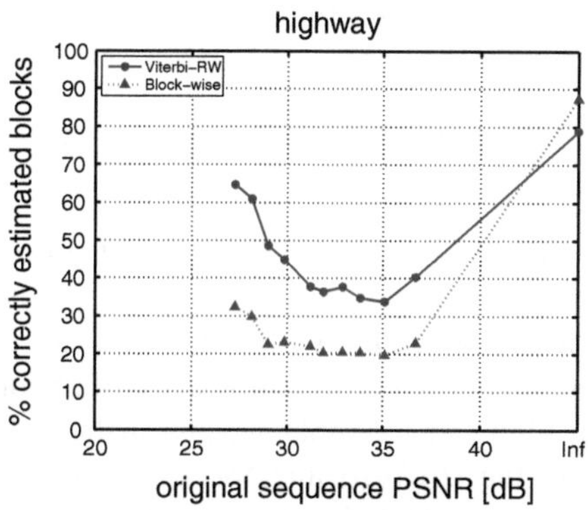

(c) *Highway* sequence.

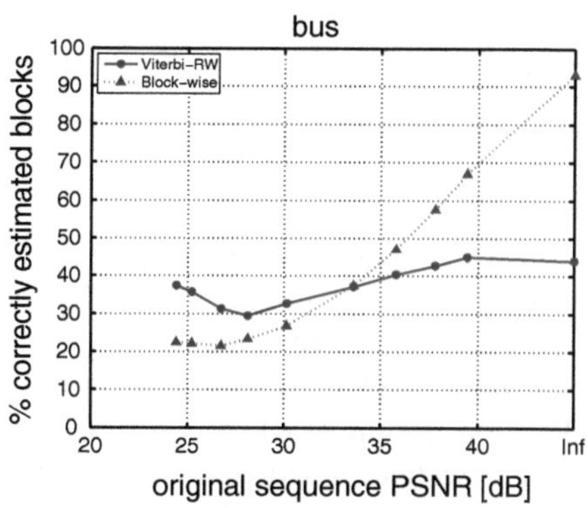

(d) *Bus* sequence.

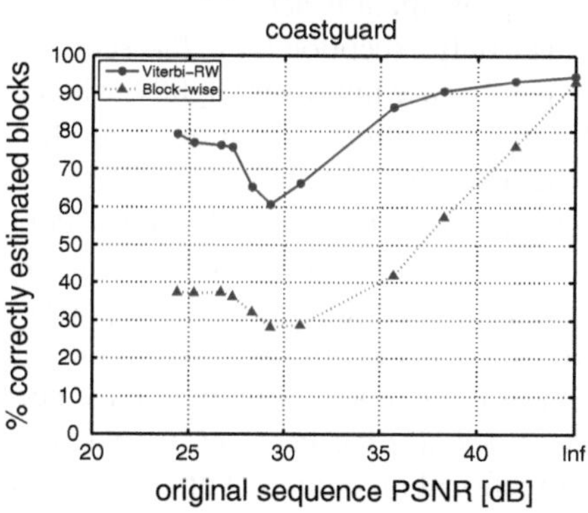

(e) *Coastguard* sequence.

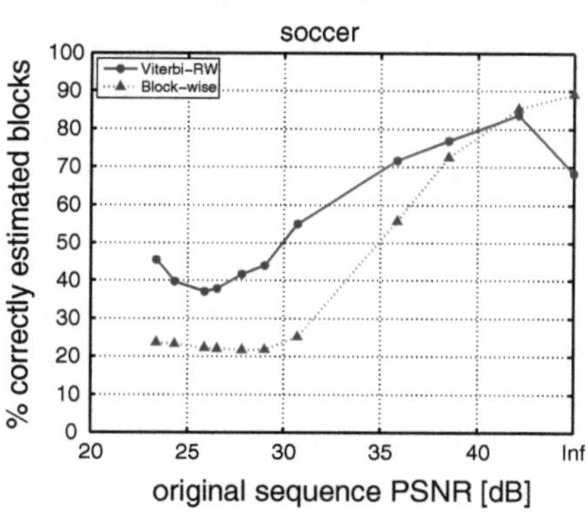

(f) *Soccer* sequence.

Fig. 16.10 Performance assessment for the row-wise (RW) and block-wise (BW) algorithm for the *Foreman*, *Mobile*, *Highway*, *Bus*, *Coastguard* and *Soccer* sequences

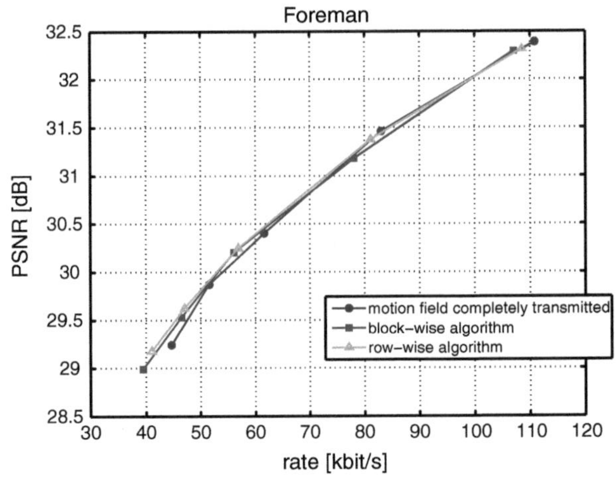

Fig. 16.11 Estimated rate-distortion curves for the *Foreman* sequence

reconstruction significantly improves the performance of the block-wise selection. The gain in terms of correct blocks depends on the sequence and on the working point, but it is substantial in general. For example, for the *Mobile* (Fig. 16.10b) sequence it goes from about 15 % more correct blocks at lower quality to a difference of about 40 % more coding blocks at high quality. Very high gains are obtained also in the case of the *Coastguard* sequence (Fig. 16.10e). For the *Highway* and *Soccer* sequences (Fig. 16.10c, f respectively) high improvement is obtained especially at low rate. Also in the case of *Foreman* (Fig. 16.10a) an improvement is achieved, even if less strong than in the cases previously discussed. The only exception is the *Bus* sequence (Fig. 16.10d): for this sequence, only the low quality case improves slightly, whereas at higher qualities the performance of the Viterbi algorithm is fixed on about 40 % correct blocks, whereas the performance of the block-wise algorithm increases.

An approximation of the rate-distortion curve for the proposed algorithm, compared with the block-wise algorithm, is also reported in Fig. 16.11, for the *Foreman* sequence. In order to give an idea of how the presented algorithm could perform in a realistic scenario, it has been applied to a lossy codec. In more detail, the rate and PSNR values for the case of transmission of the whole motion field have been obtained using a simplified H.264 codec. The block size has been set to 16 and the considered prediction mode is P, i.e., mono-directional prediction, with a single reference picture. An important remark about the rate estimation must be given: the coding efficiency in modern predictive codecs, such as the H.264 codec, depends heavily on how arithmetic coding is performed. Since our methods has not been really implemented in H.264 yet, it is impossible to measure exactly the rate savings. In order to produce a reliable estimate, the bits devoted to the motion transmission for each block have been computed, and, for the correctly predicted block, the result has been subtracted from the overall bit-rate. A signalling overhead has also been taken into account. The three plots appear to be almost overlapping: in the considered case the coding gain seems to be limited by both the signalling overhead and the fact that the motion represents a small amount of the total rate.

16.7 Conclusions

In this chapter a novel algorithm for motion estimation at the decoder side is presented. The main feature of this algorithm is that the blocks belonging to a whole region of the frame are decoded at once. The decoding relies on the spatial coherence of the current frame, i.e., on the assumption that the signal does not change abruptly at the block boundaries. In order to evaluate the spatial coherence of a macroblock composed of 2×2 blocks, a Spatial Coherence Parameter based on the energy concentration property of the DCT is introduced. The decoding consists in generating a candidate set for each block in the row and selecting the combination of candidates which minimizes the sum of the SCPs over the whole row. In order to perform the minimization, and thus to reconstruct the row, a model based on the Viterbi algorithm is introduced. Simulation results show that the proposed approach outperforms the block-wise algorithm that employs the same parameter. Moreover, the preliminary analysis of the rate-distortion performance seems to be promising.

References

1. Kamp S, Evertz M, Wien M (2008) Decoder side motion vector derivation for inter frame video coding. In: Proceeding of the ICIP08, San Diego, 12–15 Octoper 2008. pp 1120–1123
2. Klomp S, Munderloh M, Vatis Y, Ostermann J (2009) Decoder-side block motion estimation for H.264/MPEG-4 AVC based video coding. In: Proceedings of IEEE International Symposium on Circuits and Systems, Taipei, May 2009
3. Tonoli C, Migliorati P, Leonardi R (2006) Video coding with motion estimation at the decoder. In: Giusto D, Lera A, Morabito G, Atzori L (eds.) The internet of the things: 20th thyrrenian workhshop on digital communication. LNCS. Springer, Heidelberg, pp 195–204
4. Wiegand T, Sullivan GJ, Bjontegaard G (2003) Overview of the H.264/AVC video coding standard. IEEE Trans Circuits Syst Video Technol 13:560–576
5. Wiegand T, Sullivan GJ, Luthra A (2003) Draft ITU-T recommendation and final draft international standard of joint video specification. Joint Video Team Doc. JVT-G050r1 (June 2003)
6. Kappagantula S, Rao K (1985) Motion compensated interframe image prediction. IEEE Trans Commun 33:1011–1015
7. Park JW, Kim GJW, Lee SU (1997) DCT coefficients recovery-based error concealment technique and its application to the MPEG-2 Bit Stream Error. IEEE Trans Circuits Syst Video Technol 7:845–854
8. Viterbi AJ (1967) Error bounds for convolutional codes and an asymptotically optimum decoding algorithm. IEEE Trans Inf Theory 13:260–269

MIX
Papier aus verantwortungsvollen Quellen
Paper from responsible sources
FSC® C105338

If you have any concerns about our products,
you can contact us on
ProductSafety@springernature.com

In case Publisher is established outside the EU,
the EU authorized representative is:
Springer Nature Customer Service Center GmbH
Europaplatz 3, 69115 Heidelberg, Germany

Printed by Libri Plureos GmbH
in Hamburg, Germany